Biomolecular Approaches and Drugs for Neurodegeneration

Biomolecular Approaches and Drugs for Neurodegeneration

Guest Editors

Giovanni N. Roviello
Caterina Vicidomini

MDPI

Basel • Beijing • Wuhan • Barcelona • Belgrade • Novi Sad • Cluj • Manchester

Guest Editors
Giovanni N. Roviello
Institute of Biostructures and Bioimaging (IBB)
Italian National Council for Research (CNR)
Naples
Italy

Caterina Vicidomini
Institute of Biostructures and Bioimaging (IBB)
Italian National Council for Research (CNR)
Naples
Italy

Editorial Office
MDPI AG
Grosspeteranlage 5
4052 Basel, Switzerland

This is a reprint of the Special Issue, published open access by the journal *Biomolecules* (ISSN 2218-273X), freely accessible at: https://www.mdpi.com/journal/biomolecules/special_issues/O40SG4E9N5.

For citation purposes, cite each article independently as indicated on the article page online and as indicated below:

Lastname, A.A.; Lastname, B.B. Article Title. *Journal Name* **Year**, *Volume Number*, Page Range.

ISBN 978-3-7258-6193-4 (Hbk)
ISBN 978-3-7258-6194-1 (PDF)
https://doi.org/10.3390/books978-3-7258-6194-1

Contents

About the Editors

Giovanni N. Roviello

Giovanni N. Roviello holds a master's degree with honors in chemistry and a Ph.D. in biotechnology. He works as a senior researcher at the Institute of Biostructure and Bioimaging (IBB) of the Italian National Council for Research (CNR) in Naples, Italy. He has been a visiting researcher in the US (Coastal Carolina University, Gupta Science College, November 2025), Germany (IMB Jena, Georg-August University, FAU University Erlangen), the UK (University of Greenwich), and Ireland (UCD Dublin), specializing in bioorganic chemistry. He has strong academic ties with AMU University of Poznan (Poland), where he lectured on nucleic acids for Ph.D. students in 2021, and with YSU (Armenia) and Geomedi University (Georgia), where he was appointed honorary professor in medicinal chemistry in 2012. Dr. Roviello is the principal investigator for CNR on various international projects, including EU- and the Royal Society (UK)-funded projects. He serves as an academic editor for several international journals and has authored more than 140 scientific papers (Hindex in Scopus is 32). Since 2021, he has been consistently listed among the top 2% of scientists in Stanford University's global author rankings, including the 2024 Author Database (Ioannidis, J. P. A., Updated Science-Wide Author Databases, 2025).

Caterina Vicidomini

Caterina Vicidomini has built a multidisciplinary scientific career at the interface between molecular sciences and advanced medical imaging. Trained initially in the field of biostructures, she gained solid experience in the design, synthesis, and characterization of novel biologically active molecules. This expertise was later complemented by professional development in radiochemistry and in the qualitative and quantitative analysis of neuroimaging data. This combined background enabled her to contribute actively to European-funded translational research projects, including work involving the synthesis of molecular precursors, radiolabeling procedures, and in vivo microPET/CT image analysis in preclinical models. Through this experience, Dr. Vicidomini has integrated chemistry, radiopharmaceutical development, and imaging analytics within a single research pathway. Her current work involves research on innovative molecular systems and contributions to projects developing and applying multimodal imaging protocols. Her publication record spans molecular and radiochemical studies, along with high-impact diagnostic research in neurological disorders, reflecting a broad and interdisciplinary scientific vision.

Preface

Neurodegenerative diseases represent one of the most pressing health challenges of our time. Their increasing prevalence, coupled with the absence of definitive therapies, shows the urgent need for innovative scientific approaches that can address both the complexity of their pathophysiology and the limitations of current treatments. This reprint presents a collection of recent contributions that collectively aim to shed light on these challenges, offering new perspectives and strategies for neurotherapeutic intervention. The subject of this reprint is broad: it encompasses the multifaceted nature of neurodegeneration, ranging from molecular mechanisms to translational applications. In particular, readers will explore therapeutic convergence through natural products, drug repurposing, and biomolecular targets. These approaches are complemented by investigations into specific aspects of disease pathology, such as the selective vulnerability of neuronal populations in Alzheimer's disease (AD), cholesterol metabolism in microglia, and innovative methods for prion inhibition. The aim and purpose of this work are twofold: First, it seeks to highlight the breadth of strategies currently under development in the fight against neurodegenerative diseases, emphasizing how molecular insights and pharmacological innovation can converge to create new therapeutic opportunities. Second, it intends to foster dialogue and collaboration among researchers and clinicians by presenting a curated set of studies that illustrate both the challenges and the promise of ongoing research. In doing so, the reprint aspires to serve as a resource that not only informs but also inspires further exploration and discovery. The motivation for compiling this reprint arises from the recognition that neurodegeneration is not a single disease entity but rather a spectrum of disorders, like AD, Parkinson's disease (PD), prion diseases, multiple sclerosis, spinocerebellar ataxia, and others, that share overlapping mechanisms yet also exhibit unique features. Progress in understanding these conditions requires a multidisciplinary approach that integrates biochemistry, pharmacology, molecular biology, and clinical science. Several reasons further justify the creation of this collection, including the growing interest in natural products and nutraceuticals as potential neuroprotective agents. Plant-derived molecules, such as those investigated for their effects in PD models or their capacity to modulate adrenergic receptor signaling in AD, represent promising avenues for therapy. Another reason is the recognition that drug repurposing, i.e., finding new applications for existing compounds, can accelerate the development of effective treatments by taking advantage of the known pharmacological profiles. This reprint is addressed to a wide readership. For researchers, it offers a comprehensive overview of current advances and emerging directions, providing both detailed data and conceptual frameworks that may inform future investigations. For clinicians, it highlights translational aspects of research that could eventually impact patient care, from novel therapeutic strategies to insights into disease mechanisms. For early career scientists and students, it serves as an educational resource that illustrates the richness and diversity of neurodegeneration research, while also demonstrating the importance of critical thinking and interdisciplinary collaboration.

Giovanni N. Roviello and Caterina Vicidomini
Guest Editors

biomolecules

MDPI

Editorial

Therapeutic Convergence in Neurodegeneration: Natural Products, Drug Repurposing, and Biomolecular Targets

Caterina Vicidomini and Giovanni N. Roviello *

Institute of Biostructures and Bioimaging, Italian National Council for Research (IBB-CNR), Area di Ricerca Site and Headquarters, Via Pietro Castellino 111, 80131 Naples, Italy; caterina.vicidomini@ibb.cnr.it
* Correspondence: giovanni.roviello@cnr.it

Neurodegenerative diseases pose an escalating global health burden, caused by their intricate pathophysiological mechanisms, and consequently, a persistent lack of curative therapies. A detailed exploration of these concepts, and of those presented below in this work, can be found in the Special Issue "Biomolecular Approaches and Drugs for Neurodegeneration", available at https://www.mdpi.com/journal/biomolecules/special_issues/O40SG4E9N5 (accessed on 30 August 2025). Importantly, recent advances in molecular neuroscience have elucidated key processes underlying disorders such as Alzheimer's disease, Parkinson's disease, prion disorders, multiple sclerosis, and spinocerebellar ataxia type 1 [1–3]. These conditions share common features including selective neuronal vulnerability, dysregulated lipid metabolism, mitochondrial dysfunction, neuroinflammation, and protein misfolding. Emerging prophylactic [4] and therapeutic strategies increasingly target these converging pathways, with the latter leveraging both natural compounds and synthetic agents to restore cellular homeostasis and mitigate neurodegeneration [5–9]. Notably, neuropeptides play a multifaceted role in the pathophysiology of Alzheimer's disease [10]. Alterations in their expression and receptor activity have been observed in affected brain regions, particularly the hippocampus and entorhinal cortex. These molecules contribute to neuroprotection by modulating synaptic plasticity, reducing beta-amyloid accumulation, enhancing glucose metabolism, and regulating stress responses such as endoplasmic reticulum stress and autophagy. Their involvement in key signaling pathways suggests that neuropeptides may serve not only as biomarkers of disease progression, but also as promising therapeutic targets for intervention in Alzheimer's disease [10,11]. Complementing these molecular insights, recent findings have highlighted the role of microglial responses to myelin debris in aging brains, which contribute to cholesterol ester accumulation, a process implicated in neurodegenerative progression. This accumulation can be attenuated by ACAT1 inhibition, which promotes ABCA1-mediated cholesterol efflux via the LXR pathway [12,13]. Interestingly, these findings underscore the therapeutic relevance of lipid regulation in neurodegenerative contexts. Technological innovations such as protein misfolding cyclic amplification (PMCA) [14] have enabled cross-species screening of anti-prion compounds, identifying methylene blue as a potent inhibitor of prion replication [14,15] and validating PMCA's utility in drug selectivity and structure–activity profiling. Moreover, in Parkinson's disease models, high-throughput screening of marine and plant-derived fractions has uncovered bioactive molecules capable of alleviating mitochondrial stress, highlighting the untapped potential of natural products [16–18]. Among these, dehydrozingerone and its dimeric form have demonstrated notable neuroprotective effects in animal models, preserving dopaminergic neurons and motor function [19].

Similarly, ladostigil (*N*-propargyl-(3R)-aminoindan-5-yl)-*N*-propylcarbamate) [20,21] has shown promise in aging rats by modulating immune regulators such as TNFAIP3 and EGR1, thereby reducing memory decline and neuroinflammation [22]. Within the expanding realm of plant-derived therapeutics, *Syzygium aromaticum* (clove) [23,24] has emerged as a compelling candidate for intervention in Alzheimer's disease, with its phytochemicals and amino acids contributing to antioxidant defense, anti-inflammatory activity, and neurotransmitter modulation [25,26]. Across various neurodegenerative conditions, research into adrenergic receptors suggests that repurposing existing drugs may offer new therapeutic avenues for Alzheimer's disease [27], despite their incomplete mechanistic clarity. Other drug repurposing studies have highlighted the potential of atomoxetine [28] as part of neuroprotective strategies, including in the context of Alzheimer's disease. Notably, a phase II clinical trial demonstrated excellent safety, tolerability, and target engagement in individuals with mild cognitive impairment, all advantages of repurposing a well-characterized FDA-approved medication [29]. Remarkably, in multiple sclerosis, immunoglobulin G and complement activation have been implicated in demyelination and neuronal damage, reinforcing the role of immune dysregulation in disease progression [30,31]. In parallel, spinocerebellar ataxia type 1 has inspired diverse therapeutic approaches, including genetic, pharmacological, and cellular investigations, aimed at preserving Purkinje cells and restoring motor coordination [32,33]. Meanwhile, nuclear medicine continues to refine TSPO radiotracers [34] for dementia imaging [35–37], with newer compounds addressing limitations in sensitivity and specificity (Table 1).

Table 1. Overview of the pathological features, therapeutic strategies, and technological innovations herein discussed for some of the main neurodegenerative diseases.

Disease/Condition	Key Pathological Features	Therapeutic Strategies/Compounds
Alzheimer's disease	Neuronal vulnerability, metabolic stress, proteotoxicity, protein misfolding	*Syzygium aromaticum* (clove), adrenergic receptor-targeting drugs, antioxidant and neurotransmitter support
Parkinson's disease	Mitochondrial dysfunction, dopaminergic neuron loss	Dehydrozingerone and its dimer (in *Drosophila*), marine/plant-derived bioactives
Prion disorders	Protein misfolding, cross-species infectivity	Methylene blue
Multiple sclerosis	Myelin debris, cholesterol ester accumulation, immune dysregulation	ACAT1 inhibition, ABCA1-mediated cholesterol efflux via LXR pathway
Spinocerebellar ataxia type 1	Purkinje cell degeneration, motor coordination loss	Genetic, pharmacological, and cellular therapies
Aging brain	Lipid metabolism dysregulation, neuroinflammation, mitochondrial stress, protein misfolding	Natural compounds, synthetic agents, drug repurposing
Alzheimer's disease	Cholesterol accumulation, immune activation	Ladostigil (modulates TNFAIP3 and EGR1), ACAT1 inhibition
	Neuronal vulnerability, metabolic stress, proteotoxicity, protein misfolding	*Syzygium aromaticum*, adrenergic receptor-targeting drugs, antioxidant and neurotransmitter support

For further details, readers are invited to consult the Special Issue "Biomolecular Approaches and Drugs for Neurodegeneration", available at https://www.mdpi.com/journal/biomolecules/special_issues/O40SG4E9N5 (accessed on 30 August 2025).

Altogether, these multidisciplinary efforts reflect a rapidly evolving landscape in neurodegenerative disease research. By integrating molecular insights, pharmacological innovation, and diagnostic precision, the field is advancing toward targeted interventions that address both etiological mechanisms and clinical manifestations. The convergence of natural product discovery, drug repurposing, and imaging technologies offers renewed promise for reducing disease burden and improving patient outcomes.

Conflicts of Interest: The authors declare no conflicts of interest.

Abbreviations

The following abbreviations are used in this manuscript:

AD	Alzheimer's disease
PD	Parkinson's disease
MS	Multiple sclerosis
SCA1	Spinocerebellar ataxia type 1
ACAT1	Acyl-CoA:cholesterol acyltransferase 1
ABCA1	ATP-binding cassette transporter A1
LXR	Liver X receptor
PMCA	Protein misfolding cyclic amplification
TNFAIP3	Tumor necrosis factor alpha-induced protein 3
EGR1	Early growth response protein 1
TSPO	Translocator protein
IgG	Immunoglobulin G

References

1. Pathak, N.; Vimal, S.K.; Tandon, I.; Agrawal, L.; Hongyi, C.; Bhattacharyya, S. Neurodegenerative disorders of alzheimer, parkinsonism, amyotrophic lateral sclerosis and multiple sclerosis: An early diagnostic approach for precision treatment. *Metab. Brain Dis.* **2022**, *37*, 67–104. [CrossRef]
2. Scheckel, C.; Aguzzi, A. Prions, prionoids and protein misfolding disorders. *Nat. Rev. Genet.* **2018**, *19*, 405–418. [CrossRef]
3. Matilla-Dueñas, A.; Goold, R.; Giunti, P. Clinical, genetic, molecular, and pathophysiological insights into spinocerebellar ataxia type 1. *Cerebellum* **2008**, *7*, 106–114. [CrossRef]
4. Vicidomini, C.; Borbone, N.; Roviello, V.; Roviello, G.N.; Oliviero, G. Summary of the Current Status of DNA Vaccination for Alzheimer Disease. *Vaccines* **2023**, *11*, 1706. [CrossRef]
5. Miller, J.H.; Das, V. Potential for treatment of neurodegenerative diseases with natural products or synthetic compounds that stabilize microtubules. *Curr. Pharm. Des.* **2020**, *26*, 4362–4372. [CrossRef]
6. Rasool, M.; Malik, A.; Qureshi, M.S.; Manan, A.; Pushparaj, P.N.; Asif, M.; Qazi, M.H.; Qazi, A.M.; Kamal, M.A.; Gan, S.H. Recent updates in the treatment of neurodegenerative disorders using natural compounds. *Evid. Based Complement. Altern. Med.* **2014**, *2014*, 979730. [CrossRef]
7. Leuci, R.; Brunetti, L.; Poliseno, V.; Laghezza, A.; Loiodice, F.; Tortorella, P.; Piemontese, L. Natural compounds for the prevention and treatment of cardiovascular and neurodegenerative diseases. *Foods* **2020**, *10*, 29. [CrossRef]
8. Falanga, A.P.; Piccialli, I.; Greco, F.; D'Errico, S.; Nolli, M.G.; Borbone, N.; Oliviero, G.; Roviello, G.N. Nanostructural Modulation of G-Quadruplex DNA in Neurodegeneration: Orotate Interaction Revealed Through Experimental and Computational Approaches. *J. Neurochem.* **2025**, *169*, e16296. [CrossRef]
9. Vicidomini, C.; Cioffi, F.; Broersen, K.; Roviello, V.; Riccardi, C.; Montesarchio, D.; Capasso, D.; Gaetano, S.D.; Musumeci, D.; Roviello, G.N. Benzodifurans for biomedical applications: BZ4, a selective anti-proliferative and anti-amyloid lead compound. *Future Med. Chem.* **2019**, *11*, 285–302. [CrossRef]
10. Rahman, M.M.; Islam, M.R.; Supti, F.A.; Dhar, P.S.; Shohag, S.; Ferdous, J.; Shuvo, S.K.; Akter, A.; Hossain, M.S.; Sharma, R. Exploring the therapeutic effect of neurotrophins and neuropeptides in neurodegenerative diseases: At a glance. *Mol. Neurobiol.* **2023**, *60*, 4206–4231. [CrossRef]
11. Chen, X.-Y.; Du, Y.-F.; Chen, L. Neuropeptides exert neuroprotective effects in Alzheimer's disease. *Front. Mol. Neurosci.* **2019**, *11*, 493. [CrossRef]

12. Huynh Krumeich, T.N.P. Targeting the Cholesterol Storage Enzyme ACAT1/SOAT1 in Brain Cells and Mouse Model: A Novel Approach to Address Aging and APOE4-Related Neuroinflammation. Ph.D. Thesis, Dartmouth College, Hanover, NH, USA, 2024.
13. Zhou, Y.; Miles, J.R.; Tavori, H.; Lin, M.; Khoshbouei, H.; Borchelt, D.R.; Bazick, H.; Landreth, G.E.; Lee, S.; Fazio, S. PMP22 regulates cholesterol trafficking and ABCA1-mediated cholesterol efflux. *J. Neurosci.* **2019**, *39*, 5404–5418. [CrossRef] [PubMed]
14. Saá, P.; Cervenakova, L. Protein misfolding cyclic amplification (PMCA): Current status and future directions. *Virus Res.* **2015**, *207*, 47–61. [CrossRef] [PubMed]
15. Cavaliere, P.; Torrent, J.; Prigent, S.; Granata, V.; Pauwels, K.; Pastore, A.; Rezaei, H.; Zagari, A. Binding of methylene blue to a surface cleft inhibits the oligomerization and fibrillization of prion protein. *Biochim. Biophys. Acta (BBA) Mol. Basis Dis.* **2013**, *1832*, 20–28. [CrossRef] [PubMed]
16. Silva, J.; Alves, C.; Soledade, F.; Martins, A.; Pinteus, S.; Gaspar, H.; Alfonso, A.; Pedrosa, R. Marine-derived components: Can they be a potential therapeutic approach to Parkinson's disease? *Mar. Drugs* **2023**, *21*, 451. [CrossRef]
17. Huang, C.; Zhang, Z.; Cui, W. Marine-derived natural compounds for the treatment of Parkinson's disease. *Mar. Drugs* **2019**, *17*, 221. [CrossRef]
18. Fakhri, S.; Abdian, S.; Zarneshan, S.N.; Akkol, E.K.; Farzaei, M.H.; Sobarzo-Sánchez, E. Targeting mitochondria by plant secondary metabolites: A promising strategy in combating Parkinson's disease. *Int. J. Mol. Sci.* **2021**, *22*, 12570. [CrossRef]
19. Kesharwani, A.; Sree, B.K.; Singh, N.; Gajbhiye, R.L.; Murti, K.; Peraman, R.; Pandey, K.; Limoli, C.L.; Velayutham, R.; Parihar, V.K. Dehydrozingerone Improves Mood and Memory in Diabetic Mice via Modulating Core Neuroimmune Genes and Their Associated Proteins. *ACS Pharmacol. Transl. Sci.* **2025**, *8*, 1694–1710. [CrossRef]
20. Weinstock, M. Role of Oxidative Stress and Neuroinflammation in the Etiology of Alzheimer's Disease: Therapeutic Options. *Antioxidants* **2025**, *14*, 769. [CrossRef]
21. Weinreb, O.; Amit, T.; Bar-Am, O.; BH Youdim, M. Ladostigil: A novel multimodal neuroprotective drug with cholinesterase and brain-selective monoamine oxidase inhibitory activities for Alzheimer's disease treatment. *Curr. Drug Targets* **2012**, *13*, 483–494. [CrossRef]
22. Zohar, K.; Lezmi, E.; Reichert, F.; Eliyahu, T.; Rotshenker, S.; Weinstock, M.; Linial, M. Temporal shifts in microRNAs signify the inflammatory state of primary murine microglial cells. *Int. J. Mol. Sci.* **2025**, *26*, 5677. [CrossRef] [PubMed]
23. Cortés-Rojas, D.F.; de Souza, C.R.F.; Oliveira, W.P. Clove (Syzygium aromaticum): A precious spice. *Asian Pac. J. Trop. Biomed.* **2014**, *4*, 90–96. [CrossRef] [PubMed]
24. Valarezo, E.; Ledesma-Monteros, G.; Jaramillo-Fierro, X.; Radice, M.; Meneses, M.A. Antioxidant Application of Clove (Syzygium aromaticum) Essential Oil in Meat and Meat Products: A Systematic Review. *Plants* **2025**, *14*, 1958. [CrossRef]
25. Sargsyan, T.; Simonyan, H.M.; Stepanyan, L.; Tsaturyan, A.; Vicidomini, C.; Pastore, R.; Guerra, G.; Roviello, G.N. Neuroprotective Properties of Clove (Syzygium aromaticum): State of the Art and Future Pharmaceutical Applications for Alzheimer's Disease. *Biomolecules* **2025**, *15*, 452. [CrossRef]
26. Uchewa, O.O.; Egwuagu, C.B.; Ibegbu, A.O. Clove oil as a neuromodulator in environmental cadmium cognitive impairment on the prefrontal cortex of Wistar rats. *J. Trace Elem. Miner.* **2025**, *11*, 100212. [CrossRef]
27. Wang, A.; Since, M.; Dallemagne, P.; Rochais, C. Implication of Central β2 Adrenergic Receptor for the Development of Novel Drugs Against Alzheimer's Disease. *Arch. Der Pharm.* **2025**, *358*, e2400750. [CrossRef]
28. Roviello, G.; Cioffi, C.; Moccia, M.; Gillick-Healy, M.W.; Kelly, B.G.; Adamo, M.F.A. Synthesis of active pharmaceutical ingredient atomoxetine via desulfurative halogenation. *Tetrahedron Lett.* **2025**, *171–172*, 155800. [CrossRef]
29. Levey, A.I.; Qiu, D.; Zhao, L.; Hu, W.T.; Duong, D.M.; Higginbotham, L.; Dammer, E.B.; Seyfried, N.T.; Wingo, T.S.; Hales, C.M. A phase II study repurposing atomoxetine for neuroprotection in mild cognitive impairment. *Brain* **2022**, *145*, 1924–1938. [CrossRef]
30. Saez-Calveras, N.; Stuve, O. The role of the complement system in Multiple Sclerosis: A review. *Front. Immunol.* **2022**, *13*, 970486. [CrossRef]
31. Garton, T.; Gadani, S.P.; Gill, A.J.; Calabresi, P.A. Neurodegeneration and demyelination in multiple sclerosis. *Neuron* **2024**, *112*, 3231–3251. [CrossRef] [PubMed]
32. Ripolone, M.; Lucchini, V.; Ronchi, D.; Fagiolari, G.; Bordoni, A.; Fortunato, F.; Mondello, S.; Bonato, S.; Meregalli, M.; Torrente, Y. Purkinje cell COX deficiency and mtDNA depletion in an animal model of spinocerebellar ataxia type 1. *J. Neurosci. Res.* **2018**, *96*, 1576–1585. [CrossRef]
33. Tejwani, L.; Lim, J. Pathogenic mechanisms underlying spinocerebellar ataxia type 1. *Cell. Mol. Life Sci.* **2020**, *77*, 4015–4029. [CrossRef]
34. Pulagam, K.R.; Colás, L.; Padro, D.; Plaza-García, S.; Gómez-Vallejo, V.; Higuchi, M.; Llop, J.; Martín, A. Evaluation of the novel TSPO radiotracer [18F] VUIIS1008 in a preclinical model of cerebral ischemia in rats. *EJNMMI Res.* **2017**, *7*, 93. [CrossRef]
35. Salerno, S.; Viviano, M.; Baglini, E.; Poggetti, V.; Giorgini, D.; Castagnoli, J.; Barresi, E.; Castellano, S.; Da Settimo, F.; Taliani, S. TSPO radioligands for neuroinflammation: An overview. *Molecules* **2024**, *29*, 4212. [CrossRef]

36. Uzuegbunam, B.C.; Rummel, C.; Librizzi, D.; Culmsee, C.; Hooshyar Yousefi, B. Radiotracers for Imaging of Inflammatory Biomarkers TSPO and COX-2 in the Brain and in the Periphery. *Int. J. Mol. Sci.* **2023**, *24*, 17419. [CrossRef]
37. Herholz, K.; Carter, S.; Jones, M. Positron emission tomography imaging in dementia. *Br. J. Radiol.* **2007**, *80*, S160–S167. [CrossRef]

biomolecules

MDPI

Article

Multifaceted Role of Specialized Neuropeptide-Intensive Neurons on the Selective Vulnerability to Alzheimer's Disease in the Human Brain

Manci Li [1,2,*], Nicole Flack [2,3] and Peter A. Larsen [2,3]

1 Department of Electrical and Computer Engineering, College of Science and Engineering, University of Minnesota, Minneapolis, MN 55455, USA
2 Department of Veterinary and Biomedical Sciences, College of Veterinary Medicine, University of Minnesota, St. Paul, MN 55108, USA
3 Minnesota Center for Prion Research and Outreach, College of Veterinary Medicine, University of Minnesota, St. Paul, MN 55108, USA
* Correspondence: li000021@umn.edu

Abstract: Regarding Alzheimer's disease (AD), specific neuronal populations and brain regions exhibit selective vulnerability. Understanding the basis of this selective neuronal and regional vulnerability is essential to elucidate the molecular mechanisms underlying AD pathology. However, progress in this area is currently hindered by the incomplete understanding of the intricate functional and spatial diversity of neuronal subtypes in the human brain. Previous studies have demonstrated that neuronal subpopulations with high neuropeptide (NP) co-expression are disproportionately absent in the entorhinal cortex of AD brains at the single-cell level, and there is a significant decline in hippocampal NP expression in naturally aging human brains. Given the role of NPs in neuroprotection and the maintenance of microenvironments, we hypothesize that neurons expressing higher levels of NPs (HNP neurons) possess unique functional characteristics that predispose them to cellular abnormalities, which can manifest as degeneration in AD with aging. To test this hypothesis, multiscale and spatiotemporal transcriptome data from ~1900 human brain samples were analyzed using publicly available datasets. The results indicate that HNP neurons experienced greater metabolic burden and were more prone to protein misfolding. The observed decrease in neuronal abundance during stages associated with a higher risk of AD, coupled with the age-related decline in the expression of AD-associated neuropeptides (ADNPs), provides temporal evidence supporting the role of NPs in the progression of AD. Additionally, the localization of ADNP-producing HNP neurons in AD-associated brain regions provides neuroanatomical support for the concept that cellular/neuronal composition is a key factor in regional AD vulnerability. This study offers novel insights into the molecular and cellular basis of selective neuronal and regional vulnerability to AD in human brains.

Keywords: single-cell gene expression analysis; RNA-seq; dementia; Alzheimer disease; neuropeptides; neurons

1. Introduction

Alzheimer's disease (AD) is a neurodegenerative disorder characterized by progressive cognitive decline and memory loss [1]. The neuropathological features of AD encompass both "positive"—Aβ plaques and tau tangles, glial responses, and cerebral amyloid angiopathy—and "negative" lesions, such as the loss of neurons and synapses [2]. The progression of AD pathology in the brain follows a stereotypical pattern (Braak stages): In the early stages of AD, the transentorhinal regions are primarily affected, followed by the spread of pathology to neocortical regions in later stages [3]. The majority of AD cases occur sporadically, with no clear understanding of their cause or pathogenesis [1]. However, aging is the biggest risk factor for developing AD. The decline in cognitive function during natural aging bears resemblance to that seen in the early stages of AD [4,5]. Mild cognitive impairment (MCI) is defined as a stage between normal age-related cognitive changes and pathological cognitive impairments, and early AD is frequently preceded by MCI [6].

"Epicenters" are described as the sites exhibiting the peak pathological changes or atrophy within the brain, often considered to coincide with the initial site of disease onset by the network-based degeneration/spread hypothesis for AD [7,8]. Such regions, including the entorhinal cortex (EC), could also be interpreted as brain regions that are selectively vulnerable to AD [9]. Despite extensive research, the mechanisms underlying the selective vulnerability of neuronal subtypes and brain regions to cellular dysfunction and protein misfolding are unknown [9,10]. Investigating "epicenters" from the perspectives of cellular processes, the temporal progression of AD and aging, and the spatial vulnerability of brain regions to AD could provide valuable insights into this enigma.

The susceptibility of a particular brain region to certain diseases may be influenced by the inherent vulnerabilities of the specific cell types and states found within that region [9,11]. Single-cell sequencing technology has revolutionized our understanding of neuronal diversity by revealing a large number of neuronal subtypes that extend beyond the previously established categories [12], which were generally defined using molecular markers in combination with morphology and other cellular characteristics [13–15]. The expression of neuropeptides (NPs) has played a pivotal role in neuronal heterogeneity by assisting single-cell transcriptomic neurotaxonomy—an approach first introduced by Tasic and then applied by Smith et al. in their study of mouse brains as a proof of concept [16,17]. Subsequently, the comprehensive single-cell transcriptomic investigation of adult human brains also found that neuronal subpopulations can have complex and combinatorial NP co-expression networks, many of which are uniquely localized to specific brain regions [12]. Therefore, single-cell transcriptomic investigation of NPs offers a unique opportunity to elucidate the vulnerability of brain regions based on the susceptibility of the specific neuronal populations residing within them, particularly in the context of AD.

Beyond the role of neuronal identity, the neuroprotective and homeostatic functions of NPs may be of even greater value in understanding and treating AD. Adding to previous reports on NP dysfunction in AD [18], we recently reported a widespread disruption of NP networks and a disproportionate absence of neurons with high NP expression in the entorhinal cortex of AD brains [19]. These findings were further corroborated by subsequent research, highlighting the involvement of NPs in AD neuropathology and neurodegeneration during aging [20,21]. Given the crucial role of NPs in intercellular communication and neuronal health [18], they could have a significant impact on the earlier stages and potentially the etiology of AD. As NPs act through G protein-coupled receptors (GPCRs), which are among the most druggable targets for treating diseases in the central nervous system [22], investigating the role of NPs in the selective vulnerability of AD could be fruitful for developing preventative strategies and targeted interventions for AD, especially considering that GPCRs did not exhibit changes as significant as those of NPs in AD brains [19].

Considering the metabolic alterations in early AD and the potential energetic burden imposed on neurons with high levels of NP production (HNP neurons) [23,24], we hypothesize that regional dysfunction in AD may originate from these specialized neurons serving as focal points in the "epicenters" of cascading cellular dysfunction, including metabolic stress and protein misfolding (Figure 1) [8–10]. Cells expressing Alzheimer's-associated NPs (ADNPs, Table S1) are expected to be particularly vulnerable. To test this hypothesis from the perspectives of cellular mechanisms (function), the continuum of AD progression and aging (time), and regional brain vulnerability to AD (space), we analyzed publicly available single-cell and spatiotemporal RNA-seq datasets, encompassing ~1900 human brain samples. We expect that: (1) HNP neurons will express enhanced functional networks, such as increased metabolic demands and protein misfolding vulnerability, that can contribute to AD development; (2) the abundance of HNP neurons, especially those co-expressing ADNPs, will decrease with AD progression; (3) the decrease in ADNP expression with aging will be more pronounced in early AD-impacted brain regions compared to regions affected later in the disease; and (4) HNP neurons co-expressing ADNPs will be preferentially distributed in the "epicenters" of AD, and their spatial pattern will coincide with the regional progression of AD pathology. If any of these expectations are not observed, our hypothesis should be revised.

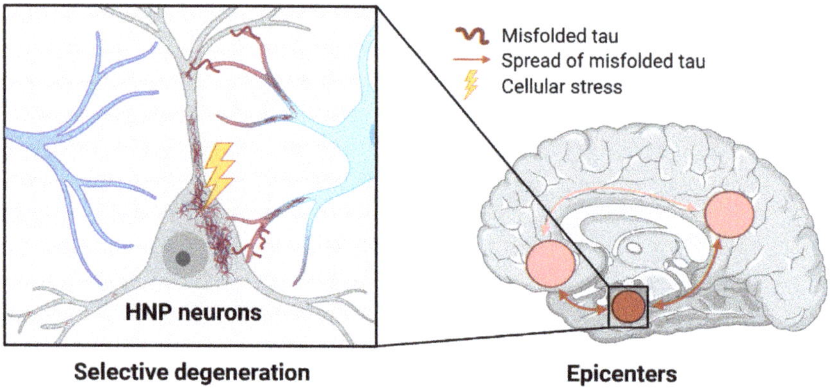

Figure 1. Hypothetical model illustrating the potential mechanisms underlying the selective vulnerability of high-neuropeptide-producing (HNP) cells and their contribution to region-specific emergence of Alzheimer's disease (AD). The model proposes that the unique functions of HNP neurons make them more prone to stress and protein misfolding, and this susceptibility becomes more evident with advancing age. The regional vulnerability observed in Alzheimer's disease can be attributed to the distribution and density of these cells, as well as their excretory functions and interactions with other brain networks. Specifically, it predicts that (1) temporal, limbic, and prefrontal cortical regions have a higher density of HNP neurons expressing ADNPs; (2) disruption of cellular processes in HNP neurons leads to a decrease in ADNPs during aging and localized formation of misfolded proteins in AD, causing various degrees of cognitive decline and selective degeneration of these neurons; and (3) the dynamic paracrine and secretory activities of HNP cells facilitate the propagation of misfolded proteins and transneuronal degeneration in closely connected temporal, limbic, and prefrontal cortical regions, resulting in the widespread deposition of misfolded tau proteins in AD.

2. Methods

2.1. Overview of Datasets and Analyses

A schematic overview of the analytical workflow is presented in Figure 2. Code for the bioinformatic analyses included in this study can be found in Supplementary Materials.

Three publicly available single-cell RNA-sequencing datasets of the human EC were included for analysis in this study: Grubman et al. (the Grubman dataset) included samples from 6 control and 6 AD brains [25]. Leng et al. (the Leng dataset) focused on progression of neuropathology in AD, sampling three donors from Braak stage 0, four donors from Braak stage 2, and three donors from Braak stage 6 [26]. The EC dataset associated with Mathys et al.—generated by the MIT ROSMAP Single-Nucleus Multiomics Study (the MIT ROSMAP Multiomics dataset)—centered on the cognitive status of donors [21,27]; we included 8 control, 8 MCI, and 8 AD samples. To establish that a higher number of co-expressed NPs can serve as a proxy marker for HNP neurons, based on the premise that higher NP co-expression is indicative of greater NP transcript levels [19], correlational analyses were performed using three single-cell RNA-sequencing datasets. The method was motivated by the following reasons: (1) the dataset from the Siletti et al. study (detailed below) represents a robust test of the hypothesis at the spatial level, providing insights into the selective vulnerability of neuronal subpopulations and brain regions in AD based on "combinatorial neuropeptide co-expression" [12]; (2) the presence or absence of a gene is likely more conserved across different studies than exact transcript counts; and (3) this approach facilitates future empirical studies investigating HNP neurons.

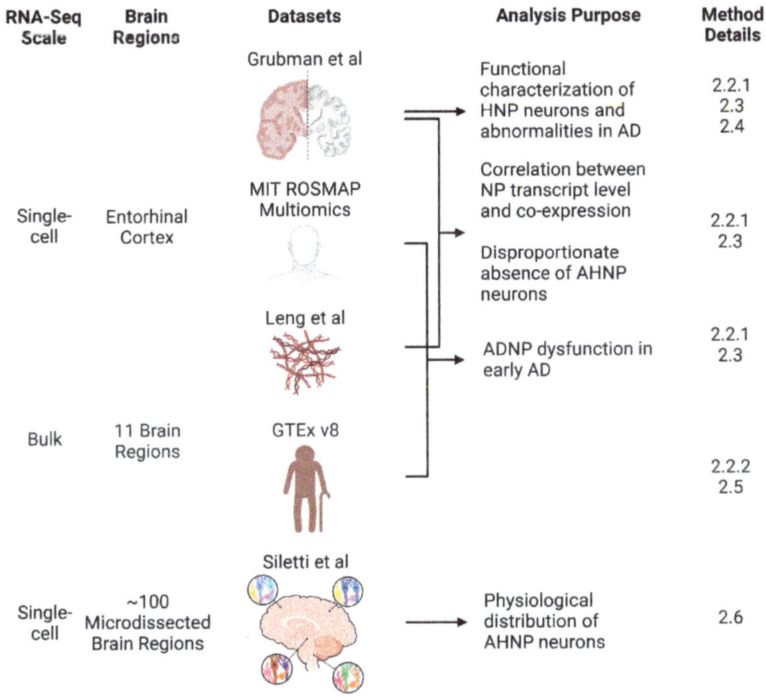

Figure 2. Schematic overview of datasets and analyses. Three single-cell RNA-sequencing datasets of the human entorhinal cortex (Grubman [25], Leng [26], and MIT ROSMAP Multiomics) were used to investigate the relationship between neuropeptide (NP) co-expression and Alzheimer's disease (AD) progression. The Grubman dataset was used for mechanistic analyses, while the MIT ROSMAP Multiomics and Leng datasets were included in demonstrating the early involvement of Alzheimer's disease-associated neuropeptides (ADNPs) during AD development/progression over time from the aspect of cognitive status and neuropathology. Bulk transcriptomic data from the GTEx project were used to study the expression of ADNPs during aging. A comprehensive description of neurons co-expressing high levels of ADNPs (AHNP neurons) across microdissected brain regions was performed using the Siletti et al. [12] single-cell dataset. The method details indicate where the analysis information can be found in the Methods section. Sing-cell, single-cell RNA-sequencing; Bulk, bulk RNA-sequencing; HNP neurons, high NP-expressing neurons. The arrows and brackets indicate the datasets used for each purpose.

The distribution of neurons based on neuropeptide (NP) co-expression was assessed for each dataset. Only the Grubman et al. dataset exhibited a distribution similar to those previously reported in high-quality mouse RNA-seq datasets [17], making it suitable for the mechanistic portion of this study—see below for detailed methods for differential gene expression, functional enrichment analysis, and regression analyses, as well as a hypergeometric test. Both the Leng and MIT ROSMAP Multiomics datasets had over 70% of neurons not expressing any NPs in the NP list compiled previously (Figure S1), which is significantly lower than expected [12,17,19]. Therefore, the ADNP list previously identified from the Grubman dataset was used throughout this manuscript [17,19].

Despite the limitation of NP expression, the Leng and MIT ROSMAP Multiomics datasets were included in the analysis of AD development and progression (time), as they characterize changes of neuropathology and cognitive functions during AD pathogenesis. The Genotype-Tissue Expression (GTEx) project's v8 data release included bulk transcriptomic data derived from tissues sampled from donors aged 20 to 79 years [28], providing a valuable resource for studying gene expression changes during aging [29–31]. All brain regions from the GTEx v8 data were included for this study, subject to further exclusion criteria applied to ensure data quality (See below).

A comprehensive analysis of ADNP occurrences and cell counts across microdissected brain regions was performed using the single-cell transcriptomic dataset generated by Siletti et al. [12]. The dataset comprised samples from 3 postmortem human brain donors, encompassing approximately 100 microdissected regions, 2 million neurons, and 461 clustered neuronal subpopulations. The dataset was accompanied by an annotated cell cluster (cluster_annotation) file that tagged clustered neuronal subpopulations with co-expressed NPs [12], which was used in this study.

2.2. Acquisition and Preprocessing of Transcriptome Data

2.2.1. Single-Cell RNA-Seq Data from the Human Entorhinal Cortex (EC)

Data generated by Grubman et al. from 6 control (CT) and 6 AD donors (12 total) were obtained [25]. The detailed documentation of data acquisition, preprocessing steps, and downstream analyses (dimension reduction and cell identification) can be found in Li and Larsen (2023) [19]. Briefly, cells were filtered based on gene expression and mitochondrial content, retaining those with 200–2500 expressed genes and less than 5% mitochondrial reads. Data normalization followed Seurat's (version 4.1.1; R 4.2.3 unless specified) guidelines using a count per million (CPM) matrix [32]. Using cell identification methods from *BRETIGEA* and Grubman et al., six primary cell types were identified: astrocytes, microglia, neurons, oligodendrocytes, oligodendrocyte precursor cells, and endothelial cells. Cells were labeled by their highest association score, but some were categorized as unidentified or hybrid based on specific criteria described in detail by Grubman et al. as well as Li and Larsen [19,25]. Only cells identified as neurons were used in the downstream analyses.

Preprocessed and annotated single-cell expression matrices and metadata from the Leng et al. study were obtained from CZ CELLxGENE [33]. Cells annotated as excitatory and inhibitory neurons from the entorhinal cortex were included for downstream analyses.

Preprocessed and annotated single-cell expression matrices and metadata from the MIT ROSMAP Multiomics study were downloaded from ADKnowledge portal [21,34]. Three diagnostic categories from the ROSMAP study were included in the presented study: control (CT or NCI: no cognitive impairment), mild cognitive impairment (MCI: no other condition contributing to CI), and AD (Alzheimer's dementia: no other condition contributing to CI (NINCDS/ADRDA Probable AD)) [27]. The clinical study design and detailed diagnostic criteria were described in the original ROSMAP manuscript and deposited on the ADKnowledge portal [6,34]. Individuals with inconsistent clinical diagnosis, clinical cognitive diagnosis summary, and final consensus cognitive diagnosis documented in the ROSMAP study were excluded from the analysis. Methods for isolation of nuclei from frozen post-mortem brain tissue, droplet-based snRNA-seq, and snRNA-seq data preprocessing are available in detail on the ADKnowledge portal [27]. De-identified metadata for individuals and experiments included in this study were detailed in supplementary materials (Table S2). To provide equal representation of each condition, we randomly selected eight AD and CT samples (set.seed = 123) to match the eight available MCI samples. Overall, 41,373 neurons (n_neuron = 41,373) from eight AD (n_neuron = 16,214), MCI (n_neuron = 11,732), and CT (n_neuron = 13,427) EC regions were included in the final analysis.

2.2.2. Spatiotemporal Bulk RNA-Seq Data from Human Brains

RNA-seq transcript matrices (transcript per million (TPM)) of 11 human brain regions from a cohort of individuals from the general population were obtained from the GTEx portal (Table S3) [35]. Samples from individuals that lacked complete metadata regarding age, sex, or death classification were excluded, as were those that scored 3 or 4 on the Hardy scale that indicates intermediate or slow death. Only samples with an RNA integrity number (RIN) larger than 6 were included in the analysis. The results were visualized with cerebroViz (version 1.0; R 3.6.3) [36] and assembled in BioRender.

2.3. *Single-Cell Correlational Analysis of NP Transcripts and NP Co-Expression*

The Spearman rank correlation was utilized to assess the relationship between the transcript level of NP and the number of co-expressed NPs for all human single-cell transcriptome datasets (i.e., the Grubman, Leng, and MIT ROSMAP Multiomics datasets) using *cor.test* in R [21,25,26,34]. The correlation coefficient (rho) and *p*-values were reported. The significance cut-off was set at 0.05.

2.4. *Downstream Analyses for Single-Cell Transcriptome Data from Human EC*

2.4.1. Differential Gene Expression (DGE) Analysis

As such, DGE analysis was performed for the single-cell dataset generated by Grubman et al. [25]. The number of co-expressed NPs was used as a proxy for transcript levels of NPs; neurons were divided into low (0–1), medium (2–5), and high (6+) NP-producing groups (LNP, MNP, and HNP) based on the number of co-expressed NPs in both conditions. DGE analysis was implemented between LNP and HNP in control neurons as well as MNP groups in control and AD using the *FindMarkers* function in *Seurat* [32]. Default parameters for DGE in *Seurat* were used (Wilcoxon rank sum test); the statistical cut-off was set at 0.05 for a false discovery rate (FDR) adjusted by the Benjamin–Hochberg (BH) method. Pseudo-bulk methods were not applied, as they would lead to reduced statistical power [37]; given the relatively small sample size and number of cells in the Grubman dataset, using pseudo-bulk would limit our ability to detect biologically meaningful differences in gene expression.

2.4.2. Functional Enrichment Analysis

The output of the DGE analysis from *Seurat* was used as input to the STRING database (version 11.5) [38]. Key enrichment output from STRING analysis was visualized using the *Enrichplot* package (version 1.18.4) [39]. STRING uses the Bonferroni method to correct for multiple comparisons and provides adjusted *p*-values [38]. The significance cut-off was set at 0.05 for FDR.

2.4.3. Regression Analysis of NP Transcripts and ADNPs Co-Expression

Regression analysis was used to discern the relationship between gene transcript levels and the presence of NP. Utilizing the *glm* function in R, general linear models were constructed for each differentially expressed gene (DEG), with transcript levels as the response variable and the number of co-expressed NPs as the explanatory variable. The BH method was used to adjust for multiple comparisons for all *p*-values; 0.05 was used as the significance cut-off for adjusted *p*-values. The goal was to identify genes whose expression is notably influenced by the increased number of co-expressed NPs.

2.4.4. Hypergeometric Test of Cell Abundance and ADNP Co-Expression

The hypergeometric test was employed using the *phyper* function in R to assess the overlap of genes with decreased expression in AD compared to those that are functionally enriched in HNP neurons [40]. Specifically, the following were defined: (1) the number of overlapped genes as "successes" in our sample ($x = 25$); (2) genes with significantly decreased expression in AD MNP neurons as "successes" in the population ($m = 91$); (3) the total number of unique genes expressed by AD MNP and control HNP neurons, minus those with decreased expression in AD MNP neurons, as "failures" in the population (n = 19,430); and (4) genes with significantly increased expression in HNP neurons as the sample size (k = 307). Because *phyper* is a cumulative distribution function, the conduction of a one-tailed such analysis (*lower.tail = FALSE*) would calculate a probability of observing as extreme and more extreme results in the direction of higher values (*p*-value) [40]. The significance cut-off was set at 0.05.

2.5. Spatiotemporal Correlation Analysis Between ADNP Gene Expression and Age in the Human Brain

In this study, aging is proxied by pseudo-aging, defined as a cross-sectional approach that simulates the effects of aging by analyzing samples from individuals of varying ages at the time of death. This method enables the assessment of age-related changes at a single point in time, addressing ethical concerns and the practical impossibility of longitudinally collecting brain tissues from multiple regions in human subjects.

Gene lists of NPs and ADNPs were downloaded from existing publications [19,41]. The TPM count matrix of spatiotemporal RNA-seq data and metadata of human brains were downloaded from GTEx [35]. All TPM counts were log-transformed. The following exclusion criteria were applied to ensure data quality for the analysis. First, individuals lacking complete metadata, including age, sex, and death classification, were removed from the dataset. Second, to ensure only high-quality RNA samples were used, subjects with Hardy scores of 3 or 4, indicating intermediate or slow death, and those with RNA integrity numbers (RIN) lower than 6, were excluded from the analysis. Finally, as brain development is known to continue throughout the early 20s [42], subjects aged 20 to 29 years were excluded to focus on age-related changes in the mature brain. The relationship between the expression of individual ADNPs and their cumulative expression with respect to age was investigated using the *PResiduals* package (*megabot* function; version 1.0-1) [43], adjusting for RIN and Hardy scale. The significance cut-off was set at 0.05. Analyzing the cumulative expression of ADNPs, in addition to individual NPs, provides a biologically relevant (overall burden and decreased population of neurons/cells co-expressing them) and robust measure (combined effect of multiple ADNPs) of their changes collectively during aging, while also improving the signal-to-noise ratio.

2.6. Examination of ADNP-Co-Expressing HNP (AHNP) Neurons Across Microdissected Brain Regions

The file used in this study was downloaded from the GitHub link provided by Siletti et al. [12]. Firstly, the presence of ADNPs from the file was quantified, and counts for NPs were generated further to calculate non-ADNPs. Based on these counts, we defined ADNP-HNP (AHNP) neurons as those tagged with 6+ ADNPs and <3 non-ANDPs, accounting for the difference in the input NPs [12,19]. Concurrently, we estimated the number of cells in different brain regions and dissections based on percentage data extracted from the *cluster_annotation* file. These estimates were summed across unique regions and dissections to provide a granular view of cell distribution. Additional analyses were conducted on specific regions for MEC, where the dataset was grouped by various attributes such as neurotransmitter, subtype, and MTG label [44], and the number of cells in each group was summed and visualized.

3. Results

3.1. Alterations of HNP Neuronal Abundance and Functions in AD: Overlap of HNP Dysfunction and Molecular Signature of AD

Analyzing the single-cell dataset by Grubman et al. [19,25,45], we showed a very strong correlation (>96%) between transcript abundance and the number of co-expressed NPs generally exists for neurons in both control and AD groups (Figure 3A). This effect was also observed (>90%) in both the Leng and MIT ROSMAP Multiomics datasets (Figure S2). Applying the number of co-expressed NPs as a proxy for transcript levels of NPs, we stratified neurons into low (0–1), medium (2–5), and high (6+) NP groups. We observed a similar absence of HNP neurons in AD, which was also corroborated by the Leng and MIT ROSMAP Multiomics datasets (Figure 3B; Table S4; Figure S3) [19]. As explained in the Methods section, only the Grubman dataset was used for the following analyses.

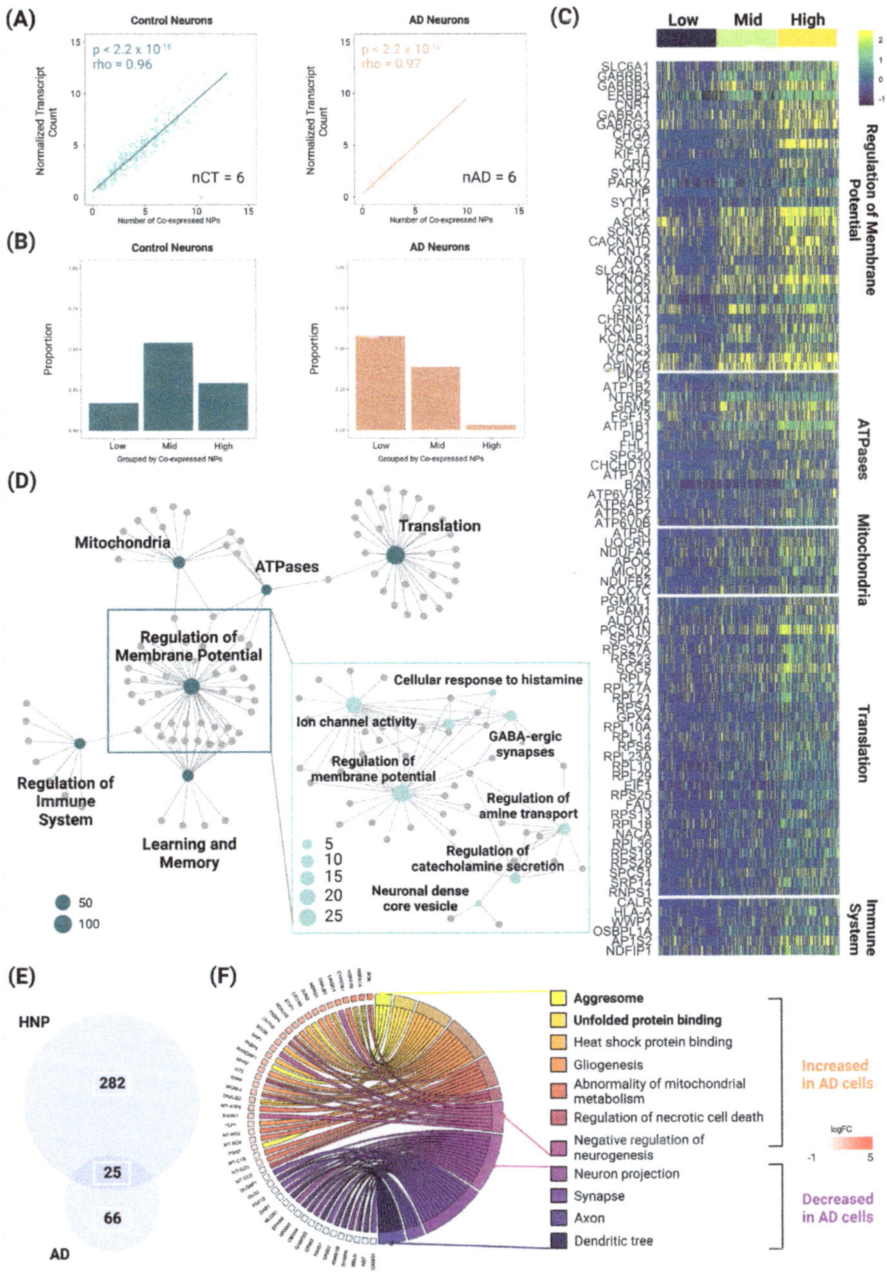

Figure 3. Changes in high-neuropeptide-producing (HNP) neuronal abundance and function: overlap of dysfunction and AD molecular signatures. (**A**) The relationship between transcript abundance and the number of co-expressed neuropeptides (NPs) in neurons from both control and AD EC. (**B**) The

distribution of neurons based on the number of co-expressed NPs: low: 0–1; middle (mid): 2–5, and high: 6+. Proportion = in-group neuron counts in the condition/total neuron count in the condition. (**C**) Heat plot showing differentially expressed genes (increased) in HNP neurons (neurons in high NP co-expression group). (**D**) Gene network plot showing results of functional enrichment for HNP neurons. (**E**) Venn diagram showing the overlap between gene expressions higher in HNP neurons (neurons in high NP co-expression group that express 6+ NPs) and significantly decreased in AD neurons co-expressing 2–5 NPs (MNP neurons). The hypergeometric test was applied to evaluate the overrepresentation of genes upregulated in control HNP neurons but notably reduced in AD MNP neurons; $p < 0.00001$; $\alpha = 0.05$. (**F**) Molecular signatures differentially increased and decreased in AD MNP neurons. nCT, number of healthy donors; nAD, number of AD donors.

As we hypothesized that HNP neurons would experience greater energetic demands and higher metabolic stress, we predicted that HNP neurons would show more metabolic activity. To test this, differentially expressed genes (DEG) were examined between high and low groups of control neurons followed by enrichment analysis [19,38]. We found that all the NPs expressed at significantly higher levels in HNP neurons were ADNPs [19], and genes required for NP transportation, translation, and metabolic processes were significantly increased in HNP neurons in comparison to LNP neurons (Figure 3C,D; Tables S1, S5 and S6). Although it is widely known that GABA is often co-expressed with NPs [46], we observed that HNP neurons also participated in other chemical communications, such as using histamine and catecholamines, to regulate membrane potential (Figure 3D; Table S5). In addition, ~36% of DEGs regulating membrane potentials were functionally enriched for learning and memory (Figure 3D; Table S5). Surprisingly, the regulation of innate immune response was increased in HNP cells (Figure 3C,D; Table S5).

To elucidate the increase observed across the LNP, MNP, and HNP groups, we sought to identify genes whose expression was significantly influenced by the level of NP co-expression, as a proxy for the abundance of NP transcripts. We utilized regression models to examine the relationship between gene expression and NP co-expression and then ranked the increased DEGs in HNP neurons by coefficients and R^2. Excluding NP components, genes ranked in the top 10 for coefficients or R^2 included sodium/potassium-transporting ATPase [47] and intracellular transport vesicles [48], supporting the hypothesized functional enhancement of HNP neurons (Table S7). Notably, *ERBB4*, the protein products of which induce tau hyperphosphorylation [49], was among the top genes related to NP co-expression (Table S7; Figure S4).

Because we discovered that HNP neurons exhibited higher performance in several expected cellular functions—including transportation, translation, and metabolic processes—and participated more heavily in other chemical communications, regulation of innate immune response, and circadian rhythm, we wondered whether dysregulation of processes in these functions may lead to the loss of neuronal functions and the accumulation of tau pathology, which are hallmarks of several neurodegenerative disorders including AD.

Speculating that disrupted functions of neurons expressing more NPs are associated with protein misfolding, we examined the DEGs for neurons stratified by the number of co-expressed NPs and analyzed the neurons in the medium group (as HNP neurons were virtually absent in AD brains). Enrichment analysis revealed that these cells displayed molecular characteristics related to both "positive" and "negative" neuropathology in AD [2]. We note the decreased DEGs in AD cells were significantly enriched for those functionally increased in HNP cells ($p < 0.00001$; Figure 3E; Tables S6 and S8), indicating that loss of HNP functions participates in the molecular pathogenesis of AD. Genes with protein products showing significantly decreased expression included those with functional roles in axons, synapses, and dendrites (Figure 3F; Table S8). Increased molecular processes included those known to be disturbed in AD, such as negative regulation of neurogenesis, gliogenesis, and abnormal mitochondrial metabolism (Figure 3F; Table S9). Notably, genes involved in forming aggresomes and unfolded protein binding were highlighted [50], indicating the active occurrence of protein misfolding in AD cells that co-express NPs (Figure 3F; Table S9).

3.2. ADNP Dysfunction Observed in Early Pathogenesis of AD

To test the hypothesis that changes in the abundance of HNP neurons expressing AD-NPs contribute to the selective vulnerability of brain regions and neuronal subpopulations in AD, we systematically analyzed three distinct datasets encompassing different stages and aspects of AD development. These datasets included: (1) the progression of cognitive impairment from normal aging to mild cognitive impairment (MCI) and finally to AD (the MIT ROSMAP Multiomics dataset) [21,51], (2) the advancement of AD neuropathology through different Braak stages (the Leng dataset) [26,33], and (3) the effects of aging, a major risk factor for AD, on ADNP expression in various brain regions (the GTEx v8 dataset) (Figure 4A) [28]. By integrating findings from these diverse datasets, we aimed to provide a comprehensive understanding of how changes in ADNP expression in HNP neurons relate to the spatiotemporal progression of AD pathogenesis. Given the disproportionate absence of neurons expressing ADNPs in the EC of AD brains and the higher level of ADNPs physiologically expressed by HNP neurons, we specifically focused on ADNPs in the following analyses.

Analysis of the MIT ROSMAP Multiomics dataset revealed that, despite the majority of neurons not expressing any NPs (Figure S1), patterns of the ADNP expression by neurons were similar to those observed in the Grubman dataset when considering cognitive status and AD. Specifically, both the MCI and AD groups had significantly more neurons co-expressing 0–1 ADNPs (CT vs. MCI p-value = 0.025; CT vs. AD p-value = 0.025) and significantly fewer neurons expressing 6+ ADNPs compared to cognitively normal individuals (CT vs. MCI p-value = 0.032; CT vs. AD p-value = 0.014) (Figure 4B). But they were not statistically different from each other at any point (Low MCI vs. AD p-value = 0.36; Mid MCI vs. AD p-value = 0.40; High MCI vs. AD p-value = 0.60) (Figure 4B). We note that most of the MCI donors (6/8) had a Braak stage of 1–2 (Table S2). Similarly, although the Leng dataset had even fewer neurons expressing NPs (Figure S2), results from visualizing the ADNP co-expression and transcript levels were consistent with the observations from the MIT ROSMAP Multiomics dataset, clearly demonstrating a depletion of HNP neurons expressing ADNPs in donors with Braak stages 2 and 6 (Figure 4C). These results provided further evidence that the loss of HNP neurons expressing ADNPs occurs early in the AD neuropathological process. Overall, it was surprising to find that as early as Braak stage 2, there was already a stark contrast between control and Braak 2 donors in ADNP expression patterns similar to those observed late in AD.

To investigate even earlier in the disease process and study the spatiotemporal changes of ADNPs in the context of aging, a major risk factor for AD, we analyzed bulk RNA-sequencing data generated by the GTEx consortium [28]. We expected that if HNP neurons expressing ADNPs diminish during the aging process, a corresponding decrease in ADNP expression levels would be observed in this dataset, particularly in brain regions affected early in AD. Our recent report indicated that ADNP expression decreased with age in the hippocampus [19]; however, we do not yet know if the decline of NPs with aging is ADNP- and brain region-specific. If our hypothesis were to hold, only brain regions affected by early AD should show age-related changes in ADNP expression, and only the accumulative expression of ADNPs, but not other NPs, should decrease with age in the human brain. We first examined the expression of ADNPs during aging among 11 brain regions selected from GTEx (Table 1) [19,35]. As expected, we found that only the hippocampus, frontal cortex, anterior cingulate gyrus, and amygdala—all of which are brain regions affected by early AD [52–57]—showed a significant decrease of ADNP transcription during aging among the 11 brain regions selected from GTEx (Table 1; Figure 4D) [3,19,35].

We also examined the expression of NPs in these brain regions and whether the expression of individual NPs was correlated with age. We found that NPs demonstrating a significant decrease in expression with age in the aforementioned brain regions consisted mostly of ADNPs (Tables S1 and S10) [19]. To support the specificity of the observed change in ADNP expression, we analyzed the transcript count of all non-ADNP NPs in the brain regions and found no age-related changes specific to AD-related regions in their overall levels (Table S11). We confirmed that the larger decrease of NP expression in these brain regions with aging was not attributed to their intrinsic capacity to express NPs: (1) Using this dataset and the NP list compiled previously, ~80 NPs were expressed by each brain region (Table S12); (2) the differences in the total NP transcript count among the surveyed brain regions do not explain the AD-specific patterns observed in Table 1 (Table S12). In short, the spatiotemporal transcriptomic analysis presented here showed that the decrease of NPs in the aging human brain was specific to brain regions and NPs implicated in AD, supporting the idea that age-related cognitive decline shares mechanisms with AD and may be mediated by loss of ADNP expression during aging.

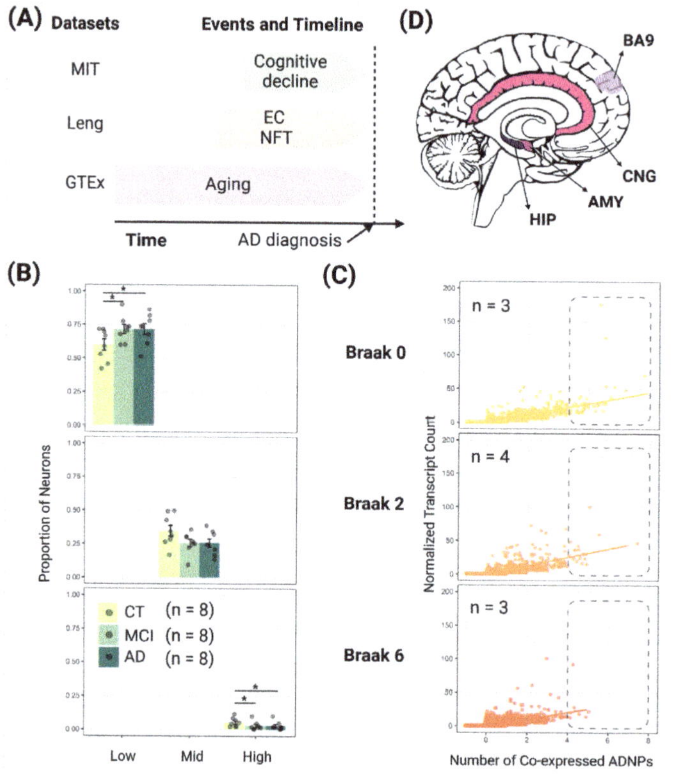

Figure 4. Decreased expression of Alzheimer's disease-associated neuropeptides (ADNPs) across brain regions during disease progression and aging. (**A**) Timeline of datasets relative to typical diagnosis of AD. (**B**) Comparison of proportions of neurons from the entorhinal cortex (EC) of control (CT), mild cognitive impairment (MCI), and AD donor brains (MIT, MIT ROSMAP Multiomics dataset), stratified by the number of co-expressed ADNPs (Low: 0–1, Mid: 2–5, High, 6+). One-tailed Wilcoxon rank sum test was used. $\alpha = 0.05$. * $p < 0.05$. (**C**) Scatterplot showing the relationship between transcript abundance and the number of ADNPs in neurons from donor brains classified as Braak stages 0, 2, and 6. The reduced abundance of neurons co-expressing higher levels of ADNPs during AD progression is lighted. (**D**) Brain regions (medial sagittal view) showing significant decrease in ADNPs with aging. AMY, amygdala; CNG, anterior cingulate cortex; HIP, hippocampus; BA9, Brodmann area 9; n, sample size.

Table 1. Significantly decreased expression of ADNPs with aging only occurred in early AD-impacted regions.

Brain Region	Sample Size	Correlation	p-Value	Significance
Hippocampus	103	−0.30	0.00	*
Frontal cortex (BA9)	116	−0.34	0.00	*
Anterior cingulate cortex (BA24)	88	−0.30	0.01	*
Amygdala	73	−0.30	0.02	*
Hypothalamus	109	0.02	0.85	
Caudate basal ganglia	146	−0.08	0.28	
Nucleus accumbens basal ganglia	143	0.00	0.96	
Putamen basal ganglia	126	0.11	0.25	
Substantia nigra	66	−0.23	0.08	
Cerebellar hemisphere	141	−0.03	0.74	
Cerebellum	156	−0.08	0.34	

ADNP, Alzheimer's disease (AD)-associated neuropeptides; BA, Brodmann area; *, p-value < 0.05. Conditional test for association [43], $\alpha = 0.05$.

3.3. Physiological Distribution of AHNP Neurons May Mediate Brain Region Vulnerability to AD

"Epicenters" have been described as brain regions showing the most significant pathological alterations, hypothesized to be the initial site of disease onset [7,8]. Stressed "nodes" are known as brain regions with high network traffic, also referred to as "hubs", that experience activity-induced deterioration that can lead to or exacerbate diseases [10,58]. While both concepts are instrumental in theories of AD etiology devised by connectome and network-based studies [7,8,10,58], the cell types underlying these "epicenters" and stressed "nodes" are unclear. Based on the regions that display reduced ADNP expression with aging, we propose that AHNP neurons serve as one of the cellular components of the "hubs" and "epicenters" leading to the onset and/or progression of AD. To test this, we analyzed the distribution of AHNP neurons across various regions of the human brain. Two potential observations and implications exist: (1) AHNP neurons ubiquitously exist in all brain regions, but those in early-AD-impacted regions are, regardless of the underlying cause, more susceptible to dysfunction than others, or (2) AHNP neurons are preferentially distributed in early-AD-impacted brain regions to physiologically perform cognitive functions, but they are more prone to dysfunction than other cell types, therefore mediating the regional vulnerability to AD with aging-related cell dysfunctions. We predict that a single-cell transcriptomic survey of brain regions would show that AHNP neurons are more abundantly distributed in early-AD-affected brain regions that engage extensively in memory and executive functions, such as the entorhinal cortex, hippocampus, and basal forebrain [59–61]. However, if the first scenario were true, our hypothesis could be negated altogether.

A recent study published single-cell transcriptome data from ~100 dissections across the forebrain, midbrain, and hindbrain of human donors and classified brain cells into 461 clusters [12]. The authors also compared their cell clusters with previous publications that used NP diversity to classify neurons [12,44]. We first investigated the top brain regions where AHNP neurons existed using cluster annotations provided by Siletti et al. and calculated the number of neurons in each of the top three brain regions and microdissections (see Section 2) [12]. As anticipated, the amygdala and hippocampus—where the age-associated decrease of ADNP expressions was observed [19]—were among the top five brain regions (Figure 5A; Table 2). While previous studies identified the hypothalamus as considerably relevant in AD [62,63], the hypothalamus did not show an age-related decrease in ADNPs. We also found that the cerebral cortex contained the most AHNP neurons, but this was likely due to the number of dissections assigned to the cerebral cortex. To overcome this issue, we set out to identify the distribution of AHNP neurons among cortical regions separately by ranking the cortical dissections based on the number of AHNP neurons. Again, we found that medial and lateral EC (MEC and LEC) ranked among the top five cortical regions (Figure 5A; Table 3). In particular, the MEC harbored the highest number of AHNP neurons among all micro-dissected regions (Table S13). In contrast, very few AHNP cells were found in the cerebellum, spinal cord, and medulla, which are generally thought to be spared by AD neuropathology (Table S14), further supporting our hypothesis that dysfunction of AHNP cells mediates the brain region-specific vulnerability to AD.

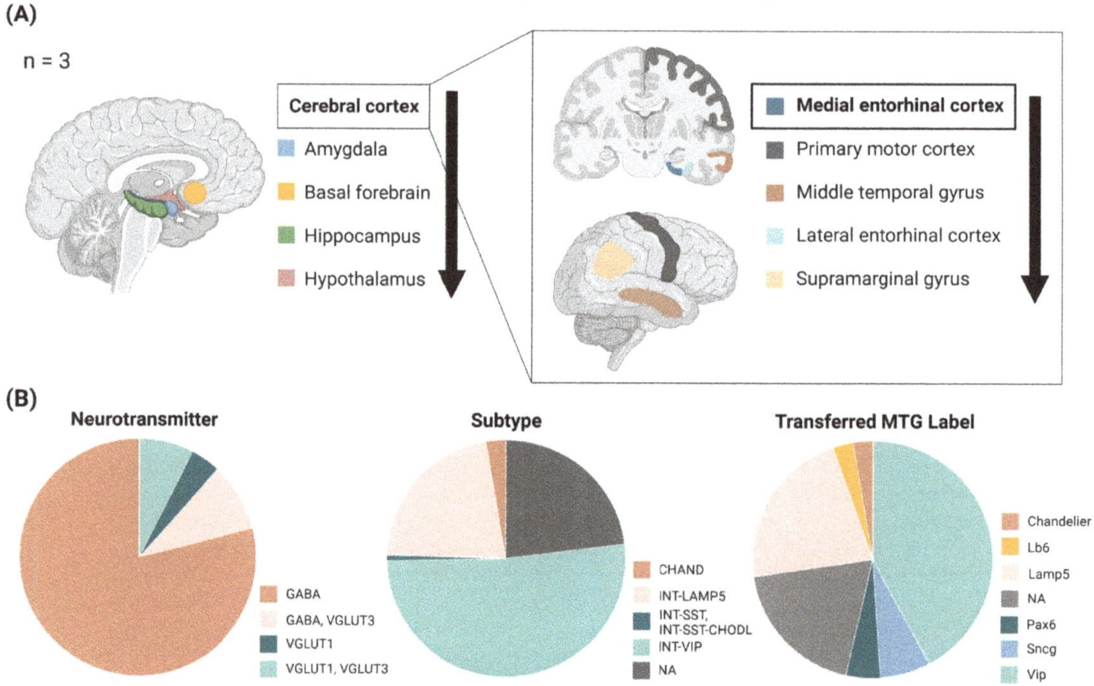

Figure 5. Local density of neurons co-expressing high levels of Alzheimer's disease-associated neuropeptides (AHNP neurons) may govern brain vulnerability to AD. (**A**) Top five brain and cortical regions ranked by AHNP neuron abundance are visualized. n, sample size. Arrows indicate the decrease of abundance. (**B**) Categorization of AHNP neurons predominantly found in EC by neurotransmitter, subtype, and transferred MTG label (common cell type nomenclatures for the medial temporal gyrus of the mammalian brain) [12].

Table 2. Top five brain regions ranked by AHNP neuron abundance.

Brain Regions	AHNP Neuronal Count
Cerebral cortex	218,145
Amygdala	72,256
Basal forebrain	45,559
Hippocampus	40,870
Hypothalamus	23,217

Table 3. Top five cortical regions ranked by AHNP neuron abundance.

Cortical Regions	Full Cortical Description	AHNP Neuronal Count
MEC	Anterior parahippocampal gyrus, posterior part (APH)—Medial entorhinal cortex	19,206
M1C	Precentral gyrus (PrCG)—Primary motor cortex	11,821
MTG	Middle Temporal Gyrus	10,654
LEC	Anterior parahippocampal gyrus (AG)—Lateral entorhinal cortex	2625
A40	Supramarginal gyrus (SMG)—A40	1732

As the EC—the earliest and most heavily affected region in AD neuropathology—harbored the most significant population of AHNP neurons, we sought to investigate the specific cell types, as previous investigation has suggested that most ADNP co-expressing neurons are GABA-ergic [19]. Specifically, we aimed to understand how they align with established neuronal cell types. We selected neurons that had MEC in the top three dissections and examined the proportions of neuron types, including based on the neurotransmitter, sub types, and transferred MTG labels (common cell type nomenclatures for the MTG regions of the mammalian brain) [12]. We found that most of the AHNP neurons were indeed GABA-ergic interneurons; however, about a quarter of the neurons have not been described before (Figure 5B).

4. Discussion

Why some ubiquitously expressed proteins (e.g., tau and Aβ) exhibit selective accumulation in particular regions of the brain and cells, yet spare their comparable neighbors, is a fundamental question of AD research. Examining multiscale and spatiotemporal RNA-seq data from 1890 human brain samples, we aimed to gain a comprehensive understanding of the potential mechanistic roles that NP-intensive neurons that co-express high levels of NPs (HNP neurons) play in mediating the selective vulnerability of brains to AD.

To address our previous hypothesis that HNP neurons, given their secretory and peptidal signaling functions, would demand more translation and transportation, leading to increased metabolic vulnerability [19], we investigated the specific characteristics of HNP neurons and the link between HNP neuron dysfunction and the hallmark molecular indicators of AD. We note that the criteria for HNP neurons in this study were primarily based on the co-expression of AD-associated NPs (ADNPs), which may be subject to change with advancements in sequencing technology, the availability of more datasets, and the discovery of additional NPs. Due to the significant variability in the proportion of neurons expressing NPs across different datasets (see Section 2), we utilized the NP list compiled in our previous study to maintain consistency in the follow-up analysis on the Grubman dataset [19,25]. To facilitate the incorporation of more NPs and additional datasets as more data are generated and datasets with better NP characterization become available, we have made the source code used in this study available for reanalysis. In addition, using co-expressed ADNPs to label neuronal subpopulations was of timely value for AD research because many neuronal subpopulations have complex combinatorial expression of NPs, and many of them exist outside of characterized cell populations [12,64].

Acknowledging these complexities, we found that (1) HNP neurons were more metabolically active and had gene expression profiles suggesting higher connectivity; (2) MEC AD neurons co-expressing higher levels of NPs showed the molecular signature of AD, including protein misfolding; and (3) the deficiency of AD cells was linked to loss of function of HNP cells. While we anticipated a greater metabolic burden as a source of neuronal vulnerability for HNP neurons, it was surprising to discover that HNP neurons can be predisposed to tau hyperphosphorylation and misfolding [49,65]. In addition to

ERBB4, recent research has revealed that the impairment of NMD, an elevated process observed in HNP cells and suppressed by cellular stress, mediates tau-induced neurotoxicity [66]. Therefore, we posit that the disruption of cellular processes elevated in HNP neurons could occur more readily/earlier in these cells and contribute to the formation of misfolded proteins within them, subsequently leading to their selective degeneration.

To test the idea that the loss of ADNPs expressed by HNP neurons could participate in early AD development and/or progression, we considered the perspectives of neuropathology, cognitive status, and spatial progression during aging. Although we anticipated a decrease in ADNP expression with age and AD progression, it was surprising to observe this phenomenon in donors with neuropathological changes as early as Braak stage 2. While the majority of MCI donors in the MIT ROSMAP Multiomics dataset presented neuropathology falling in Braak stages 1 and 2, deviating from the typical distribution where most MCI patients fall between stages 3–4 [67,68], the significant decline in HNP neuron abundance among MCI brains in this dataset further substantiates the early alteration of ADNP expressions associated with AD progression. Despite both datasets being unsuitable for the functional characterization of HNP neurons (as detailed in the Section 2), our findings indicate that the ADNPs identified in the Grubman dataset, when analyzed alongside other single-cell sequencing datasets, highlight HNP neurons relevant to Alzheimer's disease (AD). This suggests that the loss of ADNPs may represent an early molecular alteration in the brains of individuals at risk for developing AD. We note that while the disproportionate absence of HNP/ADNP co-expressing HNP (AHNP) neurons was described as selective degeneration in this manuscript, with the increased proportion of cells not expressing any NPs observed in multiple datasets, the degeneration could be interpreted as neuronal death and/or loss of functions. The definitive interpretation will need longitudinal tracking of neurodegeneration at the single-neuron level.

If AHNP neurons were indeed important for early AD, their presence in physiological conditions should help explain the regional vulnerability seen in AD. The preferential localization of AHNP neurons in regions of the brain that are susceptible to AD, such as the medial entorhinal cortex, amygdala, basal forebrain, and hippocampus, further underscores their potential role as key cellular contributors to the disease's pathology. Two brain regions from the analyses were surprising to us—the hypothalamus and primary cortex (M1). Firstly, while the hypothalamus was highlighted in the regional distribution of AHNP neurons and demonstrated atrophy in AD [69], its expression of ADNPs did not decrease significantly with age. This difference may be attributed to AD-specific changes in hypothalamus being more disease-specific than age-related. The initial surprise at finding AHNP neurons concentrated in the M1 stems from the fact that motor deficits typically manifest later in AD [70]. However, the absence of overt motor symptoms in AD can be attributed to compensatory neural rewiring and hyperexcitability in the motor regions, rather than the non-existence of neuropathology [71,72]—this is not mutually exclusive with the proposed selective degeneration, as the term could be interpreted as loss of function rather than cell death, as discussed above. Notably, variant AD with abnormal tau accumulation in the M1 has been reported [73]. As AHNP neurons likely play a role in maintaining M1 microenvironment homeostasis, their dysfunction could exacerbate regional vulnerabilities, potentially contributing to upper motor neuron dysfunction and other neurodegenerative diseases, such as amyotrophic lateral sclerosis (ALS). The presence of neurofibrillary tangles in ALS [1,74] and the molecular signatures of AHNP neurons predisposing them to tau misfolding suggest that these neurons may be relevant in understanding mixed pathologies, common susceptibilities, and overlapping etiologies across neurodegenerative disorders. Considering the changes of AHNP neurons in MCI donors and the alterations in expression of ADNPs within cognitive regions during aging, an alternative to our hypothesis could be that these neurons increase the risk of cognitive decline and tau-related neurodegenerative diseases, rather than being AD-specific. While this perspective necessitates a re-evaluation of our original hypothesis, it may be of greater importance, as it could potentially address the prevention of a spectrum of neurodegenerative diseases, including AD.

Our findings contribute to the broader framework of AD theories and our understanding of the transition from brain function to dysfunction. Macroscopically, they support the "nodal stress" hypothesis [9,10,58], which posits that brain regions with strong anatomical connectivity are particularly vulnerable to damage due to their heightened susceptibility to cytotoxic events. This hypothesis aligns with the understanding that certain brain regions, which accumulate pathologically associated proteins, are also those that contain the most vulnerable cell types and are often the earliest affected in the disease's progression [7–9]. Microscopically, HNP neurons with enhanced connectivity and high metabolic activity may be especially susceptible to the deleterious effects of cytotoxic events and the presence of misfolded proteins [3,58]. The observed decline in ADNP expression with age, coupled with the physiological distribution of HNP neurons in regions vulnerable to Alzheimer's disease, suggests a critical link between macroscopic brain vulnerability and specific microscopic cellular components. This connection implies that while regional vulnerabilities can be observed at a larger scale, they may be driven by the dysfunction of particular cell types, such as HNP neurons. Furthermore, cellular dysfunctions, marked by distinct molecular pathways and biochemical properties, play a crucial role in the onset and progression of AD [9]. Extensive research over the past decade has illuminated the connectomic landscape of neurodegenerative diseases, enhancing our understanding of the brain regions and neural circuits that are susceptible to AD [7,8,58,75]. However, the specific cell types and cellular mechanisms that underpin these connectomic findings remain largely unexplored. Our findings potentially bridge this gap by highlighting the importance of HNP neurons in the context of AD, emphasizing their potential role in both the physiological and pathological processes that characterize the disease.

ADNPs are known to play important roles in cellular processes crucial to the pathogenesis of neurodegeneration, such as mitochondrial dysfunction, persistent neuroinflammation, and disrupted circadian rhythm (briefly summarized by Li and Larsen) [19]. Together with the mechanistic considerations, the alterations in the abundance and diversity of ADNP-producing neurons during different aspects of AD progression as well as the spatially specific reduced expression of ADNPs during aging identified in this study indicate that decreased expression of ADNPs by neurons may accelerate, or even drive, the progression of protein misfolding, cognitive decline, and neurodegeneration in AD.

Given the limited regenerative capacity of neurons, early detection and treatment of AD are paramount. Our research highlights the relevance of ADNPs as combinatorial and longitudinal biomarkers to evaluate the risk and progression of AD development. For instance, an analysis of the CSF proteome identified that CHGA and VGF exhibited significant differences in abundance among the CT, MCI, and AD groups [76]. As such, CSF levels of ADNPs could serve as direct biomarkers for AD. As aging progresses, peripheral tissues could also be valuable for tracking these changes. Specifically, monitoring ADNP levels in more accessible tissues such as blood, skin, or saliva could provide a less invasive method for longitudinal tracking of AD progression and response to treatments. Even though ADNPs may be secreted by other tissue types, there could be proxy peripheral blood biomarkers reflecting changes of ADNPs in the brain. This potential connection can be explored through longitudinal studies correlating CSF ADNP levels with blood biomarker identification, such as gene expression panels. Such an approach could lead to the development of less invasive and more accessible diagnostic and monitoring tools for AD. Additionally, further studies validating these findings and investigating the underlying mechanisms responsible for the observed decline in AHNP neuron abundance are needed. Evaluating whether interventions aimed at mimicking and/or preserving the ADNP signaling network could impede the progression of AD is also valuable, especially considering the subtle differences in GPCR expression observed between CT and AD [19]. Collectively, these observations suggest that deficiencies in ADNPs with aging contribute to AD development and progression and that these deficits could be a consequence of losing AHNP neurons during aging, which underlies the shared cognitive decline during aging and in early AD.

It is crucial to acknowledge that the interpretation and generalizability of these results may be limited by the lack of diversity in the dataset. The sole dataset available for this analysis was derived exclusively from brain samples of white male individuals. It is well-established that gene expression varies both within and between different ethnic groups [41,77] and sexes [78–80]; future research collecting more inclusive datasets is needed to ensure that knowledge gained from data analysis benefits the wider population. As NPs and their signaling activities are highly influenced by sex hormones [78–80], such datasets could help understand the epidemiology of AD and the differential accumulation of tau proteins between sexes during the aging process [1,81]. Other potential confounding variables, such as transcription factors and epigenetic mechanisms, warrant careful consideration. Recent research has made tremendous advancements on providing a comprehensive multiomic brain atlas for physiological conditions [12,82,83]. Our findings underscore the importance of extending these efforts to investigate the molecular landscapes of brains in various disease states. By understanding how dysfunction in HNP neurons occurs with aging and its detrimental effects on interconnected behavioral domains, we can gain valuable insights into the cell types and dysfunctions at the early stages of AD to develop effective disease models as well as cell-type-specific targeting strategies for prevention and therapy.

5. Conclusions

NPs play essential roles in cellular communication and homeostasis but can confer metabolic burdens due to their synthesis. This study demonstrates that NP-intensive neurons display multifaceted properties associated with AD vulnerability: (1) characterized by heightened metabolic activity and susceptibility to tau hyperphosphorylation; (2) exhibiting disproportionate and site-specific depletion during early AD progression, manifesting in both cognitive and neuropathological alterations; and (3) showing spatial distribution that correlates with AD-vulnerable brain regions under physiological conditions. We conclude that NP-intensive neurons likely participate in AD development and early progression. While causality studies are warranted to substantiate these findings, this avenue of research is promising, as understanding the roles of NPs in neuronal and cellular vulnerability of AD could facilitate earlier detection and intervention.

Supplementary Materials: The following supporting information can be downloaded at: https://www.mdpi.com/article/10.3390/biom14121518/s1, Figure S1: Neurons from Leng and MIT ROSMAP Multiomics datasets showed insufficient neuronal populations expressing neuropeptides (NPs); Figure S2: The transcript level of neuropeptides (NPs) is highly correlated with the number of co-expressed NPs; Figure S3: Disproportionate absence of neurons in the high neuropeptides (NPs)-producing group where 6+ NPs are co-expressed; Figure S4: *ERBB4* expression shares a significant positive correlation with the number of co-expressed neuropeptides (NPs) in neurons; Table S1: List of Alzheimer's disease-associated neuropeptides (ADNPs); Table S2: De-identified metadata for individuals and experiments included from the MIT ROSMAP Multiomics dataset; Table S3: De-identified metadata for individuals and experiments included from the GTEx dataset; Table S4: Cell count for different cell types co-expressing various levels of neuropeptides (NP) in Alzheimer's disease (AD) and control (ct) donor brains from the Grubman dataset; Table S5: The molecular process increased in control HNP neurons in comparison to control LNP neurons; Table S6: Genes with significantly increased expression in HNP neurons; Table S7: Summarization of generalized linear regression analysis of gene expression and number of co-expressed neuropeptides (NPs); Table S8: Genes showing significantly decreased expression in AD MNP neurons; Table S9: Molecular process increased in Alzheimer's disease MNP neurons in comparison to control MNP neurons; Table S10: Alzheimer's disease-associated neuropeptides (ADNPs) show decreased expression with aging in early AD-impacted regions; Table S11: The expression of neuropeptides (NPs) that are not Alzheimer's disease-associated NPs (ADNP) across brain regions with aging are not specific to early AD regions; Table S12: Neuropeptide (NP) expression and their change with age in each brain region; Table S13: Distribution of AHNP neurons across micro-dissected brain regions; Table S14: Distribution of AHNP neurons across brain regions.

Author Contributions: M.L. conceived and designed the study. M.L. wrote the manuscript. M.L. developed and performed bioinformatic analyses and data visualization. N.F. and M.L. organized the code used for bioinformatics and data analysis. N.F. validated the bioinformatic analysis and verified the reproducibility of results and outputs. M.L., N.F. and P.A.L. discussed the results. M.L., N.F. and P.A.L. revised the manuscript. All authors have read and agreed to the published version of the manuscript.

Funding: M.L. was supported by P.A.L.'s discretionary funds and the Doctoral Dissertation Fellowship from the Graduate School Fellowship Office at the University of Minnesota.

Institutional Review Board Statement: Not applicable.

Informed Consent Statement: Not applicable.

Data Availability Statement: Datasets utilized in this manuscript are publicly available as stated in the availability of data and materials. Three publicly available single-cell RNA-sequencing datasets of the human EC were included for analysis in this study: Grubman et al. (http://adsn.ddnetbio.com, accessed on 31 May 2022), Leng et al. (https://cellxgene.cziscience.com/datasets, accessed on 21 May 2024), and the MIT ROSMAP Single-Nucleus Multiomics Study (https://adknowledgeportal.synapse.org/Explore/Studies/DetailsPage/StudyDetails?Study=syn52293417, accessed on 2 October 2023). GTEx v8 data can be downloaded from the GTEx portal (https://gtexportal.org/home/downloads/adult-gtex/bulk_tissue_expression, accessed on 3 October 2022). Single-cell sequencing of microdissected brain regions by Siletti et al. can be found on GitHub (https://github.com/linnarsson-lab/adult-human-brain, accessed on 18 May 2024). Code to reproduce findings can be found on GitHub (https://github.com/mancili/HNP/).

Acknowledgments: Alice Larson and Roxanne Larsen provided helpful comments and suggestions that strengthened this manuscript. The Genotype-Tissue Expression (GTEx) Project was supported by the Common Fund of the Office of the Director of the National Institutes of Health and by NCI, NHGRI, NHLBI, NIDA, NIMH, and NINDS. The results published here are in whole or in part based on data obtained from the AD Knowledge Portal. The data available in the AD Knowledge Portal would not be possible without the participation of research volunteers and the contribution of data by collaborating researchers. The results published here are in whole or in part based on data obtained from the AD Knowledge Portal (https://adknowledgeportal.org/). Study data were generated from postmortem brain tissue provided by the Religious Orders Study and Rush Memory and Aging Project (ROSMAP) cohort at Rush Alzheimer's Disease Center, Rush University Medical Center, Chicago. This work was supported in part by the Cure Alzheimer's Fund, NIH grants AG058002, AG062377, NS110453, NS115064, AG062335, AG074003, NS127187, MH119509, HG008155 (M.K.), RF1AG062377, RF1 AG054321, and RO1 AG054012 (L.-H.T.), and the NIH training grant GM087237 (to C.A.B.). ROSMAP is supported by P30AG10161, P30AG72975, R01AG15819, R01AG17917. U01AG46152, and U01AG61356. M.L. was supported by the Doctoral Dissertation Fellowship granted through the Graduate School Fellowship Office at the University of Minnesota. P.A.L. provided discretionary funds supporting M.L. Figures were organized using BioRender (biorender.com).

Conflicts of Interest: All authors declare no conflict of interest.

References

1. Better, M.A. Alzheimer's disease facts and figures. *Alzheimers Dement.* **2023**, *19*, 1598–1695.
2. Serrano-Pozo, A.; Frosch, M.P.; Masliah, E.; Hyman, B.T. Neuropathological alterations in Alzheimer disease. *Cold Spring Harb. Perspect. Med.* **2011**, *1*, a006189. [CrossRef] [PubMed]
3. Braak, H.; Braak, E. Neuropathological stageing of Alzheimer-related changes. *Acta Neuropathol.* **1991**, *82*, 239–259. [CrossRef] [PubMed]
4. Small, S.A.; Nava, A.S.; Perera, G.M.; Delapaz, R.; Stern, Y. Evaluating the function of hippocampal subregions with high-resolution MRI in Alzheimer's disease and aging. *Microsc. Res. Tech.* **2000**, *51*, 101–108. [CrossRef]
5. Mitchell, T.W.; Mufson, E.J.; Schneider, J.A.; Cochran, E.J.; Nissanov, J.; Han, L.-Y.; Bienias, J.L.; Lee, V.M.-Y.; Trojanowski, J.Q.; Bennett, D.A.; et al. Parahippocampal tau pathology in healthy aging, mild cognitive impairment, and early Alzheimer's disease. *Ann. Neurol.* **2002**, *51*, 182–189. [CrossRef]
6. Bennett, D.A.; Schneider, J.A.; Arvanitakis, Z.; Wilson, R.S. Overview and findings from the religious orders study. *Curr. Alzheimer Res.* **2012**, *9*, 628–645. [CrossRef]
7. Zhou, J.; Gennatas, E.D.; Kramer, J.H.; Miller, B.L.; Seeley, W.W. Predicting regional neurodegeneration from the healthy brain functional connectome. *Neuron* **2012**, *73*, 1216–1227. [CrossRef]

8. Vogel, J.W.; Corriveau-Lecavalier, N.; Franzmeier, N.; Pereira, J.B.; Brown, J.A.; Maass, A.; Botha, H.; Seeley, W.W.; Bassett, D.S.; Jones, D.T.; et al. Connectome-based modelling of neurodegenerative diseases: Towards precision medicine and mechanistic insight. *Nat. Rev. Neurosci.* **2023**, *24*, 620–639. [CrossRef]
9. Fu, H.; Hardy, J.; Duff, K.E. Selective vulnerability in neurodegenerative diseases. *Nat. Neurosci.* **2018**, *21*, 1350–1358. [CrossRef]
10. Saxena, S.; Caroni, P. Selective neuronal vulnerability in neurodegenerative diseases: From stressor thresholds to degeneration. *Neuron* **2011**, *71*, 35–48. [CrossRef]
11. Muratore, C.R.; Zhou, C.; Liao, M.; Fernandez, M.A.; Taylor, W.M.; Lagomarsino, V.N.; Pearse II, R.V.; Rice, H.C.; Negri, J.M.; He, A.; et al. Cell-type Dependent Alzheimer's Disease Phenotypes: Probing the Biology of Selective Neuronal Vulnerability. *Stem Cell Rep.* **2017**, *9*, 1868–1884. [CrossRef] [PubMed]
12. Siletti, K.; Hodge, R.; Albiach, A.M.; Lee, K.W.; Ding, S.-L.; Hu, L.; Lönnerberg, P.; Casper, T.; Clark, M.; Dee, N.; et al. Transcriptomic diversity of cell types across the adult human brain. *bioRxiv* **2022**. [CrossRef] [PubMed]
13. Stevens, C.F. Neuronal diversity: Too many cell types for comfort? *Curr. Biol.* **1998**, *8*, R708–R710. [CrossRef] [PubMed]
14. Masland, R.H. Neuronal cell types. *Curr. Biol.* **2004**, *14*, R497–R500. [CrossRef]
15. Zeng, H.; Sanes, J.R. Neuronal cell-type classification: Challenges, opportunities and the path forward. *Nat. Rev. Neurosci.* **2017**, *18*, 530–546. [CrossRef]
16. Tasic, B. Single cell transcriptomics in neuroscience: Cell classification and beyond. *Curr. Opin. Neurobiol.* **2018**, *50*, 242–249. [CrossRef]
17. Smith, S.J.; Sümbül, U.; Graybuck, L.T.; Collman, F.; Seshamani, S.; Gala, R.; Gliko, O.; Elabbady, L.; Miller, J.A.; Brakken, T.E.; et al. Single-cell transcriptomic evidence for dense intracortical neuropeptide networks. *eLife* **2019**, *8*, e47889. [CrossRef]
18. Chen, X.Y.; Du, Y.F.; Chen, L. Neuropeptides Exert Neuroprotective Effects in Alzheimer's Disease. *Front. Mol. Neurosci.* **2018**, *11*, 493. [CrossRef]
19. Li, M.; Larsen, P.A. Single-cell sequencing of entorhinal cortex reveals widespread disruption of neuropeptide networks in Alzheimer's disease. *Alzheimers Dement.* **2023**, *19*, 3575–3592. [CrossRef]
20. Beckmann, N.D.; Lin, W.J.; Wang, M.; Cohain, A.T.; Wang, P.; Ma, W.; Wang, Y.-C.; Jiang, C.; Audrain, M.; Comella, P.H.; et al. Multiscale causal networks identify VGF as a key regulator of Alzheimer's disease. *Nat. Commun.* **2020**, *11*, 3942. [CrossRef]
21. Mathys, H.; Peng, Z.; Boix, C.A.; Victor, B.M.; Leary, N.; Babu, S.; Abdelhady, G.; Jiang, X.; Ng, A.P.; Ghafari, K.; et al. Single-cell atlas reveals correlates of high cognitive function, dementia, and resilience to Alzheimer's disease pathology. *Cell* **2023**, *186*, 4365–4385.e27. [CrossRef] [PubMed]
22. Hauser, A.S.; Attwood, M.M.; Rask-Andersen, M.; Schiöth, H.B.; Gloriam, D.E. Trends in GPCR drug discovery: New agents, targets and indications. *Nat. Rev. Drug Discov.* **2017**, *16*, 829–842. [CrossRef] [PubMed]
23. Kaleta, C.; Schäuble, S.; Rinas, U.; Schuster, S. Metabolic costs of amino acid and protein production in *Escherichia coli*. *Biotechnol. J.* **2013**, *8*, 1105–1114. [CrossRef]
24. Pulido, C.; Ryan, T.A. Synaptic vesicle pools are a major hidden resting metabolic burden of nerve terminals. *Sci. Adv.* **2021**, *7*, eabi9027. [CrossRef] [PubMed]
25. Grubman, A.; Chew, G.; Ouyang, J.F.; Sun, G.; Choo, X.Y.; McLean, C.; Simmons, R.K.; Buckberry, S.; Vargas-Landin, D.B.; Poppe, D.; et al. A single-cell atlas of entorhinal cortex from individuals with Alzheimer's disease reveals cell-type-specific gene expression regulation. *Nat. Neurosci.* **2019**, *22*, 2087–2097. [CrossRef] [PubMed]
26. Leng, K.; Li, E.; Eser, R.; Piergies, A.; Sit, R.; Tan, M.; Neff, N.; Li, S.H.; Rodriguez, R.D.; Suemoto, C.K.; et al. Molecular characterization of selectively vulnerable neurons in Alzheimer's disease. *Nat. Neurosci.* **2021**, *24*, 276–287. [CrossRef]
27. Sage Bionetworks. Synapse. Available online: https://www.synapse.org/ (accessed on 13 October 2023).
28. GTEx Consortium. The GTEx Consortium atlas of genetic regulatory effects across human tissues. *Science* **2020**, *369*, 1318–1330. [CrossRef]
29. Yamamoto, R.; Chung, R.; Vazquez, J.M.; Sheng, H.; Steinberg, P.L.; Loannidis, N.M.; Sudmant, P.H. Tissue-specific impacts of aging and genetics on gene expression patterns in humans. *Nat. Commun.* **2022**, *13*, 5803. [CrossRef]
30. Guzzi, P.H.; Lomoio, U.; Veltri, P. GTExVisualizer: A web platform for supporting ageing studies. *Bioinformatics* **2023**, *39*, btad303. [CrossRef]
31. Schneider, A.L.; Martins-Silva, R.; Kaizeler, A.; Saraiva-Agostinho, N.; Barbosa-Morais, N.L. voyAGEr, a free web interface for the analysis of age-related gene expression alterations in human tissues. *eLife* **2024**, *12*, RP88623. [CrossRef]
32. Butlet, A.; Hoffman, P.; Smibert, P.; Papalexi, E.; Satija, R. Integrating single-cell transcriptomic data across different conditions, technologies, and species. *Nat. Biotechnol.* **2018**, *36*, 411–420. [CrossRef] [PubMed]
33. CZI Single-Cell Biology Program; Abdulla, S.; Aevermann, B.; Assis, P.; Badajoz, S.; Bell, S.M.; Bezzi, E.; Cakir, B.; Chaffer, J.; Chambers, S.; et al. CZ CELL × GENE Discover: A single-cell data platform for scalable exploration, analysis and modeling of aggregated data. *bioRxiv* **2023**. [CrossRef]
34. AD Knowledge Portal. Available online: https://adknowledgeportal.synapse.org/Explore/Studies/DetailsPage/StudyDetails?Study=syn52293417 (accessed on 2 July 2024).
35. GTEx Portal. Available online: https://www.gtexportal.org (accessed on 1 November 2022).
36. Bahl, E.; Koomar, T.; Michaelson, J.J. cerebroViz: An R package for anatomical visualization of spatiotemporal brain data. *Bioinformatics* **2017**, *33*, 762–763. [CrossRef] [PubMed]

37. Zimmerman, K.D.; Espeland, M.A.; Langefeld, C.D. A practical solution to pseudoreplication bias in single-cell studies. *Nat. Commun.* **2021**, *12*, 738. [CrossRef]
38. Szklarczyk, D.; Gable, A.L.; Lyon, D.; Junge, A.; Wyder, S.; Huerta-Cepas, J.; Simonovic, M.; Doncheva, N.Y.; Morris, J.H.; Bork, P.; et al. STRING v11: Protein–protein association networks with increased coverage, supporting functional discovery in genome-wide experimental datasets. *Nucleic Acids Res.* **2019**, *47*, D607–D613. [CrossRef]
39. Yu, G. enrichplot: Visualization of Functional Enrichment Result. R Package Version 1.26.2. 2024. Available online: https://yulab-smu.top/biomedical-knowledge-mining-book/ (accessed on 20 November 2024).
40. Johnson, N.L.; Kotz, S.; Kemp, A.W. *Univariate Discrete Distributions*; Wiley: Hoboken, NJ, USA, 1992.
41. Tai, K.Y.; Wong, K.; Aghakhanian, F.; Parhar, I.S.; Dhaliwal, J.; Ayub, Q. Selected neuropeptide genes show genetic differentiation between Africans and non-Africans. *BMC Genet.* **2020**, *21*, 31. [CrossRef]
42. Gavin, L.; MacKay, A.P.; Brown, K.; Harrier, S.; Ventura, S.J.; Kann, L.; Rangel, M.; Berman, S.; Dittus, P.; Liddon, N.; et al. Sexual and reproductive health of persons aged 10–24 years—United States, 2002–2007. *MMWR Surveill. Summ.* **2009**, *58*, 1–58. [PubMed]
43. Liu, Q.; Shepherd, B.; Li, C. PResiduals: An R Package for Residual Analysis Using Probability-Scale Residuals. *J. Stat. Softw.* **2020**, *94*, 1–27. [CrossRef]
44. Jorstad, N.L.; Song, J.H.T.; Exposito-Alonso, D.; Suresh, H.; Castro-Pacheco, N.; Krienen, F.M.; Yanny, A.M.; Close, J.; Gelfand, E.; Long, B.; et al. Comparative transcriptomics reveals human-specific cortical features. *bioRxiv* **2022**. [CrossRef]
45. Hao, Y.; Hao, S.; Andersen-Nissen, E.; Mauck, W.M., 3rd; Zheng, S.; Butler, A.; Lee, J.M.; Wilk, A.J.; Darby, C.; Zager, M.; et al. Integrated analysis of multimodal single-cell data. *Cell* **2021**, *184*, 3573–3587.e29. [CrossRef]
46. Lizbinski, K.M.; Marsat, G.; Dacks, A.M. Systematic Analysis of Transmitter Coexpression Reveals Organizing Principles of Local Interneuron Heterogeneity. *eNeuro* **2018**, *5*, 1–20. [CrossRef] [PubMed]
47. ATP1B1 ATPase Na+/K+ Transporting Subunit Beta 1 [Homo Sapiens (Human)]—Gene—NCBI. Available online: https://www.ncbi.nlm.nih.gov/gene/481 (accessed on 20 June 2024).
48. Boehm, M.; Bonifacino, J.S. Adaptins: The final recount. *Mol. Biol. Cell* **2001**, *12*, 2907–2920. [CrossRef]
49. Nie, S.D.; Li, X.; Tang, C.E.; Min, F.-Y.; Shi, X.-J.; Wu, L.-Y.; Zhou, S.-L.; Chen, Z.; Wu, J.; Song, T.; et al. High glucose forces a positive feedback loop connecting ErbB4 expression and mTOR/S6K pathway to aggravate the formation of tau hyperphosphorylation in differentiated SH-SY5Y cells. *Neurobiol. Aging* **2018**, *67*, 171–180. [CrossRef] [PubMed]
50. Albaret, M.A.; Textoris, J.; Dalzon, B.; Lambert, J.; Linard, M.; Helmer, C.; Hacot, S.; Ghayad, S.E.; Ferréol, M.; Mertani, H.C.; et al. HSV-1 cellular model reveals links between aggresome formation and early step of Alzheimer's disease. *Transl. Psychiatry* **2023**, *13*, 86. [CrossRef]
51. Greenwood, A.K.; Montgomery, K.S.; Kauer, N.; Woo, K.H.; Leanza, Z.J.; Poehlman, W.L.; Gockley, J.; Sieberts, S.K.; Bradic, L.; Logsdon, B.A.; et al. The AD Knowledge Portal: A Repository for Multi-Omic Data on Alzheimer's Disease and Aging. *Curr. Protoc. Hum. Genet.* **2020**, *108*, e105. [CrossRef]
52. Laakso, M.P.; Soininen, H.; Partanen, K.; Helkala, E.L.; Hartikainen, P.; Vainio, P.; Hallikainen, M.; Hänninen, T.; Riekkinen, P.J., Sr. Volumes of hippocampus, amygdala and frontal lobes in the MRI-based diagnosis of early Alzheimer's disease: Correlation with memory functions. *J. Neural Transm. Park. Dis. Dement. Sect.* **1995**, *9*, 73–86. [CrossRef] [PubMed]
53. Albert, M.S.; Moss, M.B. Early Features of Alzheimer's Disease. In *Cerebral Cortex: Neurodegenerative and Age-Related Changes in Structure and Function of Cerebral Cortex*; Peters, A., Morrison, J.H., Eds.; Springer: New York, NY, USA, 1999; pp. 461–474.
54. Jones, B.F.; Barnes, J.; Uylings, H.B.M.; Fox, N.C.; Frost, C.; Witter, M.P.; Scheltens, P. Differential regional atrophy of the cingulate gyrus in Alzheimer disease: A volumetric MRI study. *Cereb. Cortex* **2006**, *16*, 1701–1708. [CrossRef]
55. Amanzio, M.; Torta, D.M.E.; Sacco, K.; Cauda, F.; D'Agata, F.; Duca, S.; Leotta, D.; Palermo, S.; Geminiani, G.C. Unawareness of deficits in Alzheimer's disease: Role of the cingulate cortex. *Brain* **2011**, *134*, 1061–1076. [CrossRef]
56. Spalletta, G.; Piras, F.; Piras, F.; Sancesario, G.; Lorio, M.; Fratangeli, C.; Cacciari, C.; Caltagirone, C.; Orfei, M. Neuroanatomical correlates of awareness of illness in patients with amnestic mild cognitive impairment who will or will not convert to Alzheimer's disease. *Cortex* **2014**, *61*, 183–195. [CrossRef]
57. Poulin, S.P.; Dautoff, R.; Morris, J.C.; Barrett, L.F.; Dickerson, B.C. Alzheimer's Disease Neuroimaging Initiative. Amygdala atrophy is prominent in early Alzheimer's disease and relates to symptom severity. *Psychiatry Res.* **2011**, *194*, 7–13. [CrossRef]
58. Buckner, R.L.; Sepulcre, J.; Talukdar, T.; Krienen, F.M.; Liu, H.; Hedden, T.; Andrews-Hanna, J.R.; Sperling, R.A.; Johnson, K.A. Cortical hubs revealed by intrinsic functional connectivity: Mapping, assessment of stability, and relation to Alzheimer's disease. *J. Neurosci.* **2009**, *29*, 1860–1873. [CrossRef] [PubMed]
59. Pennanen, C.; Kivipelto, M.; Tuomainen, S.; Hartikainen, P.; Hänninen, T.; Laakso, M.P.; Hallikainen, M.; Vanhanen, M.; Nissinen, A.; Helkala, E.L.; et al. Hippocampus and entorhinal cortex in mild cognitive impairment and early AD. *Neurobiol. Aging* **2004**, *25*, 303–310. [CrossRef] [PubMed]
60. Fernández-Cabello, S.; Kronbichler, M.; Van Dijk, K.R.A.; Goodman, J.A.; Spreng, N.R.; Schmitz, T.W.; Alzheimer's Disease Neuroimaging Initiative. Basal forebrain volume reliably predicts the cortical spread of Alzheimer's degeneration. *Brain* **2020**, *143*, 993–1009. [CrossRef] [PubMed]
61. Grothe, M.; Heinsen, H.; Teipel, S.J. Atrophy of the cholinergic Basal forebrain over the adult age range and in early stages of Alzheimer's disease. *Biol. Psychiatry* **2012**, *71*, 805–813. [CrossRef]

62. Zhou, J.N.; Hofman, M.A.; Swaab, D.F. VIP neurons in the human SCN in relation to sex, age, and Alzheimer's disease. *Neurobiol. Aging* **1995**, *16*, 571–576. [CrossRef]
63. Cedernaes, J.; Osorio, R.S.; Varga, A.W.; Kam, K.; Schiöth, H.B.; Benedict, C. Candidate mechanisms underlying the association between sleep-wake disruptions and Alzheimer's disease. *Sleep Med. Rev.* **2017**, *31*, 102–111. [CrossRef]
64. Suresh, H.; Crow, M.; Jorstad, N.; Hodge, R.; Lein, E.; Dobin, A.; Bakken, T.; Gillis, J. Comparative single-cell transcriptomic analysis of primate brains highlights human-specific regulatory evolution. *Nat. Ecol. Evol.* **2023**, *7*, 1930–1943. [CrossRef]
65. Woo, R.S.; Lee, J.H.; Yu, H.N.; Song, D.Y.; Baik, T.K. Expression of ErbB4 in the apoptotic neurons of Alzheimer's disease brain. *Anat. Cell Biol.* **2010**, *43*, 332–339. [CrossRef]
66. Zuniga, G.; Levy, S.; Ramirez, P.; De Mange, J.; Gonzalez, E.; Gamez, M.; Frost, B. Tau-induced deficits in nonsense-mediated mRNA decay contribute to neurodegeneration. *Alzheimer's Dement.* **2023**, *19*, 405–420. [CrossRef]
67. Markesbery, W.R. Neuropathologic alterations in mild cognitive impairment: A review. *J. Alzheimers Dis.* **2010**, *19*, 221–228. [CrossRef]
68. Bennett, D.A.; Schneider, J.A.; Bienias, J.L.; Evans, D.A.; Wilson, R.S. Mild cognitive impairment is related to Alzheimer disease pathology and cerebral infarctions. *Neurology* **2005**, *64*, 834–841. [CrossRef] [PubMed]
69. Huang, W.C.; Peng, Z.; Murdock, M.H.; Liu, L.; Mathys, H.; Davila-Velderrain, J.; Jiang, X.; Chen, M.; NG, A.P.; Kim, T.; et al. Lateral mammillary body neurons in mouse brain are disproportionately vulnerable in Alzheimer's disease. *Sci. Transl. Med.* **2023**, *15*, eabq1019. [CrossRef] [PubMed]
70. Suvà, D.; Favre, I.; Kraftsik, R.; Esteban, M.; Lobrinus, A.; Miklossy, J. Primary motor cortex involvement in Alzheimer disease. *J. Neuropathol. Exp. Neurol.* **1999**, *58*, 1125–1134. [CrossRef]
71. Zadey, S.; Buss, S.S.; McDonald, K.; Press, D.Z.; Pascual-Leone, A.; Fried, P.J. Higher motor cortical excitability linked to greater cognitive dysfunction in Alzheimer's disease: Results from two independent cohorts. *Neurobiol. Aging* **2021**, *108*, 24–33. [CrossRef]
72. Ferreri, F.; Pauri, F.; Pasqualetti, P.; Fini, R.; Dal Forno, G.; Rossini, P.M. Motor cortex excitability in Alzheimer's disease: A transcranial magnetic stimulation study. *Ann. Neurol.* **2003**, *53*, 102–108. [CrossRef]
73. Lyoo, C.H.; Cho, H.; Choi, J.Y.; Hwang, M.S.; Hong, S.K.; Kim, Y.J.; Ryu, Y.H.; Lee, M.S. Tau Accumulation in Primary Motor Cortex of Variant Alzheimer's Disease with Spastic Paraparesis. *J. Alzheimers Dis.* **2016**, *51*, 671–675. [CrossRef]
74. Gao, Y.L.; Wang, N.; Sun, F.R.; Cao, X.P.; Zhang, W.; Yu, J.T. Tau in neurodegenerative disease. *Ann. Transl. Med.* **2018**, *6*, 175. [CrossRef] [PubMed]
75. Yu, M.; Sporns, O.; Saykin, A.J. The human connectome in Alzheimer disease—Relationship to biomarkers and genetics. *Nat. Rev. Neurol.* **2021**, *17*, 545–563. [CrossRef]
76. Quinn, J.P.; Ethier, E.C.; Novielli, A.; Malone, A.; Ramirez, C.E.; Salloum, L.; Trombetta, B.A.; Kivisäkk, P.; Bremang, M.; Selzer, S.; et al. Cerebrospinal Fluid and Brain Proteoforms of the Granin Neuropeptide Family in Alzheimer's Disease. *J. Am. Soc. Mass Spectrom.* **2023**, *34*, 649–667. [CrossRef]
77. Hicks, C.; Miele, L.; Koganti, T.; Young-Gaylor, L.; Rogers, D.; Vijayakumar, V.; Megason, G. Analysis of patterns of gene expression variation within and between ethnic populations in pediatric B-ALL. *Cancer Inform.* **2013**, *12*, 155–173. [CrossRef]
78. Genazzani, A.R.; Stomati, M.; Morittu, A.; Bernadi, F.; Monteleone, P.; Casarosa, E.; Gallo, R.; Salvestroni, C.; Luisi, M. Progesterone, progestagens and the central nervous system. *Hum. Reprod.* **2000**, *15* (Suppl. S1), 14–27. [CrossRef] [PubMed]
79. Yamashita, J.; Nishiike, Y.; Fleming, T.; Kayo, D.; Okubo, K. Estrogen mediates sex differences in preoptic neuropeptide and pituitary hormone production in medaka. *Commun. Biol.* **2021**, *4*, 948. [CrossRef] [PubMed]
80. Cetinkaya, A.; Kilinc, E.; Camsari, C.; Ogun, M.N. Effects of estrogen and progesterone on the neurogenic inflammatory neuropeptides: Implications for gender differences in migraine. *Exp. Brain Res.* **2020**, *238*, 2625–2639. [CrossRef] [PubMed]
81. Buckley, R.F.; Mormino, E.C.; Rabin, J.S.; Hohman, T.J.; Landau, S.; Hanseeuw, B.J.; Jacobs, H.I.L.; Papp, K.V.; Amariglio, R.E.; Properzi, M.J.; et al. Sex Differences in the Association of Global Amyloid and Regional Tau Deposition Measured by Positron Emission Tomography in Clinically Normal Older Adults. *JAMA Neurol.* **2019**, *76*, 542–551. [CrossRef]
82. Tian, W.; Zhou, J.; Bartlett, A.; Zeng, Q.; Liu, H.; Castanon, R.G.; Kenworthy, M.; Altshul, J.; Valadon, C.; Aldridge, A.; et al. Single-cell DNA methylation and 3D genome architecture in the human brain. *Science* **2023**, *382*, eadf5357. [CrossRef]
83. Li, Y.E.; Preissl, S.; Miller, M.; Johnson, N.D.; Wang, Z.; Jiao, H.; Zhu, C.; Wang, Z.; Xie, Y.; Poirion, O.; et al. A comparative atlas of single-cell chromatin accessibility in the human brain. *Science* **2023**, *382*, eadf7044. [CrossRef]

biomolecules
MDPI

Article

Inhibiting the Cholesterol Storage Enzyme ACAT1/SOAT1 in Myelin Debris-Treated Microglial Cell Lines Activates the Gene Expression of Cholesterol Efflux Transporter ABCA1

Thao N. Huynh [1], Matthew C. Havrda [2], George J. Zanazzi [3], Catherine C. Y. Chang [1,*] and Ta Yuan Chang [1,*]

[1] Department of Biochemistry and Cell Biology, Geisel School of Medicine at Dartmouth, Hanover, NH 03755, USA; thao.n.huynh.gr@dartmouth.edu
[2] Department of Molecular and System Biology, Geisel School of Medicine at Dartmouth, Hanover, NH 03755, USA; matthew.c.havrda@dartmouth.edu
[3] Department of Pathology and Laboratory Medicine, Dartmouth–Hitchcock Medical Center, Lebanon, NH 03766, USA; george.j.zanazzi@dartmouth.edu
* Correspondence: catherine.chang@dartmouth.edu (C.C.Y.C.); ta.yuan.chang@dartmouth.edu (T.Y.C.)

Abstract: Aging is the major risk factor for Alzheimer's disease (AD). In the aged brain, myelin debris accumulates and is cleared by microglia. Phagocytosed myelin debris increases neutral lipid droplet content in microglia. Neutral lipids include cholesteryl esters (CE) and triacylglycerol (TAG). To examine the effects of myelin debris on neutral lipid content in microglia, we added myelin debris to human HMC3 and mouse N9 cells. The results obtained when using ^{3}H-oleate as a precursor in intact cells reveal that myelin debris significantly increases the biosynthesis of CE but not TAG. Mass analyses have shown that myelin debris increases both CE and TAG. The increase in CE biosynthesis was abolished using inhibitors of the cholesterol storage enzyme acyl-CoA:cholesterol acyltransferase 1 (ACAT1/SOAT1). ACAT1 inhibitors are promising drug candidates for AD treatment. In myelin debris-loaded microglia, treatment with two different ACAT1 inhibitors, K604 and F12511, increased the mRNA and protein content of ATP-binding cassette subfamily A1 (ABCA1), a protein that is located at the plasma membrane and which controls cellular cholesterol disposal. The effect of the ACAT1 inhibitor on ABCA1 was abolished by preincubating cells with the liver X receptor (LXR) antagonist GSK2033. We conclude that ACAT1 inhibitors prevent the accumulation of cholesterol and CE in myelin debris-treated microglia by activating ABCA1 gene expression via the LXR pathway.

Keywords: Alzheimer's disease; myelin debris; microglia; cholesterol; cholesteryl esters; acyl-CoA: cholesterol acyltransferase; sterol O-acyltransferase; ACAT inhibitor; ATP-binding cassette subfamily A member 1; aging; foam cell; liver X receptor

1. Introduction

Myelin is a cholesterol-rich material that forms the insulating sheath around nerve cells [1]. Myelin contains up to 70–80% of the total brain cholesterol in the adult brain [2]. Proper myelination is crucial for synaptic transmission and metabolic support in neuronal cells [1]. Myelin debris results from the breakdown of myelin [3] and accumulates in the aging brain and in certain neurodegenerative diseases [4,5]. Under normal conditions, microglia and other phagocytes clear myelin debris to enable proper remyelination [4]. However, in aging mouse models, myelin debris accumulates inside the microglia [4]. When microglia become overloaded with myelin debris, the cells experience a pro-inflammatory response, which may further contribute to cognitive decline in aging and neurodegenerative diseases [4,6]. Myelin debris-overloaded microglia often exhibit a "foamy" phenotype, characterized by the accumulation of neutral lipid droplets [7–9]. Neutral lipid droplets consist mainly of cholesteryl esters (CE) and triacylglycerol (TAG) [10].

Aging is the major risk factor for Alzheimer's disease (AD). While animal models for aging are becoming available [11], cell models for aging are sparse. Microglia play important roles in the pathogenesis of AD [9,12–15]. Myelin debris-treated microglia have been considered as cell models for mammalian cell aging. For example, a previous study by Nugent et al. showed that treatment of myelin debris in human iPSC microglia and in mouse primary microglia caused CE accumulation, while inhibition of the cholesterol storage enzyme acyl-CoA:cholesterol acyltransferase, also known as sterol O-acyltransferase 1 (ACAT1/SOAT1), reduced CE accumulation, demonstrating that myelin debris stimulates CE biosynthesis [7]. However, the downstream effects of inhibiting ACAT1/SOAT1 in myelin debris-treated microglia have not been well examined. Altering ACAT1 activity in these cells is likely to affect cholesterol homeostasis through its interacting with other key players involved in regulating cellular lipid metabolism [16,17]. Liver X receptors (LXRs) are nuclear receptors that regulate cholesterol homeostasis, lipid metabolism and membrane phospholipid composition and they play important roles in immune responses [18–20]. The natural ligands for LXR are oxysterols [21]. In a mouse model for tauopathy, Litvinchuk et al. [22] have shown that a synthetic LXR agonist, GW3965, increases gene expression of ABCA1, which is the major cholesterol efflux protein that controls cellular cholesterol disposal [23]. Additional results by Litvinchuk et al. show that overexpression of the ABCA1 gene also reduces CEs and suppresses tauopathy [22]. This study suggests that interventions to stimulate cholesterol efflux from microglia via activating ABCA1 to defend against AD warrant further investigation. Gouna et al. have also shown that treating wild-type mouse microglia with myelin debris causes CE accumulation [24]. In these studies, whether myelin debris also affects TAG biosynthesis is unclear. To address these issues, in the current work, we use two cell lines, mouse N9 and human HMC3 microglia, as our cell models. These cell lines have been used successfully by other investigators as microglia cell models for AD research [25–31]. We prepared myelin debris from both mouse and human brains, and then added them to these cells to monitor their effects on CE and TAG biosynthesis. We also studied the downstream effects of the ACAT1 inhibitor in these cells.

2. Materials and Methods

2.1. Animals

Wild-type (WT) mice in C57B6/J background were obtained from Jackson Laboratory (Bar Harbor, ME, USA). Mice that were 3–6 months old were used for myelin debris isolation. All mouse procedures were approved by the Dartmouth Institutional Animal Care and Use Committee.

2.2. Human Brain Samples

Human brain samples were donated by the Anatomy Gifts Registry (Hanover, MD, USA) and the Department of Pathology and Laboratory Medicine at Dartmouth–Hitchcock Medical Center (Lebanon, NH, USA). All brain samples were received frozen prior to their use.

2.3. Cell Culture

Both the N9 and HMC3 (ATCC®CRL-3304) cell lines were obtained from ATCC. All cells were maintained at 37 °C with 5% CO_2 in a humidified incubator. Mouse N9 microglial cells were maintained in RPMI-1640 with 10% heat-inactivated serum. Human HMC3 microglial cells were maintained in MEM supplemented with 10% heat-inactivated serum and 1% non-essential amino acids. HEK-293 cells were maintained in DMEM with 10% serum.

The ACAT1/SOAT1 inhibitors K604 and F12511 were first dissolved in DMSO at 5 mM as stock solutions and then diluted into the culture medium such that the final concentration was 0.5 μM, as previously described in [16,32–34]. For LXR agonist treatment, T0901317 was diluted to a final concentration of 10 μM in complete media as described above and added to the cells for 24 h, as previously described in [32,35]. For LXR antagonist treatment, GSK2033 was diluted to a final concentration of 5 μM in complete media and added to cells for 24 h, as previously described in [36].

2.4. Myelin Isolation and Characterization of Myelin Debris

Myelin debris were isolated from mouse and human frozen brain samples following a protocol adapted from [37], with minor modifications. Briefly, brain samples were minced and homogenized in 0.3M sucrose in Tris buffer (20 mM Tris-HCl, pH 7.45, 1mM EDTA with protease inhibitor cocktail), using an automatic Dounce chamber for 20 strokes. The homogenates were layered on top of a 0.83M sucrose solution and spun at 75,000$\times$ g for 70 min with minimum acceleration and deceleration. Cloudy myelin layers at the interface were collected and subjected to two rounds of osmotic shock by adding Tris buffer, followed by centrifugation at 75,000$\times$ g for 70 min with minimum acceleration and deceleration. Myelin debris, post osmotic shock and pelleted at the bottom of the tube, were resuspended in MilliQ water and spun at 15,000$\times$ g for 20 min to obtain clean myelin debris for cell treatment.

Myelin debris protein content was determined by Lowry protein assay, and cholesterol content was measured following the manufacturer's protocol using the Wako Free Cholesterol Kit E. For both mouse and human myelin debris preparations, the ratio of total protein to cholesterol was 2:1 (µg protein to µg cholesterol). This value is consistent with the value reported by [3]. Protein content characterization of human and mice myelin debris was analyzed by running a 15% SDS-PAGE gel and staining with colloidal Coomassie blue dye.

2.5. Intact Cell ^{3}H-Oleate Pulse

^{3}H-oleate pulse in intact cells was performed according to a procedure previously described in [38,39]. ^{3}H-oleate enters the cell interior and forms oleyl coenzyme A through acyl coenzyme A synthetases [40]. Oleyl coenzyme A then serves as a common precursor for CE biosynthesis via ACATs, and for TAG biosynthesis via diacylglycerol acyltransferases (DGATs) [41,42].

Mouse N9 and human HMC3 microglia were treated for 24 h with various concentrations of either control (DMSO), myelin debris, or the ACAT1 inhibitors K604 or F12511. Following treatment, cells were pulsed for 30 min (N9) or 2 h (HMC3) at 37 °C with ^{3}H-oleate/fatty acid-free BSA. Cells were then washed 3 times with ice-cold PBS and lysed with 0.2M NaOH, under orbital shaking for 30 min. Aliquots were withdrawn for protein concentration determination using the Lowry assay. The solubilized cell slurries were neutralized with 3M HCl and 1M KH_2PO_4. Nonradioactive CE and TAG, of 40 µg each, were added per sample to aid in identification after TLC analysis. Lipid extractions were performed with $CHCl_3$:MeOH (2:1) and water. Samples were vortexed and centrifuged at 500 rpm for 10 min. The top-phase was removed, and the bottom phase was blow dried by N_2 using an N-evap apparatus. Dried samples were vortexed vigorously with 50 µL of ethyl acetate and spotted on TLC plates. The solvent system used was petroleum ether:ethyl ether:acetic acid (90:10:1). The CE bands (Rf = 0.9) and the TAG bands (Rf = 0.5) were identified after TLC by iodine staining and scraped from the TLC. Radioactivity readouts were measured by scintillation counter.

2.6. Nile Red Staining and Image Analysis

Nile Red was used to monitor lipid droplets in live cells according to procedures previously described in [43,44]. Briefly, primary microglia were plated on MaTek (Ashland, MA, USA) 35 mm dishes pre-coated with poly-L-lysine at 2 $\times$ 10^6 cells per plate overnight. Treatment was performed in serum-free DMEM, and cells were rinsed three times with HBSS (Gibco by ThermoFisher, Waltham, MA, USA). Cells were treated with 100 ng/mL Nile Red and incubated for 10 min at 37 °C, 5% CO_2, protected from light. Cells were rinsed in HBSS and imaged in serum-free MEM, with no phenol red (Gibco by ThermoFisher, Waltham, MA, USA), on the confocal fluorescence microscope. For image analysis, data were collected and analyzed using Fiji-ImageJ software version 2.1.0/1.53c. Briefly, all images were set to the same threshold, fluorescence intensities were measured and normalized against total cell area to obtain a mean fluorescence intensity.

2.7. TLC Analysis of Intracellular Cholesterol, CE and TAG Contents

For intracellular cholesterol, CE and TAG content analysis, cells were analyzed following the procedure described in [34,45]. Briefly, cells were treated as indicated in the treatment scheme and lysed by scraping in distilled water. Lipids were then extracted using $CHCl_3$:MeOH (2:1) extraction and water, then dried in a similar manner as described in Section 2.5. Dried samples were spotted on a TLC plate and separated using hexanes, ethyl ether: acetic acid (65:35:2). Bands were visualized by iodine staining and quantified on a standard curve produced in the same TLC plate.

2.8. Whole Cell Protein Isolation and Western Blot Analyses

For whole cell protein isolation, cells were harvested in 10% SDS for ACAT1 Western blots, or in RIPA buffer for the other Western blots, containing protease inhibitor cocktail (Sigma) and incubated at 4 °C for 30 min. Cell lysates were then centrifuged, and the protein concentration of the supernatant was determined using the Lowry protein assay. For mouse and human myelin debris Western blot, Laemmli sample buffer (supplemented with 350mM dithiothreitol final concentration) was directly added to the isolated myelin debris fraction. The lysates were run on a 6% and 10% SDS-PAGE gel and transferred to a 0.45 μm nitrocellulose membrane for 4 h at 300 mA. After blocking in 5% milk in TBST buffer overnight at 4 °C, the membranes were incubated overnight with anti-ABCA1 (Novus NB400-105), anti-ACAT1 (DM102) produced in the Chang lab, anti-PLIN2 (Proteintech 15294-1-AP), anti-MBP (Novus NB600-717) and anti-vinculin (Millipore 05-386) as a protein loading control. Blots were then washed and incubated with secondary antibodies from Li-Cor appropriate for the species. Western blot images were captured on the Li-Cor Odyssey CLx and analyzed on Li-Cor Image Studio. Analyzed protein bands were within the linear range of detection.

2.9. RNA Extraction

For whole cell RNA extraction, cells were lysed in Trizol. Chloroform was added (0.2 mL per ml of Trizol) and the samples were vortexed and spun at 10,000× g for 18 min. The top aqueous phase was removed and transferred to a new tube. An equal volume of 100% EtOH was added and 700 μL aliquots were loaded onto a Qiagen RNeasy column. The Qiagen RNeasy Plus Mini Kit was used for all subsequent steps according to the manufacturer's protocol.

2.10. NanoString nCounter Elements XT Assay and Analysis

These experiments were performed using facilities available at the Genomics and Molecular Biology Shared Resources at the Dartmouth Cancer Center. NanoString-recommended protocols were followed, as described in the nCounter Elements XT Assay user's manual (June 2018). Data collection was performed using an nCounter digital analyzer. An nCounter reporter library file (RLF), specific to the reporter code sets used, was uploaded prior to scanning on the digital analyzer. The nCounter cartridge was loaded into a cartridge carrier at the top of the digital analyzer, which can hold up to six cartridges at a time. The digital analyzer uses epifluorescence microscopy and a CCD camera to yield target molecule counts by imaging each of the 12 channels of the cartridges independently. Digital images were processed within the nCounter instrument, and the reporter probe counts were tabulated in comma-separated value (CSV) format for data analysis. After sample imaging, data were downloaded and imported into NanoString nSolver analysis software version 4.0.

2.11. Synthesis of ACAT1/SOAT1 Inhibitors

K604 and F12511 were custom synthesized by WuXi AppTec in China. Based on HPLC-MS and NMR profiles, the purity of K604 was 98%. F12511 was 98% pure in terms of stereospecificity. K604 is a high-affinity, selective ACAT1 with K_i = 0.45 μmol/L for ACAT1 and K_i = 102.9 μmol/L for ACAT2 [46]. F12511 is another potent ACAT inhibitor that has an affinity for both ACAT1 (K_i = 0.039 μmol/L) and ACAT2 (K_i = 0.110 μmol/L) [46]. K604 is competitive against oleyl-CoA (K_i = 0.378 μmol/L), which is the preferred fatty acyl-CoA substrate for ACAT1 [47]. F12511 is a fatty acid anilide derivative [48,49].

2.12. Statistical Analysis

All statistical analyses were performed using Prism10 software version 10.1.0 (GraphPad). A one-way ANOVA test with Sidak correction for multiple comparisons between treatment groups were used. Error bars indicate SEM. * $p < 0.05$; ** $p < 0.01$; *** $p < 0.001$; **** $p < 0.0001$.

3. Results

3.1. Characterization of Myelin Debris Isolated from Human and Mouse Brain Tissues

To establish a simple system by which to study myelin debris-loaded microglia, we first isolated myelin debris from human and mouse brain samples, following the procedure shown in Figure 1A. The myelin debris fraction was characterized using a 15% SDS-PAGE gel, to identify proteins typically associated with myelin debris, including proteolipid protein (PLP) and myelin basic protein (MBP). PLP provides structural stability for myelin [50], while MBP is responsible for signaling and adhesion on the myelin cytosolic surface [51]. Results in Figure 1B show a migration of dominated protein patterns that is similar to PLP and MBP. These protein patterns are consistent with isolated myelin debris in previously published literature [52]. Additionally, the protein components of myelin debris remain consistent across different batches of isolated myelin (Figure 1B). When comparing human myelin debris from the corpus callosum and mouse myelin debris from whole brain tissue, their protein band patterns on SDS-PAGE are similar. In our isolated myelin debris, PLP and MBP are the two most enriched proteins (Figure 1B). To ensure that our isolated myelin debris is enriched in crucial myelin-associated protein such as MBP, we performed Western blot analysis against MBP on our samples. The results in Figure 1C confirm that MBP is present in our isolated myelin debris fraction.

3.2. Myelin Debris Loading in Mouse and Human Microglia Activates Cholesteryl Ester (CE) Synthesis in Intact Cells

We next investigated whether myelin debris loading alters CE synthesis by incubating two microglial cell lines with ^{3}H-oleate for a short time period (described in Section 2). This method is used to measure the biosynthesis of CEs and TAGs (in addition to other lipids derived from ^{3}H-oleyl coenzyme A) in intact cells. ^{3}H-oleate uptake rate in cells is cell-line dependent, thus we exposed N9 cells and HMC3 cells to ^{3}H-oleate with different incubation time to establish a comparable incorporation baseline between the two cell lines, as well as ensuring that a raw radioactive count is within the linear detection range [34]. Raw radioactive counts for each lipid species and cell lines exposed to DMSO control treatment are reported in the legend for Figure 2. For CE, the raw radioactive count for N9 cells was 119.8 dpm/min/mg and for HMC3 this was 68.78 dpm/min/mg. The equivalent values for TAG, were 2088.08 dpm/min/mg for N9 cells and 3933.08 dpm/min/mg for HMC3. We exposed N9 cells to mouse myelin debris and HMC3 cells to human myelin debris at various dosages for 24 h and monitored the biosynthesis of CE and TAG. We tested a range of 5–25 μg/mL of myelin cholesterol based on previous work by Nugent et al., 2020 [7]. The experimental parameters are graphically demonstrated in Figure 2A.

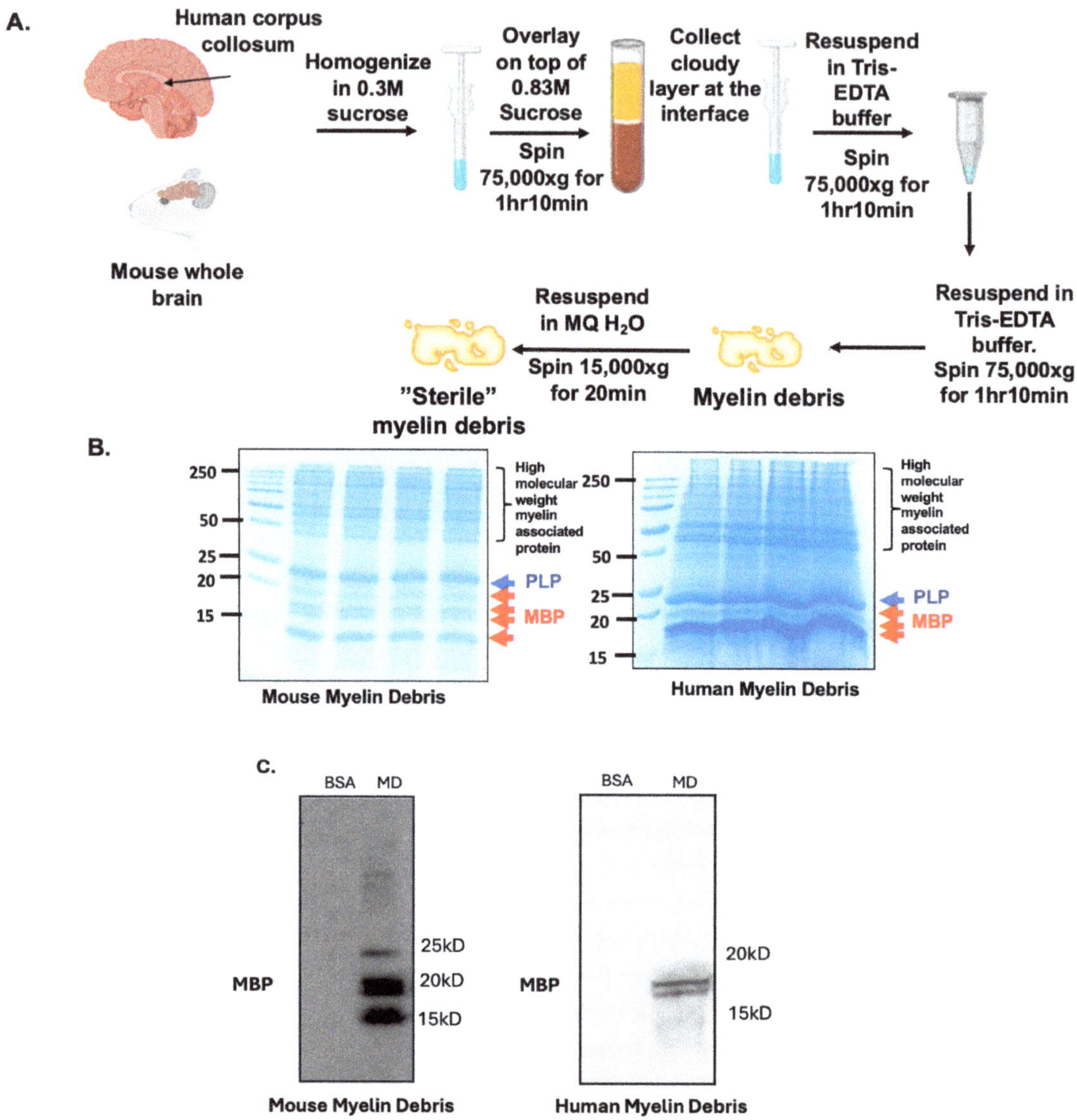

Figure 1. Characterization of mouse and human myelin debris. (**A**) Procedure for crude myelin debris isolation from frozen brain tissue as described in Section 2. Images were created using BioRender. (**B**) Representative colloidal Coomassie blue staining on a 15% SDS-PAGE gel for characterization of the crude myelin debris fraction. Signature protein associated with myelin fraction, such as proteolipid protein (PLP) or myelin basic protein (MBP) are labeled in blue and red, respectively. N = 4. (**C**) Representative Western blot for MBP in isolated human and mouse myelin debris. BSA at equal protein concentration was used as negative control. Western blot original images are in the Supplementary Materials.

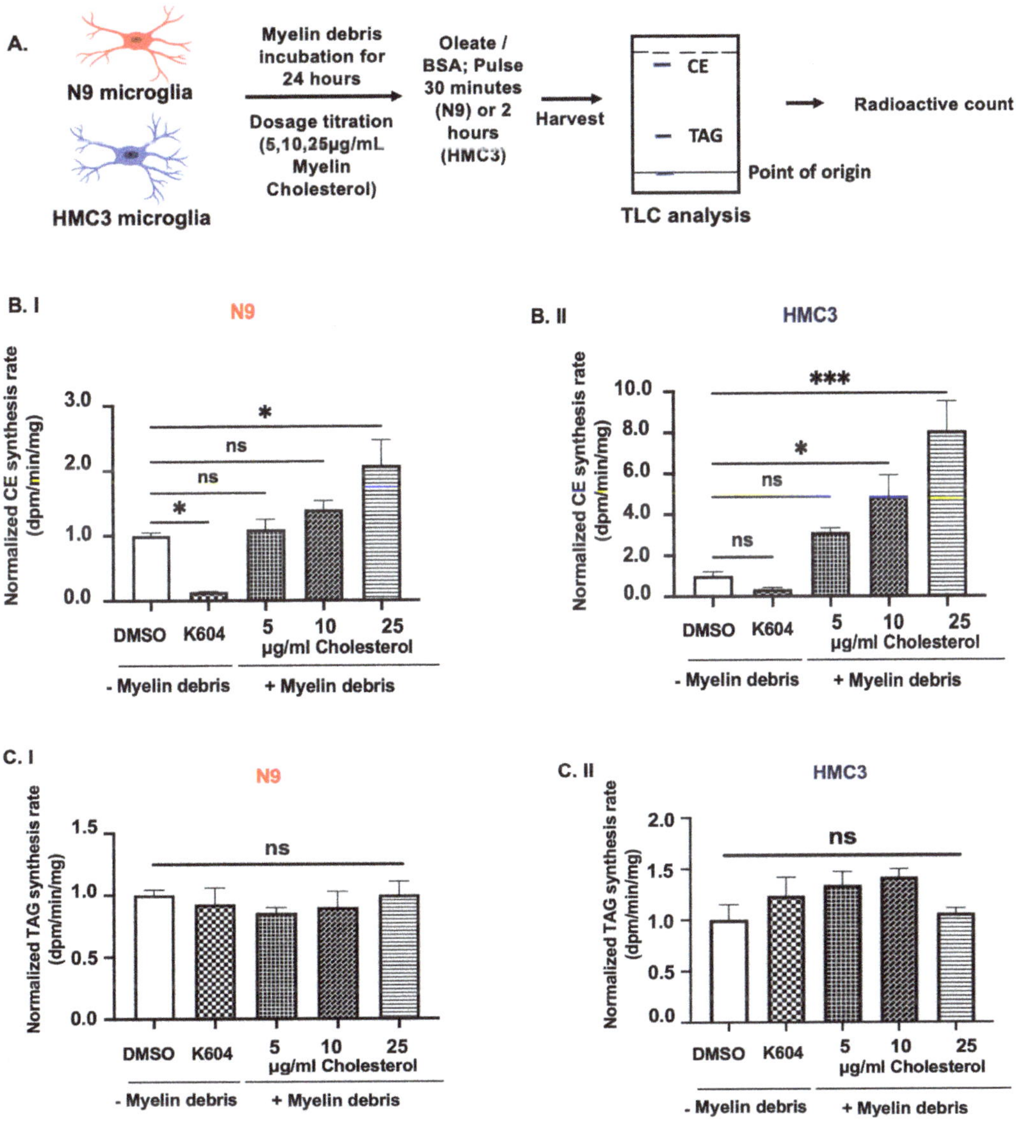

Figure 2. Myelin debris treatment in mouse and human microglia activates the synthesis of cholesteryl esters (CE), but not triacylglycerides (TAG), in a dose-dependent manner using oleic acid as substrate. (**A**) Treatment scheme: The extracted lipids underwent thin-layer chromatography (TLC) to separate individual lipids, which were identified by internal lipid standards (CEs and TAG) added during lipid extraction and as described in Section 2. Images were creared using BioRender. (**B**) CE and (**C**) TAG synthesis rates in intact (**I**) N9 cells and (**II**) HMC3 cells were monitored using a ^{3}H-oleate pulse. Cells were grown to 80% confluency and treated with DMSO alone, with the ACAT1/SOAT1 inhibitor K604, or with myelin debris at 5, 10 and 25 μg/mL cholesterol for 24 h. Afterwards, CE and

TAG biosynthesis rates in intact cells were monitored by providing ^{3}H-oleate to intact N9 cells for 30 min and to HMC3 cells for 2 h. Cholesterol content in myelin debris was measured using the Cholesterol Wako kit. Raw radioactive counts were normalized to 119.8 dpm/min/mg (CE) and 2088.08 dpm/min/mg (TAG) for N9 cells, and 68.78 dpm/min/mg (CE) and 3933.08 dpm/min/mg (TAG) for HMC3 cells. N = 3. Data are expressed as mean ± SEM. * $p < 0.05$; *** $p < 0.001$.

After their exposure to myelin debris, we monitored the rate of CE synthesis in intact cells as shown in Figure 2B.I,B.II. Our analysis revealed that, in N9 cells, the relative CE biosynthesis rate was upregulated by 9%, 39%, and 108% when exposed to 5, 10, and 25 µg/mL of myelin cholesterol, respectively. In HMC3 cells, the CE biosynthesis rate increased by 312%, 486%, and 806% at 5, 10, and 25 µg/mL of myelin cholesterol, respectively. These results indicate that, in response to myelin debris loading, both N9 mouse and human HMC3 microglial cells upregulate CE synthesis in intact cells. Our results show that the magnitude of CE synthesis in HMC3 cells in response to human myelin debris is significantly higher than in N9 cells.

We also monitored the rate of TAG biosynthesis in N9 and HMC3 intact cells using the ^{3}H-oleate pulse method, as reported in Figure 2C.I,C.II. Our analysis showed that, unlike the CE biosynthesis rate, no statistically significant increase in TAG biosynthesis was observed under the same treatment conditions. However, the data suggest a trend towards a small, dose-dependent increase in TAG biosynthesis in N9 cells, but not in HMC3 cells.

3.3. *Pharmaceutical Inhibition of ACAT1/SOAT1 Reduces CE Accumulation and Intracellular Cholesterol Content in HMC3 Treated Myelin Debris*

Cholesterol from myelin debris can be converted to CE for storage by the cholesterol storage enzyme ACAT1/SOAT1 [53]. To test this possibility, we treated cells with or without myelin debris and the small molecule ACAT1 inhibitor K604 [47], followed by monitoring the rate of CE biosynthesis in intact HMC3 cells. The experimental scheme is illustrated in Figure 3A. The results in Figure 3B demonstrate that inhibition of ACAT1 in myelin debris-loaded HMC3 microglia reduced CE biosynthesis by approximately 87%.

To further validate these data, we used Nile Red, a commonly used neutral lipid stain for lipid droplets [54], to determine the effect of ACAT1 blockage in live HMC3 cells (Figure 3C). Nile Red stains both components of lipid droplets (CE and TAG) [55]. In cells with excess cholesterol loading, cholesterol is converted into CE in the form of lipid droplets, and, when the storage process is blocked by inactivating ACAT1, CE droplets disappear. These processes can be monitored quantitatively via Nile Red staining in living cells [43,44]. Here, we switched to use F12511 instead of K604 as the small molecule ACAT1 blocker for the following two reasons: Firstly, F12511 is a more potent ACAT1 inhibitor compared with K604 [43]. K604 has a K_i = 0.45 µmol/L, while F12511 has a K_i = 0.039 µmol/L for ACAT1 [46]. Secondly, F12511 has been widely studied [56–60], and (2) our laboratory recently developed a drug delivery system using F12511 to treat a mouse model for Alzheimer's disease [59,60]. Thus, we moved forward with the use of F12511 in later experiments as these results can help future studies in vivo. As shown in Figure 3C, 24 h of myelin debris treatment increased the Nile Red signal in HMC3 cells by 71%, while ACAT1 inhibitor F12511 treatment largely reduced the Nile Red signal by 65%, These data agree with the ^{3}H-oleate pulse data in the HMC3 cell line (Figures 2B.II and 3B). We also found that, when comparing the cell population in the F12511-treated group (without myelin) to that in the F12511-treated group (with myelin), the latter population showed a higher level of Nile Red staining. (Figure 3C, comparing the second box to the fourth box). This result suggests that myelin debris-loaded HMC3 cells may increase other neutral lipids that are not CEs.

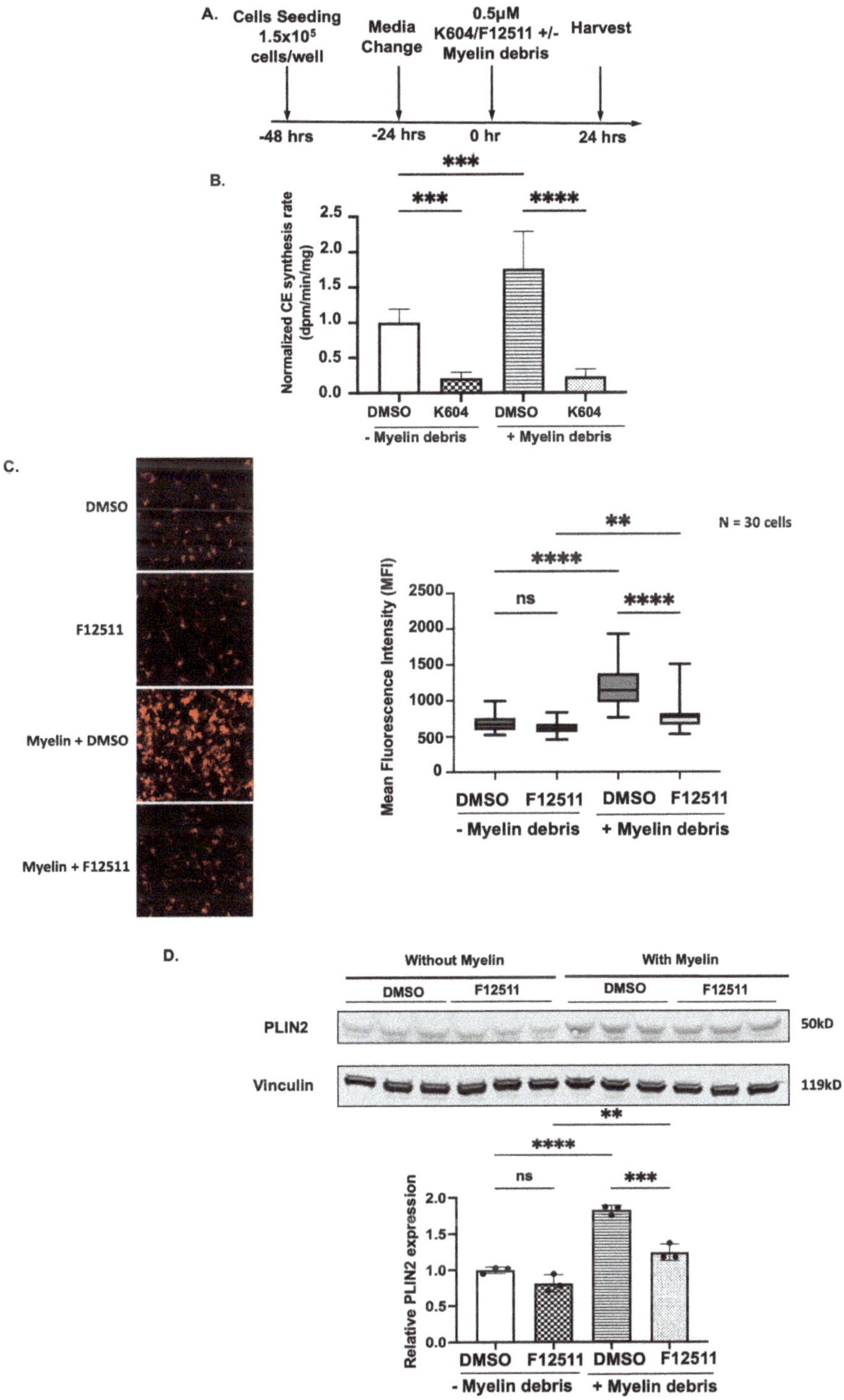

Figure 3. *Cont.*

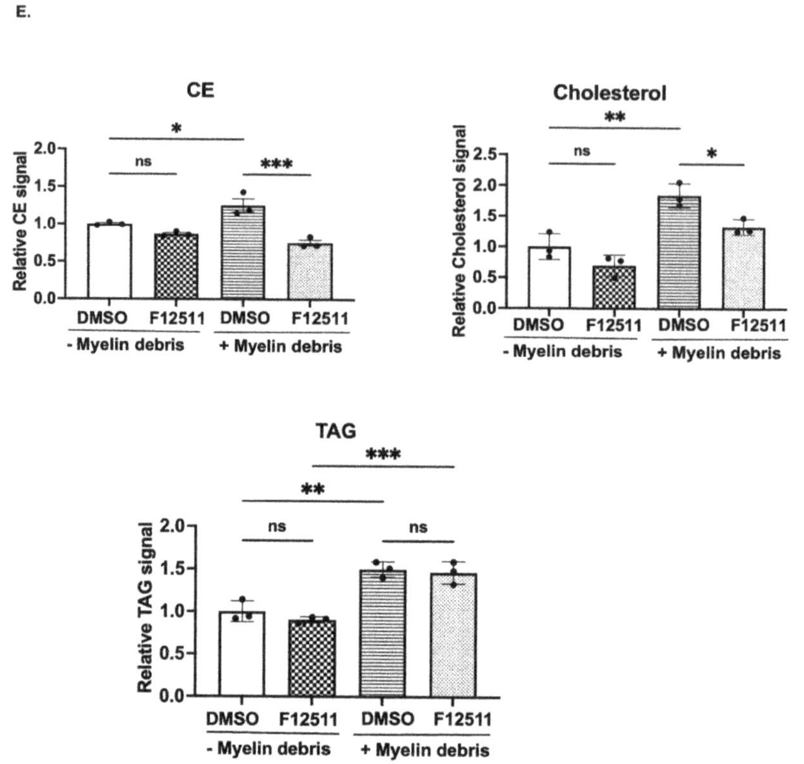

Figure 3. ACAT1 inhibition by K604 or F12511 reduces cholesteryl ester accumulation in HMC3 microglia treated with myelin debris. (**A**) Treatment scheme for all experiments. (**B**) HMC3 cells grown to 80% confluency were treated with DMSO, either with or without the ACAT1/SOAT1 inhibitor K604, or with myelin debris at 25 μg/mL cholesterol for 24 h as described in Section 2. Cholesteryl ester biosynthesis rates in intact cells were monitored by providing ^{3}H-oleate to HMC3 for 2 h. Cholesterol content in myelin debris was measured using the Cholesterol Wako kit. Raw radioactive counts were normalized to 1, yielding 464.9 or 723.4 dpm/min/mg. (**C**) Representative image and quantification of neutral lipid staining by Nile Red staining, demonstrating lipid droplet accumulation in HMC3 cells. (**D**) Representative perilipin 2 (PLIN2) Western blot and quantification in HMC3 cells. (**E**) Quantification of TLC analysis for intracellular CE and free (unesterified cholesterol) and TAG in HMC3 cells. Values were calculated based on cells without myelin or ACAT1 inhibitors (K604 or F12511) treatment, normalized to 1. N = 3–6 for all experiments. Data are expressed as mean ± SEM. * $p < 0.05$; ** $p < 0.01$; *** $p < 0.001$; **** $p < 0.0001$. Western blot original images are in the Supplementary Materials.

To validate the Nile Red staining data, we used Perilipin 2 (PLIN2), a protein marker of lipid droplets, to quantify lipid droplet levels in HMC3 cells under different treatment conditions. PLIN2 has been widely used by different laboratories as a marker to monitor lipid droplets [24,61,62]. As expected, PLIN2 Western blot data support our Nile Red staining data (Figure 3D): The PLIN2 signal in myelin debris-treated cells increased by 80% compared with DMSO-treated cells with no myelin debris (Figure 3D, third bar vs. first bar), and treatment with F12511 in myelin debris-treated cells reduced the PLIN2 signal by approximately 57% compared with myelin debris-treated cells alone (Figure 3D, fourth bar vs. third bar). Our results, shown in Figure 3C,D, are consistent with previously published findings that myelin debris treatment increases not only CEs but also other neutral lipids [7,24]. The identities of these non-CE neutral lipid species are largely unknown. As our ^{3}H-oleate pulse results for both N9 and HMC3 cells (Figure 2C) show that myelin debris does not increase biosynthesis of oleate-derived TAG, it is possible that these non-CE, neutral lipid species may not use oleyl CoA as their precursor. In other words, myelin debris may increase the biosynthesis of certain neutral lipid species that do not contain oleate.

The results described above show that the HMC3 cell line is a good model by which to study cholesterol homeostasis in myelin debris-loaded microglia. We next assessed the effect of ACAT1 inhibition on intracellular cholesterol and cholesterol ester content in HMC3 microglia. We used thin-layer chromatography (TLC) to separate and quantify cellular free cholesterol (unesterified cholesterol), CE and TAG content, based on the method described in Macala et al. [45] and Harned et al. [34]. Our data (Figure 3E, top two panels) show that, while myelin debris increased both CE by 125% and free cholesterol content by 180% in HMC3 cells, interestingly, F12511 treatment not only reduced CE content by 50%, but also reduced free cholesterol content in myelin debris-treated cells by 51%. We believe this result is likely due to ACAT1 inhibition leading to an increase in cellular cholesterol efflux (to facilitate cellular cholesterol disposal). The action of ACAT inhibitors on cellular cholesterol efflux has been previously documented in many cell lines, but not in microglia [16,63].

The results in Figure 3E, left panel, show that a significant amount of CE is present in HMC3 cells without myelin debris treatment. Treating cells with F12511 for 24 h did not cause a large reduction in this "ACAT inhibitor insensitive" CE content. The origin of these CEs is unknown. We speculate that they might be derived from materials present in serum, that they may reside in the endolysosome compartment, and that their metabolic fate is not directly linked with ACAT1.

TAG is mainly synthesized by diglyceride acyltransferase-1 and-2 (DGAT1 and DGAT2). Prakash and Manchada et al., of Chopra's lab, have recently reported that DGAT2 inhibitors could reduce lipid droplets in microglia induced by amyloid beta (Aβ) peptides [64]. However, their study did not use myelin debris-treated microglia and did not include use of ACAT inhibitors. Here, instead of testing the possible involvement of DGAT, we monitored the total TAG content upon myelin debris treatment with or without ACAT inhibitor and report the result in Figure 3E (bottom panel). The results show that myelin debris incubation did increase TAG content by 150%, and that treatment with F12511 did not impact the TAG level. Overall, these data are consistent with the data from the study of Nugent et al. in bone narrow-derived macrophages (BMDM) [7]. They have shown that both the diacylglycerides (DAG) and TAG contents are upregulated in BMDM, when incubated with myelin debris [7]. Myelin lipids contain a very high amount of saturated long chain fatty acids [65,66]. We speculate that, upon entering the cell interior, the saturated long-chain fatty acids become fatty acyl CoA to serve as a substrate for DGATs (or perhaps other enzymes) to produce TAG [40]; this process may not be easily detectable using a ^{3}H-oleate pulse. Further investigations are needed to investigate the myelin debris–TAG connection.

3.4. Treatments with Myelin Debris and/or ACAT1 Inhibitor F12511 Do Not Change ACAT1 Protein Expression in HMC3 Cells

As myelin debris treatment increased CE synthesis and cellular CE content, while F12511 reduced these levels in HMC3 cells, we next asked whether these treatments affect ACAT1 protein expression. ACAT1 is a membrane-bound enzyme located in the endoplasmic reticulum (ER) and ER-associated membrane [34]. Previous work has shown that, in many cell types examined, the primary mode of ACAT1 regulation is attributed to allosteric regulation by its substrates, cholesterol, and oxysterols (as reviewed in [67]). F12511 is a potent inhibitor of ACAT1 [48], but F12511 may or may not alter ACAT1 protein expression. We analyzed whole HMC3 cell lysates treated with and without myelin debris, with and without F12511, with the experimental plan shown in Figure 4A, a representative Western blot shown in Figure 4B, and the quantitative analysis shown in Figure 4C. Our analysis demonstrated that ACAT1 protein content does not change under any experimental condition, with or without myelin debris, and with or without F12511. As myelin debris is cholesterol-rich, the increase in CE biosynthesis, as demonstrated in Figure 2B.I,B.II and Figure 3B, is likely due to the increased availability of cholesterol as substrate to ACAT1 in the ER when cells are exposed to myelin debris.

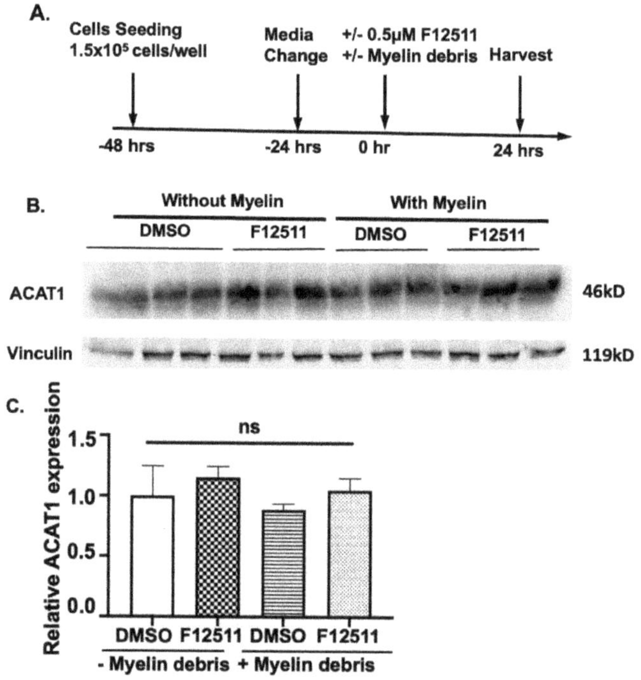

Figure 4. Treatment with myelin debris and/or the ACAT inhibitor K604 does not alter ACAT1 protein content in HMC3 cells. (**A**) Treatment scheme. HMC3 cells grown to 80% confluency were treated with DMSO, either with or without ACAT1 inhibitor F12511, or with myelin debris at 25 µg/mL cholesterol for 24 h. (**B**) Representative western blot. Cells were then harvested for Western blot analyses, with vinculin used as the loading control. (**C**) Quantification of western blot analysis. Values were calculated using the untreated cells (no myelin or K604) as 1. N = 3. Data are expressed as mean ± SEM. NS, not significant. Western blot original images are in the Supplementary Materials.

3.5. Pharmaceutical Inhibition of ACAT1 for 24 h in Myelin Debris-Loaded HMC3 Microglia Upregulates the Protein Expression of the Cholesterol Efflux Transporter ABCA1

Cholesterol efflux is an important step to re-establishing cholesterol homeostasis in microglia and restoring their proper function (reviewed in [28]). Removal of excess cholesterol in cells occurs through various ABC transporters, most prominently through ABCA1 [23]. We used two different ACAT1 inhibitors, K604 and F12511, to monitor ABCA1 protein content in the whole cell lysate of HMC3 cells loaded with myelin debris for 24 h. HEK-293 cells, which lack ABCA1, were used as a negative control [16]. ABCA1 is known to be highly upregulated by LXR agonists [68]. We used the LXR agonist T0901317 to serve as a positive control [35]. The results reported in Figure 5A,B indicate that total ABCA1 protein content is significantly upregulated with either K604 or F12511 when compared with cells treated with myelin debris alone. Quantitative analyses revealed approximately 32% and 36% increases in ABCA1 protein content in myelin debris-loaded cells treated with ACAT1 inhibitors K604 and F12511, respectively.

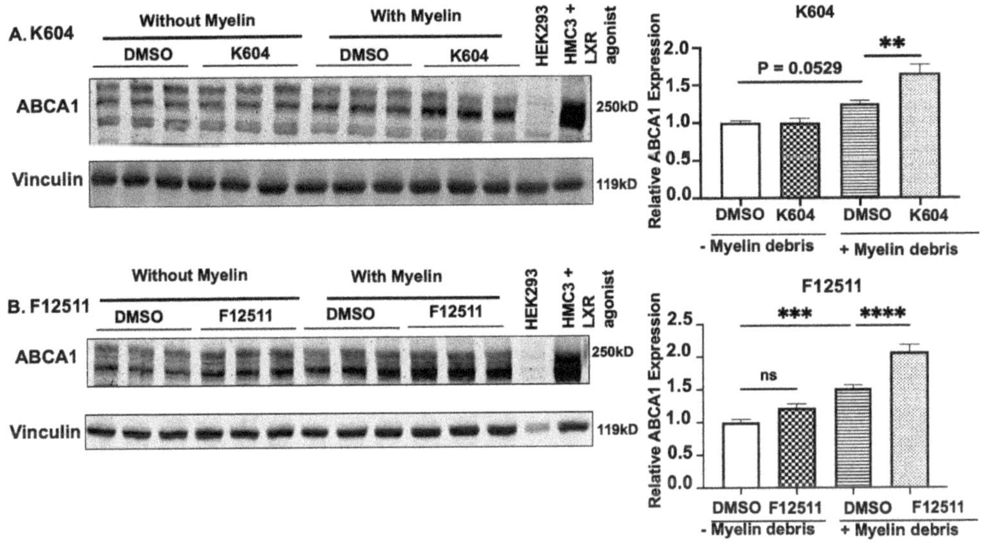

Figure 5. ABCA1 protein content increased after 24 h treatment with human myelin debris co-incubated with either K604 or F12511. Human microglial HMC3 cells grown to 80% confluency and treated with (**A**) 0.5μM K604 or (**B**) 0.5μM F12511, incubated with or without human myelin debris at 25 μg/mL cholesterol. Cells were then harvested for Western blot analyses. HEK293 cell lysate was used as a negative control for ABCA1, and HMC3 treated with 10 μM LXR agonist T0901317 for 24 h served as a positive control for ABCA1. Vinculin was the loading control. Values were calculated based on the untreated condition (no myelin and without K604 or F12511), normalized to 1. N = 3–6. Data are expressed as mean ± SEM. ** $p < 0.01$; *** $p < 0.001$; **** $p < 0.0001$. Western blot original images are in the Supplementary Materials.

We also observed that, while myelin debris loading in HMC3 cells increased ABCA1 protein content by 25–50% (Figure 5A,B), adding ACAT1 inhibitors further increased ABCA1 protein content. These results, along with the findings shown in Figure 3E, suggest that ACAT1 inhibitors enhanced cholesterol efflux activity via ABCA1 upregulation, resulting in a decrease in the intracellular cholesterol pool. We conclude that, in myelin debris-loaded microglia, ACAT1 inhibitors not only ameliorate CE accumulation but also enhance cellular cholesterol disposal through the upregulation of the ABCA1 protein content.

Previously, we had reported that, in mouse microglia N9 cells grown in medium-containing serum, adding K604 for up to 8 h had no detectable effect on cellular ABCA1 protein levels [32]. A comparison of our current results (Figure 5A,B) with previous results suggests that, in microglia, the effect of the ACAT inhibitor is dependent on cellular cholesterol loading. The magnitude of this effect is greater and easier to detect when cells are loaded with cholesterol-rich substances such as myelin debris.

3.6. F12511 Treatment in Myelin Debris-Loaded HMC3 Microglia Significantly Increases ABCA1 mRNA Expression

When ACAT1 is blocked, excess cholesterol released from the ACAT1 storage pool in the ER can participate in cholesterol trafficking to other membrane compartments [69]. This excess cholesterol can either convert to oxysterols to activate LXR, the master regulator that induces ABCA1 gene expression [68,70] or move to the plasma membrane to stabilize ABCA1 protein and decrease its turnover rate, thereby increasing cellular ABCA1 protein content [71]. Both events may occur. To determine whether ACAT1 inhibitor affects *ABCA1* gene expression, we treated HMC3 cells with or without myelin debris and with or without F12511 for 12 h, followed by RNA extraction and analysis using NanoString Elements technology (Figure 6A). NanoString Elements enables the quantification of a wide range of genes in a single reaction via hybridization and individual barcoding, without the amplification required by traditional qPCR.

Table 1. Primer sequence.

Gene	Probe A (5′-3′)	Probe B (5′-3′)
ABCA1	GGTCTGAGAG CCGGTCATCAAT CTCATGAAAGA GTTCCACAAA GGCTCCACAAT TCTGCGGGT TAGCAGGAAG GTTAGGGAAC	CGAAAGCC ATGACCTCCG ATCACTCGAA TATTTCTTCCA GGGTCGTCTC TGAGATGCCA TAACTAGAA ATGCCCA
PPARγ	GTACTCTTGAAG TTTCAGGTCAT ACTTGTAATCT GCAACCA CTGGATCTGCA TCCTCTTCTTT TCTTGGTGTT GAGAAGAT GCTC	CGAAAGCCAT GACCTCCGATC ACTCTTCTCAG AATAATAAGG TGGAGATGC AGGCTCCAC TTTGATTG CACTTTG
CH25H	ATGTCGAAGA AGCCCAAAG AAAACAGTT CCCAGACGCTC ATATACTGCGT CAAAGACGC CTATCTTCCA GTTTGATCG GGAAACT	CGAAAGCC ATGACCTCCG ATCACTCTG AGCGGGTGGC ACCCGAGCAGT GTGACGTTCATC
CYP27A1	GATGGATCGC TGCAGGCAGC CAATGCGTTTC TCGAACAGGAT GTAGCAAACT GTTGAGATTA TTGAGCTTCAT CATGACCAGAAG	CGAAAGCC ATGACCTCCGA TCACTCTTCTG GAACATTAACC CGATGGATCTG ACGAAGGTCAC GGTGTCCTCGGG
LXRα	AGGCAGCCACC AGGCCTCAGCC ATCCGGCCAAG AAAACAGAAAA TATGGGCCTCAA GACCTAAGCGA CAGCGTGACC TTGTTTCA	CGAAAGCCAT GACCTCCGATC ACTCAGGAAT GTTTGCCCTTC TCAGTCTGTT CCACTTCTAGG
ABCF1 (housekeeping)	CCAGCTTGATG TCAGATGCATT TTCTAACATGGCT TGGCGGGAGGA CATCCTTTCGGG TTATATCTATCATT TACTTGACACCCT	CGAAAGCCATG ACCTCCGATCACTCG TCTGCATTGACGAAC AGCTCCTTGCCATGA GCGGAGATGC TGAACTTCT

In addition to *ABCA1* gene expression, we also examined other related genes involved in intracellular cholesterol and lipid processing pathways that are known to be upstream of *ABCA1*, including peroxisome proliferator-activated receptor gamma (*PPARγ*) (lipid processing), liver X receptor alpha (*LXRα*) (a direct upstream regulator of ABCA1), cholesterol 25-hydroxylase (*CH25H*), and sterol 27-hydroxylase (*CYP27A1*) (oxysterol-converting enzymes). The data collected from this experiment were quantified and plotted in Figure 6B.

The results show that, firstly, and in agreement with our protein content analysis data (Figure 5), *ABCA1* gene expression is significantly upregulated in myelin debris-loaded HMC3 microglia by about 80% and is further increased by another 43% in cells treated with both myelin debris and F12511 (Figure 6B). This result suggests that ACAT1 inhibition can upregulate *ABCA1* gene expression in myelin debris-treated cells.

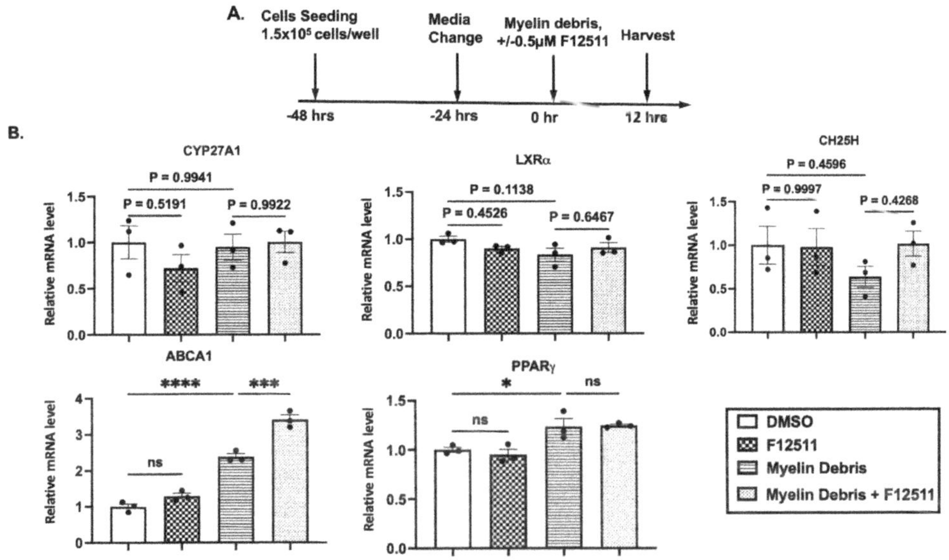

Figure 6. F12511 treatment in myelin debris-loaded HMC3 microglia significantly increases *ABCA1* mRNA expression. Myelin debris also activates *PPARγ* gene expression in HMC3 microglia. (**A**) Treatment scheme. (**B**) Treated HMC3 cells were lysed in Trizol, and RNA was purified. RNA samples were then analyzed using NanoString Elements. Bar graphs display the relative mRNA expression for key genes (*ABCA1, PPARγ, CH25H, CYP27A1* and *LXRa*) after normalization to the housekeeping gene (*ABCF1*). Gene expression in DMSO-treated cells (without myelin) was normalized to 1. N = 3. Data are expressed as mean ± SEM. * $p < 0.05$; *** $p < 0.001$; and **** $p < 0.0001$. Primers sequence are reported in Table 1.

Second, we observed an increase in *PPARγ* gene expression when cells were exposed to myelin debris. This result is consistent with previously published data in macrophages from a multiple sclerosis model [72] (Figure 6B). Most likely, phosphatidylserine (PS), a component of myelin debris, is responsible for the upregulation of *PPARγ* gene expression [72]. Treatment with F12511 did not alter *PPARγ* expression in myelin debris-treated cells (Figure 6B).

Third, regarding the two oxysterol-converting enzymes (CH25H and CYP27A1), we did not observe any significant changes across all treatment conditions, nor did we see changes in the direct upstream regulator of *ABCA1* gene expression (*LXRα*). We speculate that, in order to observe upregulation of *LXRα* and other oxysterol-converting enzyme gene expression, we may need to analyze treated cells at earlier time points, such as 4, 6, and 8 h. It is also possible that the activation of these genes is transient, and that they returned to normal levels by 12 h, while *ABCA1* gene expression remained upregulated at 12 h to induce ABCA1 protein expression for efflux at 24 h (Figures 5 and 6B).

3.7. Liver X Receptors (LXR) Antagonist GSK2033 Treatment Blocks ABCA1 Protein Expression in Myelin Debris and ACAT1 Inhibitor-Treated HMC3 Microglial Cells

As we were unable to observe *LXRα* gene upregulation at the 12 h treatment time point, we took an alternative approach to validate that the effect of the ACAT1 inhibitor on ABCA1 expression is LXR-dependent. We employed a specific LXR antagonist, GSK2033 [36,73], in our system (Figure 7A). Briefly, the LXR antagonist was added simultaneously with F12511 to these cells, and we monitored ABCA1 protein expression via Western blot (Figure 7A,B). As reported in Figure 7C, quantitative analysis of the Western blot results demonstrated that GSK2033 effectively diminished ABCA1 protein expression in both DMSO and F12511-treated cells with myelin debris, supporting our conclusion that the effect of F12511 on ABCA1 protein expression is LXR-dependent. To confirm whether diminishing ABCA1 content in myelin debris-loaded microglia treated with F12511 also abolishes its ability to reduce intracellular cholesterol, we preincubated cells with myelin debris and treated them with or without F12511 or with or without GSK2033 and followed this with TLC analysis. Our data in Figure 7D suggest that GSK2033 blocked F12511 action on intracellular cholesterol efflux, indicating that F12511 action in myelin debris-loaded HMC3 microglia is dependent on LXR.

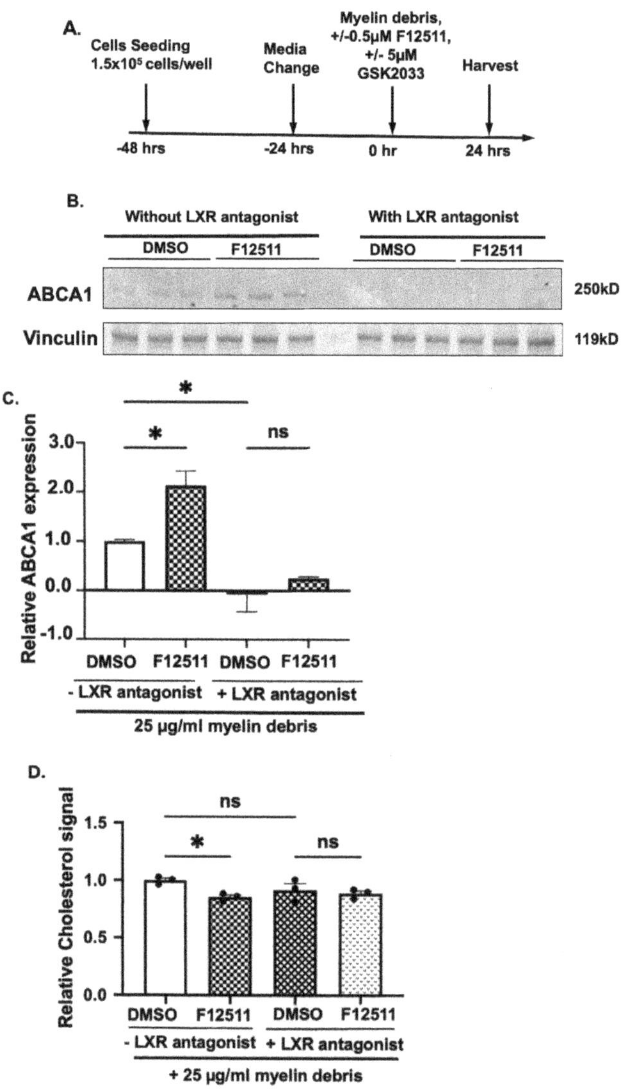

Figure 7. LXR antagonist GSK2033 treatment blocks ABCA1 protein expression in myelin debris and ACAT1 inhibitor-treated HMC3 microglial cells. (**A**) Treatment scheme. (**B**) Representative Western blot. (**C**) Quantification of the Western blot. Vinculin was used as the loading control. Values were calculated based on the myelin-treated cells without LXR agonist and F12511, normalized to 1. (**D**) Quantification of TLC analysis for intracellular free unesterified cholesterol in HMC3 cells. Values were calculated based on cells without LXR antagonist GSK2033 treatment, normalized to 1. N = 3 for all experiments. Data are expressed as mean ± SEM. * $p < 0.05$. Western blot original images are in the Supplementary Materials.

4. Discussion

Alzheimer's disease (AD) is classified as early onset (EOAD) and late onset (LOAD), with LOAD (affecting patients aged 65 or older) comprising more than 90% of all AD cases [74]. While mouse models for LOAD are becoming available [11], no simple cell culture model for LOAD is currently available. Myelin debris-loaded microglia have been considered as a cell model for aging. In our current work, we used two established microglia cell lines, mouse N9 and human HMC3, to gain a better understanding of how myelin debris-loaded microglia respond to the pharmaceutical inhibition of ACAT1.

Using ^{3}H-Oleate pulse-chase assay, we demonstrated that in both N9 and HMC3 cells, CE biosynthesis increases in a myelin debris dose-dependent manner (Figure 2). We then showed that, in myelin debris-loaded HMC3 microglia, inhibition of ACAT1 alleviates CE and cholesterol accumulation (Figure 3) without impacting the ACAT1 protein expression (Figure 4). We monitored the downstream effects of ACAT1/SOAT1 inhibition and found that treatment with two different ACAT1 inhibitors, K604 and F12511, upregulated ABCA1 protein expression (Figure 5) and ABCA1 gene expression (Figure 6). These results correlate with a decrease in cellular cholesterol content (Figure 3), suggesting that increased cholesterol efflux occurs in myelin debris-loaded HMC3 microglia upon ACAT1 inhibition. We then showed that the effect of the ACAT1 inhibitor on ABCA1 expression and on cellular cholesterol content are most likely LXR-dependent (Figure 7). We built a working model (illustrated in Figure 8) to explain the action of ACAT1 inhibition in myelin debris-loaded microglia. This model is elaborated in more detail below.

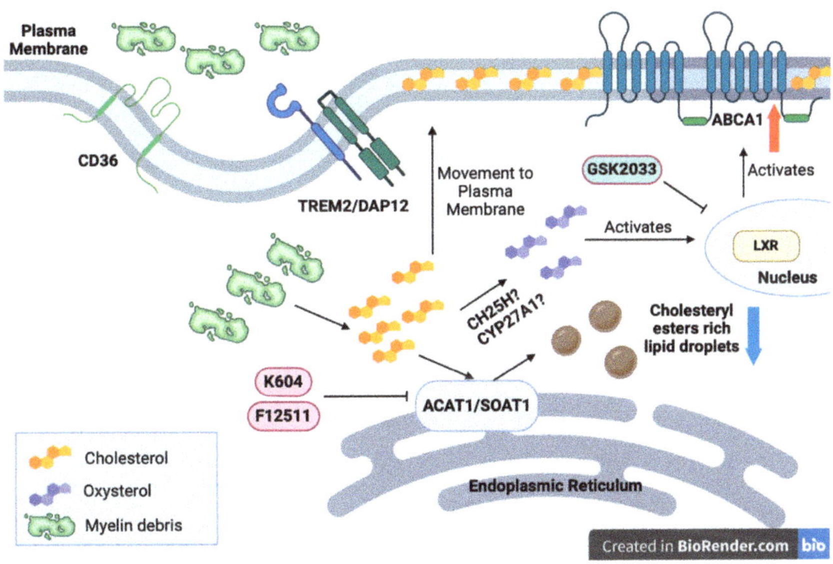

Figure 8. A working model illustrating the mechanism of ACAT1/SOAT1 inhibition in HMC3 microglia. With aging, myelin debris accumulates and is phagocytosed by microglia via receptors such as CD36, TREM2/DAP12, etc. Myelin debris is enriched in cholesterol, which can be stored as cholesteryl esters droplets by ACAT1/SOAT1 at the ER. This leads to CE lipid droplet accumulation in aging microglia. Pharmaceutical inhibition of ACAT1/SOAT1 by K604 or F12511 releases an extra pool of cholesterol to participate in cholesterol turnover. The cholesterol can then (1) migrate to the lipid–raft regions in the plasma membrane to serve as a substrate for the major lipid efflux protein and participate in cholesterol efflux, or (2) be converted into oxysterols by CH25OH, or other enzymes such as CYP27A1, which activates LXR and leads to the upregulation of ABCA1 expression for cholesterol efflux. Additionally, oxysterols can cross the blood–brain barrier into the periphery, restoring cholesterol homeostasis in the microglia and the brain.

As illustrated in Figure 8, myelin debris enters the cell surface of microglia via phagocytic receptors and unloads its cholesterol to the ER, where ACAT1 is localized. ACAT1 converts cholesterol unloaded from myelin debris into CEs for storage. Blocking ACAT1 activity diverts the cholesterol storage pool, to participate in the cholesterol translocation and turnover process. The diverted cholesterol pool can (1) migrate to the plasma membrane to serve as a substrate for efflux or (2) be converted into oxysterols, such as 27-OH cholesterol, 25-OH oxysterols, etc. Oxysterols can serve as ligands to activate LXR, which then upregulates *ABCA1* expression. ABCA1 protein is one of the most important proteins involved in the cellular cholesterol efflux process and has been linked to beneficial responses, including producing anti-inflammatory effects in various diseases (as reviewed in [75–78]). In macrophages, ABCA1 functions as an anti-inflammatory receptor [79].

The scheme described above provides a rationale by which to explain why blocking ACAT1 can restore cholesterol homeostasis in microglia. There are at least two possible mechanisms to explain ABCA1 upregulation by ACAT1 inhibition: (1) an increase in cholesterol availability at the plasma membrane as a substrate for ABCA1, and/or (2) an increase in oxysterols, perhaps by providing more cholesterol as a substrate to the enzymes such as CH25H, CYP27A1, or other oxysterol-converting enzymes. The increase in oxysterol concentration then activates the LXR pathway, which upregulates the ABCA1 protein and *ABCA1* mRNA expression [16,80].

Upregulation of ABCA1 can also be achieved through the use of synthetic LXR agonists. However, the difficulty of using synthetic LXR agonists is that they are known to also stimulate the gene expression of SREBP1-c, which overproduces fatty acids and leads to an increase in TG biosynthesis. LXR agonist is expected to lead to hypertriglyceridemia [81]. Here, we show that feeding myelin debris-treated microglia with the potent ACAT1 inhibitor F12511 inhibits CE synthesis and induces the ABCA1 gene expression. As a result, F12511 led to reductions in cellular CE and cholesterol contents, without increasing the TAG content in microglia. These results suggest that using ACAT1 inhibitor is an alternative to using synthetic LXR agonists to increase ABCA1 protein expression.

5. Conclusions

In this work, we demonstrated that ACAT1 inhibition activates ABCA1 gene expression in a LXR dependent manner in myelin debris loaded microglia. ACAT1 inhibition upregulates cholesterol efflux in cholesterol burden cells through ABCA1 activation. Our study highlights the role of ACAT1 as a promising therapeutic target for treating AD.

Besides aging, apolipoprotein E4 (*ApoE4*) is the biggest genetic risk factor for developing AD and has been linked to disruption in cholesterol homeostasis [22,62,82,83]. APOE4 expression in microglia-phagocytosed myelin debris leads to the accumulation of neutral lipids and lysosomal mass [61]. Additionally, APOE4 expression leads to aberrant cholesterol trafficking and perturbed cholesterol metabolism across various different cell types in the brain [22,62,82,83]. In APOE4 postmortem human brain tissue, CE is upregulated [82]. It remains unknown if ACAT1 inhibition would be beneficial in an aging APOE4 model, so this should be carefully investigated.

For future studies, these proposed mechanisms (Figure 8) need to be elucidated in microglia at the biochemical and cell biological levels. Additionally, oxysterols can be cleared from the brain into the periphery, which has been proposed as one of the main mechanisms for cholesterol disposal in the brain [84]. Previously, *ACAT1* genetic ablation and pharmaceutical inhibition have been reported to provide beneficial effects in various AD mouse models [17,59,85–88]. Our current work, along with the work of Nugent et al. [7], suggests that the pharmaceutical inhibition of ACAT1 may be a novel strategy by which to remove excess CE and cholesterol from the brain and attenuate the pro-inflammatory response in the brain. This possibility needs to be tested at the in vivo level.

Supplementary Materials: The following supporting information can be downloaded at: https://www.mdpi.com/article/10.3390/biom14101301/s1.

Author Contributions: T.N.H., C.C.Y.C. and T.Y.C. designed the research; T.N.H., performed the research; T.N.H., analyzed the data; T.Y.C., C.C.Y.C. and T.N.H. wrote and edited the manuscript. M.C.H. and G.J.Z. provided advice on human and mouse myelin and myelin debris. All authors have read and agreed to the published version of the manuscript.

Funding: This work was supported by NIH Grant R01 AG063544 (to T.Y.C and C.C.Y.C.) and by the Dartmouth PhD Innovation Program (to T.N.H). NIH/NIEHS Grant 2R01ES024745-06 (M.C.H), 1R01ES033462-0 (M.C.H). We acknowledge the shared facilities of the preclinical Imaging and Microscopy Resource and National Cancer Institute Cancer Center Support Grant 5P30 CA023108-37 at the Dartmouth Cancer Center, and NIH Grant P20-GM113132 to support the Institute for Biomolecular Targeting at Dartmouth. The authors acknowledge the Genomics and Molecular Biology Shared Resources at the Dartmouth Cancer Center with NCI Cancer Center Support Grant 5P30 CA023108-41. RRID: SCR_021293.

Institutional Review Board Statement: Ethics approval and consent to participants. All the experiments were performed in accordance with legal and institutional guidelines and were carried out under the ethics, consent and permissions of the Ethical Committee of Care and Use of Laboratory Animals at Geisel School of Medicine at Dartmouth (approval code: 00002020, 00002125; approval date: 13 August 2024, 9 January 2024).

Informed Consent Statement: Informed consent was obtained from all subjects involved in the study. Written informed consent has been obtained from the patients to publish this paper.

Data Availability Statement: The data presented in this study are available in this article.

Acknowledgments: We wish to thank the Anatomy Gifts Registry (Hanover, MD, USA) for their generosity supplying human brain samples. We thank Robert Hill, Christian Lytle, Haibo Li, Adrianna De La Torre, Taylor Harned and Junghoon Lee for their thoughtful feedback and support.

Conflicts of Interest: The authors declare no conflicts of interest.

References

1. Hughes, E.G.; Orthmann-Murphy, J.L.; Langseth, A.J.; Bergles, D.E. Myelin remodeling through experience-dependent oligodendrogenesis in the adult somatosensory cortex. *Nat. Neurosci.* **2018**, *21*, 696–706. [CrossRef] [PubMed]
2. Dietschy, J.M.; Turley, S.D. Thematic review series: Brain Lipids. Cholesterol metabolism in the central nervous system during early development and in the mature animal. *J. Lipid Res.* **2004**, *45*, 1375–1397. [CrossRef]
3. Hammel, G.; Zivkovic, S.; Ayazi, M.; Ren, Y. Consequences and mechanisms of myelin debris uptake and processing by cells in the central nervous system. *Cell Immunol.* **2022**, *380*, 104591. [CrossRef] [PubMed]
4. Hill, R.A.; Li, A.M.; Grutzendler, J. Lifelong cortical myelin plasticity and age-related degeneration in the live mammalian brain. *Nat. Neurosci.* **2018**, *21*, 683–695. [CrossRef]
5. Bartzokis, G.; Lu, P.H.; Tingus, K.; Mendez, M.F.; Richard, A.; Peters, D.G.; Oluwadara, B.; Barrall, K.A.; Finn, J.P.; Villablanca, P.; et al. Lifespan trajectory of myelin integrity and maximum motor speed. *Neurobiol. Aging* **2010**, *31*, 1554–1562. [CrossRef] [PubMed]
6. Williams, K.; Ulvestad, E.; Waage, A.; Antel, J.P.; McLaurin, J. Activation of adult human derived microglia by myelin phagocytosis in vitro. *J. Neurosci. Res.* **1994**, *38*, 433–443. [CrossRef]
7. Nugent, A.A.; Lin, K.; van Lengerich, B.; Lianoglou, S.; Przybyla, L.; Davis, S.S.; Llapashtica, C.; Wang, J.; Kim, D.J.; Xia, D.; et al. TREM2 Regulates Microglial Cholesterol Metabolism upon Chronic Phagocytic Challenge. *Neuron* **2020**, *105*, 837–854 e839. [CrossRef] [PubMed]
8. Cantuti-Castelvetri, L.; Fitzner, D.; Bosch-Queralt, M.; Weil, M.T.; Su, M.; Sen, P.; Ruhwedel, T.; Mitkovski, M.; Trendelenburg, G.; Lutjohann, D.; et al. Defective cholesterol clearance limits remyelination in the aged central nervous system. *Science* **2018**, *359*, 684–688. [CrossRef]
9. Marschallinger, J.; Iram, T.; Zardeneta, M.; Lee, S.E.; Lehallier, B.; Haney, M.S.; Pluvinage, J.V.; Mathur, V.; Hahn, O.; Morgens, D.W.; et al. Lipid-droplet-accumulating microglia represent a dysfunctional and proinflammatory state in the aging brain. *Nat. Neurosci.* **2020**, *23*, 194–208. [CrossRef]
10. Zadoorian, A.; Du, X.; Yang, H. Lipid droplet biogenesis and functions in health and disease. *Nat. Rev. Endocrinol.* **2023**, *19*, 443–459. [CrossRef]
11. Vitek, M.P.; Araujo, J.A.; Fossel, M.; Greenberg, B.D.; Howell, G.R.; Rizzo, S.J.S.; Seyfried, N.T.; Tenner, A.J.; Territo, P.R.; Windisch, M.; et al. Translational animal models for Alzheimer's disease: An Alzheimer's Association Business Consortium Think Tank. *Alzheimers Dement* **2020**, *6*, e12114. [CrossRef] [PubMed]

12. Cameron, B.; Landreth, G.E. Inflammation, microglia, and Alzheimer's disease. *Neurobiol. Dis.* **2010**, *37*, 503–509. [CrossRef]
13. Spangenberg, E.E.; Lee, R.J.; Najafi, A.R.; Rice, R.A.; Elmore, M.R.; Blurton-Jones, M.; West, B.L.; Green, K.N. Eliminating microglia in Alzheimer's mice prevents neuronal loss without modulating amyloid-beta pathology. *Brain* **2016**, *139*, 1265–1281. [CrossRef]
14. Long, J.M.; Holtzman, D.M. Alzheimer Disease: An Update on Pathobiology and Treatment Strategies. *Cell* **2019**, *179*, 312–339. [CrossRef]
15. Elmore, M.R.P.; Hohsfield, L.A.; Kramar, E.A.; Soreq, L.; Lee, R.J.; Pham, S.T.; Najafi, A.R.; Spangenberg, E.E.; Wood, M.A.; West, B.L.; et al. Replacement of microglia in the aged brain reverses cognitive, synaptic, and neuronal deficits in mice. *Aging Cell* **2018**, *17*, e12832. [CrossRef]
16. Yamauchi, Y.; Chang, C.C.; Hayashi, M.; Abe-Dohmae, S.; Reid, P.C.; Chang, T.Y.; Yokoyama, S. Intracellular cholesterol mobilization involved in the ABCA1/apolipoprotein-mediated assembly of high density lipoprotein in fibroblasts. *J. Lipid Res.* **2004**, *45*, 1943–1951. [CrossRef]
17. Bryleva, E.Y.; Rogers, M.A.; Chang, C.C.; Buen, F.; Harris, B.T.; Rousselet, E.; Seidah, N.G.; Oddo, S.; LaFerla, F.M.; Spencer, T.A.; et al. ACAT1 gene ablation increases 24(S)-hydroxycholesterol content in the brain and ameliorates amyloid pathology in mice with AD. *Proc. Natl. Acad. Sci. USA* **2010**, *107*, 3081–3086. [CrossRef] [PubMed]
18. Li, P.; Spann, N.J.; Kaikkonen, M.U.; Lu, M.; Oh da, Y.; Fox, J.N.; Bandyopadhyay, G.; Talukdar, S.; Xu, J.; Lagakos, W.S.; et al. NCoR Repression of LXRs Restricts Macrophage Biosynthesis of Insulin-Sensitizing Omega 3 Fatty Acids. *Cell* **2013**, *155*, 200–214. [CrossRef] [PubMed]
19. Rong, X.; Albert, C.J.; Hong, C.; Duerr, M.A.; Chamberlain, B.T.; Tarling, E.J.; Ito, A.; Gao, J.; Wang, B.; Edwards, P.A.; et al. LXRs regulate ER stress and inflammation through dynamic modulation of membrane phospholipid composition. *Cell Metab.* **2013**, *18*, 685–697. [CrossRef]
20. Sodhi, R.K.; Singh, N. Liver X receptors: Emerging therapeutic targets for Alzheimer's disease. *Pharmacol. Res.* **2013**, *72*, 45–51. [CrossRef]
21. Chen, W.; Chen, G.; Head, D.L.; Mangelsdorf, D.J.; Russell, D.W. Enzymatic reduction of oxysterols impairs LXR signaling in cultured cells and the livers of mice. *Cell Metab.* **2007**, *5*, 73–79. [CrossRef] [PubMed]
22. Litvinchuk, A.; Suh, J.H.; Guo, J.L.; Lin, K.; Davis, S.S.; Bien-Ly, N.; Tycksen, E.; Tabor, G.T.; Remolina Serrano, J.; Manis, M.; et al. Amelioration of Tau and ApoE4-linked glial lipid accumulation and neurodegeneration with an LXR agonist. *Neuron* **2024**, *112*, 384–403 e388. [CrossRef] [PubMed]
23. Oram, J.F.; Heinecke, J.W. ATP-binding cassette transporter A1: A cell cholesterol exporter that protects against cardiovascular disease. *Physiol. Rev.* **2005**, *85*, 1343–1372. [CrossRef] [PubMed]
24. Gouna, G.; Klose, C.; Bosch-Queralt, M.; Liu, L.; Gokce, O.; Schifferer, M.; Cantuti-Castelvetri, L.; Simons, M. TREM2-dependent lipid droplet biogenesis in phagocytes is required for remyelination. *J. Exp. Med.* **2021**, *218*, e20210227. [CrossRef] [PubMed]
25. Stansley, B.; Post, J.; Hensley, K. A comparative review of cell culture systems for the study of microglial biology in Alzheimer's disease. *J. Neuroinflamm.* **2012**, *9*, 115. [CrossRef]
26. Dello Russo, C.; Cappoli, N.; Coletta, I.; Mezzogori, D.; Paciello, F.; Pozzoli, G.; Navarra, P.; Battaglia, A. The human microglial HMC3 cell line: Where do we stand? A systematic literature review. *J. Neuroinflamm.* **2018**, *15*, 259. [CrossRef]
27. Akhter, R.; Shao, Y.; Formica, S.; Khrestian, M.; Bekris, L.M. TREM2 alters the phagocytic, apoptotic and inflammatory response to Abeta(42) in HMC3 cells. *Mol. Immunol.* **2021**, *131*, 171–179. [CrossRef]
28. Munoz Herrera, O.M.; Zivkovic, A.M. Microglia and Cholesterol Handling: Implications for Alzheimer's Disease. *Biomedicines* **2022**, *10*, 3105. [CrossRef]
29. Yamazaki, S.; Yamaguchi, K.; Someya, A.; Nagaoka, I.; Hayashida, M. Anti-Inflammatory Action of Dexmedetomidine on Human Microglial Cells. *Int. J. Mol. Sci.* **2022**, *23*, 10096. [CrossRef]
30. Baek, M.; Yoo, E.; Choi, H.I.; An, G.Y.; Chai, J.C.; Lee, Y.S.; Jung, K.H.; Chai, Y.G. The BET inhibitor attenuates the inflammatory response and cell migration in human microglial HMC3 cell line. *Sci. Rep.* **2021**, *11*, 8828. [CrossRef]
31. Wang, Y.; Peng, Y.; Yan, H. Commentary: Neuroinflammatory In Vitro Cell Culture Models and the Potential Applications for Neurological Disorders. *Front. Pharmacol.* **2021**, *12*, 792614. [CrossRef] [PubMed]
32. Shibuya, Y.; Chang, C.C.; Huang, L.H.; Bryleva, E.Y.; Chang, T.Y. Inhibiting ACAT1/SOAT1 in microglia stimulates autophagy-mediated lysosomal proteolysis and increases Abeta1-42 clearance. *J. Neurosci.* **2014**, *34*, 14484–14501. [CrossRef] [PubMed]
33. Li, H.; Huynh, T.N.; Duong, M.T.; Gow, J.G.; Chang, C.C.Y.; Chang, T.Y. ACAT1/SOAT1 Blockade Suppresses LPS-Mediated Neuroinflammation by Modulating the Fate of Toll-like Receptor 4 in Microglia. *Int. J. Mol. Sci.* **2023**, *24*, 5616. [CrossRef] [PubMed]
34. Harned, T.C.; Stan, R.V.; Cao, Z.; Chakrabarti, R.; Higgs, H.N.; Chang, C.C.Y.; Chang, T.Y. Acute ACAT1/SOAT1 Blockade Increases MAM Cholesterol and Strengthens ER-Mitochondria Connectivity. *Int. J. Mol. Sci.* **2023**, *24*, 5525. [CrossRef] [PubMed]
35. Schultz, J.R.; Tu, H.; Luk, A.; Repa, J.J.; Medina, J.C.; Li, L.; Schwendner, S.; Wang, S.; Thoolen, M.; Mangelsdorf, D.J.; et al. Role of LXRs in control of lipogenesis. *Genes Dev.* **2000**, *14*, 2831–2838. [CrossRef]
36. Cashikar, A.G.; Toral-Rios, D.; Timm, D.; Romero, J.; Strickland, M.; Long, J.M.; Han, X.; Holtzman, D.M.; Paul, S.M. Regulation of astrocyte lipid metabolism and ApoE secretionby the microglial oxysterol, 25-hydroxycholesterol. *J. Lipid Res.* **2023**, *64*, 100350. [CrossRef] [PubMed]

37. Rolfe, A.J.; Bosco, D.B.; Broussard, E.N.; Ren, Y. In Vitro Phagocytosis of Myelin Debris by Bone Marrow-Derived Macrophages. *J. Vis. Exp.* **2017**, *130*, 56322. [CrossRef]
38. Goldstein, J.L.; Dana, S.E.; Brown, M.S. Esterification of low density lipoprotein cholesterol in human fibroblasts and its absence in homozygous familial hypercholesterolemia. *Proc. Natl. Acad. Sci. USA* **1974**, *71*, 4288–4292. [CrossRef]
39. Chang, C.C.; Chang, T.Y. Cycloheximide sensitivity in regulation of acyl coenzyme A:cholesterol acyltransferase activity in Chinese hamster ovary cells. 2. Effect of sterol endogenously synthesized. *Biochemistry* **1986**, *25*, 1700–1706. [CrossRef]
40. Igal, R.A.; Wang, P.; Coleman, R.A. Triacsin C blocks de novo synthesis of glycerolipids and cholesterol esters but not recycling of fatty acid into phospholipid: Evidence for functionally separate pools of acyl-CoA. *Biochem. J.* **1997**, *324 Pt 2*, 529–534. [CrossRef]
41. Millar, J.S.; Stone, S.J.; Tietge, U.J.; Tow, B.; Billheimer, J.T.; Wong, J.S.; Hamilton, R.L.; Farese, R.V., Jr.; Rader, D.J. Short-term overexpression of DGAT1 or DGAT2 increases hepatic triglyceride but not VLDL triglyceride or apoB production. *J. Lipid Res.* **2006**, *47*, 2297–2305. [CrossRef] [PubMed]
42. Cheng, D.; Iqbal, J.; Devenny, J.; Chu, C.H.; Chen, L.; Dong, J.; Seethala, R.; Keim, W.J.; Azzara, A.V.; Lawrence, R.M.; et al. Acylation of acylglycerols by acyl coenzyme A:diacylglycerol acyltransferase 1 (DGAT1). Functional importance of DGAT1 in the intestinal fat absorption. *J. Biol. Chem.* **2008**, *283*, 29802–29811. [CrossRef] [PubMed]
43. Chang, C.C.; Sakashita, N.; Ornvold, K.; Lee, O.; Chang, E.T.; Dong, R.; Lin, S.; Lee, C.Y.; Strom, S.C.; Kashyap, R.; et al. Immunological quantitation and localization of ACAT-1 and ACAT-2 in human liver and small intestine. *J. Biol. Chem.* **2000**, *275*, 28083–28092. [CrossRef] [PubMed]
44. Cadigan, K.M.; Chang, C.C.; Chang, T.Y. Isolation of Chinese hamster ovary cell lines expressing human acyl-coenzyme A/cholesterol acyltransferase activity. *J. Cell Biol.* **1989**, *108*, 2201–2210. [CrossRef] [PubMed]
45. Macala, L.J.; Yu, R.K.; Ando, S. Analysis of brain lipids by high performance thin-layer chromatography and densitometry. *J. Lipid Res.* **1983**, *24*, 1243–1250. [CrossRef]
46. Chang, T.Y.; Chang, C.C.Y.; Harned, T.C.; De La Torre, A.L.; Lee, J.; Huynh, T.N.; Gow, J.G. Blocking cholesterol storage to treat Alzheimer's disease. *Explor. Neuroprotective Ther.* **2021**, *1*, 173–184. [CrossRef]
47. Ikenoya, M.; Yoshinaka, Y.; Kobayashi, H.; Kawamine, K.; Shibuya, K.; Sato, F.; Sawanobori, K.; Watanabe, T.; Miyazaki, A. A selective ACAT-1 inhibitor, K-604, suppresses fatty streak lesions in fat-fed hamsters without affecting plasma cholesterol levels. *Atherosclerosis* **2007**, *191*, 290–297. [CrossRef]
48. Lopez-Farre, A.J.; Sacristan, D.; Zamorano-Leon, J.J.; San-Martin, N.; Macaya, C. Inhibition of acyl-CoA cholesterol acyltransferase by F12511 (Eflucimibe): Could it be a new antiatherosclerotic therapeutic? *Cardiovasc. Ther.* **2008**, *26*, 65–74. [CrossRef]
49. Guan, C.; Niu, Y.; Chen, S.C.; Kang, Y.; Wu, J.X.; Nishi, K.; Chang, C.C.Y.; Chang, T.Y.; Luo, T.; Chen, L. Structural insights into the inhibition mechanism of human sterol O-acyltransferase 1 by a competitive inhibitor. *Nat. Commun.* **2020**, *11*, 2478. [CrossRef]
50. Knapp, P.E.; Benjamins, J.A.; Skoff, R.P. Epigenetic factors up-regulate expression of myelin proteins in the dysmyelinating jimpy mutant mouse. *J. Neurobiol.* **1996**, *29*, 138–150. [CrossRef]
51. Boggs, J.M. Myelin basic protein: A multifunctional protein. *Cell Mol. Life Sci.* **2006**, *63*, 1945–1961. [CrossRef] [PubMed]
52. Larocca, J.N.; Norton, W.T. Isolation of myelin. *Curr. Protoc. Cell Biol.* **2007**, *3*, Unit3 25. [CrossRef] [PubMed]
53. Chang, C.C.; Huh, H.Y.; Cadigan, K.M.; Chang, T.Y. Molecular cloning and functional expression of human acyl-coenzyme A:cholesterol acyltransferase cDNA in mutant Chinese hamster ovary cells. *J. Biol. Chem.* **1993**, *268*, 20747–20755. [CrossRef] [PubMed]
54. Fowler, S.; Brown, W.; Warfel, J.; Greenspan, P. Use of nile red for the rapid in situ quantitation of lipids on thin-layer chromatograms. *J. Lipid Res.* **1987**, *28*, 1225–1232. [CrossRef]
55. Greenspan, P.; Fowler, S.D. Spectrofluorometric studies of the lipid probe, nile red. *J. Lipid Res.* **1985**, *26*, 781–789. [CrossRef]
56. Junquero, D.; Bruniquel, F.; N'Guyen, X.; Autin, J.M.; Patoiseau, J.F.; Degryse, A.D.; Colpaert, F.C.; Delhon, A. F 12511, a novel ACAT inhibitor, and atorvastatin regulate endogenous hypercholesterolemia in a synergistic manner in New Zealand rabbits fed a casein-enriched diet. *Atherosclerosis* **2001**, *155*, 131–142. [CrossRef]
57. Junquero, D.; Oms, P.; Carilla-Durand, E.; Autin, J.; Tarayre, J.; Degryse, A.; Patoiseau, J.; Colpaert, F.C.; Delhon, A. Pharmacological profile of F 12511, (S)-2′,3′, 5′-trimethyl-4′-hydroxy-alpha-dodecylthioacetanilide a powerful and systemic acylcoenzyme A: Cholesterol acyltransferase inhibitor. *Biochem. Pharmacol.* **2001**, *61*, 97–108. [CrossRef]
58. Junquero, D.; Pilon, A.; Carilla-Durand, E.; Patoiseau, J.F.; Tarayre, J.P.; Torpier, G.; Staels, B.; Fruchart, J.C.; Colpaert, F.C.; Clavey, V.; et al. Lack of toxic effects of F 12511, a novel potent inhibitor of acyl-coenzyme A: Cholesterol O-acyltransferase, on human adrenocortical cells in culture. *Biochem. Pharmacol.* **2001**, *61*, 387–398. [CrossRef]
59. De La Torre, A.L.; Huynh, T.N.; Chang, C.C.Y.; Pooler, D.B.; Ness, D.B.; Lewis, L.D.; Pannem, S.; Feng, Y.; Samkoe, K.S.; Hickey, W.F.; et al. Stealth Liposomes Encapsulating a Potent ACAT1/SOAT1 Inhibitor F12511: Pharmacokinetic, Biodistribution, and Toxicity Studies in Wild-Type Mice and Efficacy Studies in Triple Transgenic Alzheimer's Disease Mice. *Int. J. Mol. Sci.* **2023**, *24(13)*, 11013. [CrossRef]
60. De La Torre, A.L.; Smith, C.; Granger, J.; Anderson, F.L.; Harned, T.C.; Havrda, M.C.; Chang, C.C.Y.; Chang, T.Y. Facile method to incorporate high-affinity ACAT/SOAT1 inhibitor F12511 into stealth liposome-based nanoparticle and demonstration of its efficacy in blocking cholesteryl ester biosynthesis without overt toxicity in neuronal cell culture. *J. Neurosci. Methods* **2022**, *367*, 109437. [CrossRef]

61. Machlovi, S.I.; Neuner, S.M.; Hemmer, B.M.; Khan, R.; Liu, Y.; Huang, M.; Zhu, J.D.; Castellano, J.M.; Cai, D.; Marcora, E.; et al. APOE4 confers transcriptomic and functional alterations to primary mouse microglia. *Neurobiol. Dis.* **2022**, *164*, 105615. [CrossRef] [PubMed]
62. Haney, M.S.; Palovics, R.; Munson, C.N.; Long, C.; Johansson, P.K.; Yip, O.; Dong, W.; Rawat, E.; West, E.; Schlachetzki, J.C.M.; et al. APOE4/4 is linked to damaging lipid droplets in Alzheimer's disease microglia. *Nature* **2024**, *628*, 154–161. [CrossRef] [PubMed]
63. Sugimoto, K.; Tsujita, M.; Wu, C.A.; Suzuki, K.; Yokoyama, S. An inhibitor of acylCoA: Cholesterol acyltransferase increases expression of ATP-binding cassette transporter A1 and thereby enhances the ApoA-I-mediated release of cholesterol from macrophages. *Biochim. Biophys. Acta* **2004**, *1636*, 69–76. [CrossRef] [PubMed]
64. Prakash, P.; Manchanda, P.; Paouri, E.; Bisht, K.; Sharma, K.; Wijewardhane, P.R.; Randolph, C.E.; Clark, M.G.; Fine, J.; Thayer, E.A.; et al. Amyloid beta Induces Lipid Droplet-Mediated Microglial Dysfunction in Alzheimer's Disease. *bioRxiv* **2023**, *6*, 543525. [CrossRef]
65. Poitelon, Y.; Kopec, A.M.; Belin, S. Myelin Fat Facts: An Overview of Lipids and Fatty Acid Metabolism. *Cells* **2020**, *9*, 812. [CrossRef] [PubMed]
66. Sastry, P.S. Lipids of nervous tissue: Composition and metabolism. *Prog. Lipid Res.* **1985**, *24*, 69–176. [CrossRef]
67. Rogers, M.A.; Liu, J.; Song, B.L.; Li, B.L.; Chang, C.C.; Chang, T.Y. Acyl-CoA:cholesterol acyltransferases (ACATs/SOATs): Enzymes with multiple sterols as substrates and as activators. *J. Steroid Biochem. Mol. Biol.* **2015**, *151*, 102–107. [CrossRef]
68. Tontonoz, P. Transcriptional and posttranscriptional control of cholesterol homeostasis by liver X receptors. *Cold Spring Harb. Symp. Quant. Biol.* **2011**, *76*, 129–137. [CrossRef]
69. Rogers, M.A.; Chang, C.C.Y.; Maue, R.A.; Melton, E.M.; Peden, A.A.; Garver, W.S.; Lee, J.; Schroen, P.; Huang, M.; Chang, T.Y. Acat1/Soat1 knockout extends the mutant Npc1 mouse lifespan and ameliorates functional deficiencies in multiple organelles of mutant cells. *Proc. Natl. Acad. Sci. USA* **2022**, *119*, e2201646119. [CrossRef]
70. Li, A.C.; Binder, C.J.; Gutierrez, A.; Brown, K.K.; Plotkin, C.R.; Pattison, J.W.; Valledor, A.F.; Davis, R.A.; Willson, T.M.; Witztum, J.L.; et al. Differential inhibition of macrophage foam-cell formation and atherosclerosis in mice by PPARalpha, beta/delta, and gamma. *J. Clin. Invest.* **2004**, *114*, 1564–1576. [CrossRef]
71. Hsieh, V.; Kim, M.J.; Gelissen, I.C.; Brown, A.J.; Sandoval, C.; Hallab, J.C.; Kockx, M.; Traini, M.; Jessup, W.; Kritharides, L. Cellular cholesterol regulates ubiquitination and degradation of the cholesterol export proteins ABCA1 and ABCG1. *J. Biol. Chem.* **2014**, *289*, 7524–7536. [CrossRef] [PubMed]
72. Bogie, J.F.; Jorissen, W.; Mailleux, J.; Nijland, P.G.; Zelcer, N.; Vanmierlo, T.; Van Horssen, J.; Stinissen, P.; Hellings, N.; Hendriks, J.J. Myelin alters the inflammatory phenotype of macrophages by activating PPARs. *Acta Neuropathol. Commun.* **2013**, *1*, 43. [CrossRef] [PubMed]
73. El-Gendy, B.E.M.; Goher, S.S.; Hegazy, L.S.; Arief, M.M.H.; Burris, T.P. Recent Advances in the Medicinal Chemistry of Liver X Receptors. *J. Med. Chem.* **2018**, *61*, 10935–10956. [CrossRef] [PubMed]
74. Mendez, M.F. Early-Onset Alzheimer Disease. *Neurol. Clin.* **2017**, *35*, 263–281. [CrossRef]
75. Tall, A.R.; Yvan-Charvet, L. Cholesterol, inflammation and innate immunity. *Nat. Rev. Immunol.* **2015**, *15*, 104–116. [CrossRef]
76. Bi, X.; Vitali, C.; Cuchel, M. ABCA1 and Inflammation: From Animal Models to Humans. *Arterioscler. Thromb. Vasc. Biol.* **2015**, *35*, 1551–1553. [CrossRef]
77. Ito, A.; Hong, C.; Rong, X.; Zhu, X.; Tarling, E.J.; Hedde, P.N.; Gratton, E.; Parks, J.; Tontonoz, P. LXRs link metabolism to inflammation through Abca1-dependent regulation of membrane composition and TLR signaling. *Elife* **2015**, *4*, e08009. [CrossRef]
78. He, P.; Gelissen, I.C.; Ammit, A.J. Regulation of ATP-binding cassette transporter A1 (ABCA1) expression: Cholesterol-dependent and-independent signaling pathways with relevance to inflammatory lung disease. *Respir. Res.* **2020**, *21*, 250. [CrossRef] [PubMed]
79. Tang, C.; Liu, Y.; Kessler, P.S.; Vaughan, A.M.; Oram, J.F. The macrophage cholesterol exporter ABCA1 functions as an anti-inflammatory receptor. *J. Biol. Chem.* **2009**, *284*, 32336–32343. [CrossRef]
80. Saito, H.; Tachiura, W.; Nishimura, M.; Shimizu, M.; Sato, R.; Yamauchi, Y. Hydroxylation site-specific and production-dependent effects of endogenous oxysterols on cholesterol homeostasis: Implications for SREBP-2 and LXR. *J. Biol. Chem.* **2023**, *299*, 102733. [CrossRef]
81. Repa, J.J.; Liang, G.; Ou, J.; Bashmakov, Y.; Lobaccaro, J.M.; Shimomura, I.; Shan, B.; Brown, M.S.; Goldstein, J.L.; Mangelsdorf, D.J. Regulation of mouse sterol regulatory element-binding protein-1c gene (SREBP-1c) by oxysterol receptors, LXRalpha and LXRbeta. *Genes Dev.* **2000**, *14*, 2819–2830. [CrossRef] [PubMed]
82. Blanchard, J.W.; Akay, L.A.; Davila-Velderrain, J.; von Maydell, D.; Mathys, H.; Davidson, S.M.; Effenberger, A.; Chen, C.Y.; Maner-Smith, K.; Hajjar, I.; et al. APOE4 impairs myelination via cholesterol dysregulation in oligodendrocytes. *Nature* **2022**, *611*, 769–779. [CrossRef] [PubMed]
83. Tcw, J.; Qian, L.; Pipalia, N.H.; Chao, M.J.; Liang, S.A.; Shi, Y.; Jain, B.R.; Bertelsen, S.E.; Kapoor, M.; Marcora, E.; et al. Cholesterol and matrisome pathways dysregulated in astrocytes and microglia. *Cell* **2022**, *185*, 2213–2233 e2225. [CrossRef] [PubMed]
84. Russell, D.W.; Halford, R.W.; Ramirez, D.M.; Shah, R.; Kotti, T. Cholesterol 24-hydroxylase: An enzyme of cholesterol turnover in the brain. *Annu. Rev. Biochem.* **2009**, *78*, 1017–1040. [CrossRef]

85. Hutter-Paier, B.; Huttunen, H.J.; Puglielli, L.; Eckman, C.B.; Kim, D.Y.; Hofmeister, A.; Moir, R.D.; Domnitz, S.B.; Frosch, M.P.; Windisch, M.; et al. The ACAT inhibitor CP-113,818 markedly reduces amyloid pathology in a mouse model of Alzheimer's disease. *Neuron* **2004**, *44*, 227–238. [CrossRef]
86. Huttunen, H.J.; Kovacs, D.M. ACAT as a drug target for Alzheimer's disease. *Neurodegener. Dis.* **2008**, *5*, 212–214. [CrossRef] [PubMed]
87. Puglielli, L.; Konopka, G.; Pack Chung, E.; Ingano, L.A.; Berezovska, O.; Hyman, B.T.; Chang, T.Y.; Tanzi, R.E.; Kovacs, D.M. Acyl-coenzyme A: Cholesterol acyltransferase modulates the generation of the amyloid beta-peptide. *Nat. Cell Biol.* **2001**, *3*, 905–912. [CrossRef]
88. Valencia-Olvera, A.C.; Balu, D.; Faulk, N.; Amiridis, A.; Wang, Y.; Pham, C.; Avila-Munoz, E.; York, J.M.; Thatcher, G.R.J.; LaDu, M.J. Inhibition of ACAT as a Therapeutic Target for Alzheimer's Disease Is Independent of ApoE4 Lipidation. *Neurotherapeutics* **2023**, *20*, 1120–1137. [CrossRef]

biomolecules

MDPI

Article

Screening of Anti-Prion Compounds Using the Protein Misfolding Cyclic Amplification Technology

Sandra Pritzkow [1,*], Isaac Schauer [1], Ananya Tupaki-Sreepurna [1], Rodrigo Morales [1,2] and Claudio Soto [1,*]

[1] Department of Neurology, Mitchell Center for Alzheimer's Disease and Related Brain Disorders, University of Texas Health Science Center at Houston, McGovern Medical School, Houston, TX 77030, USA; rodrigo.moralesloyola@uth.tmc.edu (R.M.)

[2] Centro Integrativo de Biologia y Quimica Aplicada (CIBQA), Universidad Bernardo O'Higgins, Santiago 8370993, Chile

* Correspondence: sandra.pritzkow@uth.tmc.edu (S.P.); claudio.soto@uth.tmc.edu (C.S.)

Abstract: Prion diseases are 100% fatal infectious neurodegenerative diseases affecting the brains of humans and other mammals. The disease is caused by the formation and replication of prions, composed exclusively of the misfolded prion protein (PrP^{Sc}). We invented and developed the protein misfolding cyclic amplification (PMCA) technology for in vitro prion replication, which allow us to replicate the infectious agent and it is commonly used for ultra-sensitive prion detection in biological fluids, tissues and environmental samples. In this article, we studied whether PMCA can be used to screen for chemical compounds that block prion replication. A small set of compounds previously shown to have anti-prion activity in various systems, mostly using cells infected with murine prions, was evaluated for their ability to prevent the replication of prions. Studies were conducted simultaneously with prions derived from 4 species, including human, cattle, cervid and mouse. Our results show that only one of these compounds (methylene blue) was able to completely inhibit prion replication in all species. Estimation of the IC50 for methylene blue inhibition of human prions causing variant Creutzfeldt-Jakob disease (vCJD) was 7.7 μM. Finally, we showed that PMCA can be used for structure-activity relationship studies of anti-prion compounds. Interestingly, some of the less efficient prion inhibitors altered the replication of prions in some species and not others, suggesting that PMCA is useful for studying the differential selectivity of potential drugs.

Keywords: prions; PMCA; Creutzfeldt-Jakob disease; therapeutic compounds; prion replication

1. Introduction

Prion diseases (PrDs) are fatal neurodegenerative disorders (NDs) affecting humans and various mammals, including sheep, goats, mink, cervids, cattle, felines and ungulates [1]. The underlying mechanism in PrDs involves the accumulation of the pathological form of the prion protein (PrP^{Sc}) leading to brain damage in the form of spongiform encephalopathy, neuronal loss, synaptic dysfunction and brain inflammation [1]. PrP^{Sc} forms by autocatalytic conversion of the host's normal prion protein (PrP^{C}) and this process can be spread infectiously between individuals [2]. Creutzfeldt-Jakob disease (CJD) is the most common PrD in humans, and it can appear in sporadic (sCJD), familial or infectious forms. Animal PrDs include bovine spongiform encephalopathy (BSE) in cattle, chronic wasting disease (CWD) in deer and elk, and scrapie in sheep. CJD is a rare disease; however, the heretical nature of the prion infectious agent, the reported transmission of the disease between cattle and humans generating variant CJD (vCJD), and the recent expansion of the number of cases and geographical location of CWD made it important to develop strategies for efficient treatment [1]. The concept that PrP^{Sc} is the only component of the infectious material and that cerebral accumulation of PrP^{Sc} leads to neurodegeneration and disease is almost universally accepted in the field [3]. The atomic resolution structures for both PrP^{C}

and PrP^{Sc} are known [4] (at least for some species) and animal models fully recapitulate all characteristics of human prion diseases [5].

Despite impressive knowledge about the molecular and cellular basis of PrDs, currently, there is not any approved treatment for inhibiting prion replication in CJD or any other prion disease [6,7]. Part of the difficulty in the development of therapeutic interventions is the lack of biologically relevant screening assays to identify candidate hit molecules. Our strategy was to use the Protein Misfolding Cyclic Amplification (PMCA) technology we previously invented and developed for in vitro prion replication [8,9] to screen for compounds capable of preventing prion formation and propagation. Over the past decade, PMCA has proven to be a great tool for studying replication of infectious prions, understanding the complex prion biology, and detecting with extremely high sensitivity tiny amounts of infectious prions [10]. The PMCA technology has enabled researchers, for the first time, to cyclically amplify the folding and biochemical properties of a protein in a manner conceptually analogous to the amplification of DNA by PCR [8,9,11]. PMCA allowed the generation of infectious prions in vitro, providing the strongest proof in favor of the prion hypothesis [10,12]. The technique has also permitted detection, for the first time, of infectious prions in the blood and urine of animals and humans, offering a great possibility for early diagnosis [13,14]. Indeed, PMCA (and a variation called RT-QuIC) are now routinely used for the diagnosis of CJD worldwide. With PMCA, it has been possible to address critical issues in the prion field, including prion strains, species barriers, and de novo generation of infectious particles [10,15–17]. The efficiency of PMCA and the faithfulness with which it reproduces prion biology (e.g., infectivity, strain diversity, species barriers) suggest that compounds interfering with PMCA amplification may represent good hits for therapeutic development [18]. On the contrary, in vitro, misfolded PrP particles generated de novo from recombinant proteins have been shown to adopt a different structure and it is not infectious [19–21]. Indeed, a study evaluating the infectivity of recPrP produced in 20,000 experiments showed that $recPrP^{Sc}$ amyloid was readily formed in a test tube but generated no infectivity [21]. The difference in our study is that we used the PMCA technology to faithfully direct the templated conversion of PrP^{C} into PrP^{Sc}. Furthermore, PMCA offers the possibility to test the effect of the compounds on multiple PrP^{Sc} strains and species, which may overcome the problem identified in previous reports that compounds can inhibit replication of prions from certain strains/species, but not others [22–24]. The main goal of this study was to provide proof-of-concept data that PMCA can be used as a rapid and biologically relevant in vitro screening assay to identify compounds able to block prion replication. The compounds tested were selected based on previously published results suggesting they can have anti-prion activity.

In this article, we used PMCA to test a small set of molecules which have been reported to have anti-prion activities in diverse experiments, including Congo red, quinacrine, curcumin, tannic acid, methylene blue, rhodanine, chlorpromazine and minocycline. Congo red is perhaps one of the oldest reported prion inhibitors. Congo red is the sodium salt of 3,3′-([1,1′-biphenyl]-4,4′-diyl)bis(4-aminonaphthalene-1-sulfonic acid) and is often used to stain for amyloid deposits. The first report of Congo red as a prion inhibitor was published in 1992 by Caughey and Race [25] and confirmed later in various studies using diverse model systems [26–29]. Quinacrine, an acridine derivative formerly used as an antimalarial drug, was shown in cellular models of prion replication to inhibit prion formation [30,31]. However, studies using models of human prion replication did not show significant activity [32] and clinical trials with this drug in CJD did not produce beneficial results [33,34]. Curcumin, a natural compound and a major component of the spice turmeric, was shown to inhibit the in vitro formation of protease-resistance PrP [35]. Tannic acid is a large polyphenolic compound, which is a specific form of tannin. Tannic acid is found in the nutgalls formed by insects on twigs of certain oak trees. Using cellular models and RT-QuIC, tannic acid was shown to prevent PrP^{Sc} formation [36,37]. Methylthioninium chloride, usually called methylene blue (MB), is a salt used as a dye and as a medication approved for the treatment of methemoglobinemia. MB is an inhibitor of nitric oxide synthase and guanylate cyclase. Experiments using prion-infected cells showed that MB efficiently blocked prion replication [31]. It was also shown by NMR that MB binds PrP at a surface cleft of a fibrillogenic region of the protein and prevents its aggregation [38]. Rhodanine is a 5-membered heterocyclic organic compound possessing a thiazolidine core and a derivative was shown to inhibit prion-induced neuroinflammation [39] as well as be able to inhibit Tau protein aggregation [40]. Chlorpromazine is an antipsychotic drug used to treat psychiatric disorders such as schizophrenia, which was shown to have a potent anti-prion activity in prion-infected cells [31,41]. Minocycline and other tetracyclines have been shown to interact with and reverse protease-resistant prion protein and intraperitoneal injection of the drug in a hamster model of prion disease showed an 81% increased survival time [42].

2. Materials and Methods

Prion-infected brain samples. As inocula to trigger prion replication we used brain homogenates from: (i) a wild-type mouse experimentally infected with the RML prion strain; (ii) a human affected by vCJD; (iii) a cow affected by BSE; and (iv) a white-tailed deer naturally infected by CWD. Ten percent weight/volume (*w*/*v*) brain homogenates were prepared in PBS and large debris were removed by centrifugation at 810× *g* at 4 °C for 1 min. The supernatants were aliquoted, snap-frozen in liquid nitrogen, and stored at −80 °C until use.

Preparation of PMCA substrates. 10% *w*/*v* brain homogenates were prepared in conversion buffer (PBS supplemented with 1% Triton X-100, 150 mM NaCl, and Complete, EDTA-free protease inhibitor). Large debris were removed by centrifugation at 810× *g* at 4 °C for 1 min. The supernatants were aliquoted, snap-frozen in liquid nitrogen, and stored at −80 °C until use. For RML prion replication, we used wild-type mouse brains. For vCJD, we used transgenic mice expressing human *PRNP* with 129M polymorphism (Tg 6815 line) kindly provided by Dr. Glenn Telling (Colorado State University). For CWD amplification, we used gene-targeted transgenic mice expressing deer *PRNP* (Tg Gt226Q), provided by Dr. Glenn Telling. For BSE prion replication, we used as substrate transgenic mice expressing bovine *PRNP* (TgBoPrP) which was also provided by Dr. Glenn Telling.

Prion replication by PMCA and screening of prion inhibitors. PMCA was performed as described previously [9]. Briefly, thin PCR tubes (Eppendorf, Cat. No. 951010022) were used to perform the experiments. A 220–250 mL volume of water was poured into the sonicator holder in every experiment. Each sonication cycle comprised 20 s of sonication at an amplitude of 13 and 29 min 40 s of incubation. The horn and converter of the sonicator were placed inside a 32 °C incubator. A total of 48 PMCA cycles (24 h) were performed. To trigger prion replication, different dilutions of prion-infected brain homogenate were added to the reaction including the respective PMCA substrate (see above). The final volume of the reaction was 100 µL. At the same time, potentially inhibitory compounds (Congo red, rhodanine, quinacrine, tannic acid, methylene blue, curcumin, chlorpromazine, minocycline, azure A and thionine acetate) were added at a final concentration of 100 µM. Stock solutions of the compounds were dissolved in DMSO at 10 mM, and diluted into the reaction to reach a 1% DMSO concentration in the tube.

Proteinase K (PK) digestion and western blotting. PMCA products and standard prion-laden brain homogenates were incubated with PK (100 µg/mL) for 1 h at 37 °C with agitation, using the conditions previously described in detail [9]. PK digestions were stopped by the addition of loading sample buffer and boiling for 10 min at 100 °C. Proteins were separated by SDS-PAGE and then transferred to 0.45 µm nitrocellulose membranes, which were blocked with 10% *w*/*v* dry milk for 1 h at RT and then probed with monoclonal antibody 6D11 unless stated otherwise.

3. Results

3.1. Screening of a Small Selection of Anti-Prion Compounds

To assess whether PMCA can be used for identifying compounds able to prevent prion replication, we tested the activity of eight compounds previously reported to inhibit prion propagation in diverse systems, mostly cells [7]. The compounds tested were rhodanine, Congo red, quinacrine, tannic acid, methylene blue, curcumin, chlorpromazine and minocycline. All compounds were dissolved in DMSO and diluted to reach a 1% v/v concentration of DMSO in the reaction. Before testing the compounds, we first studied whether the presence of 1% DMSO might interfere with PMCA efficiency. For this purpose, serial dilutions of vCJD, CWD and BSE were tested by PMCA in the presence or absence of 1% DMSO (Figure 1). The results showed that regardless of whether the reaction was conducted with or without DMSO, PMCA was successful in amplifying up to a 10^{-9} dilution of each brain-infected material, indicating that 1% DMSO does not interfere with PMCA.

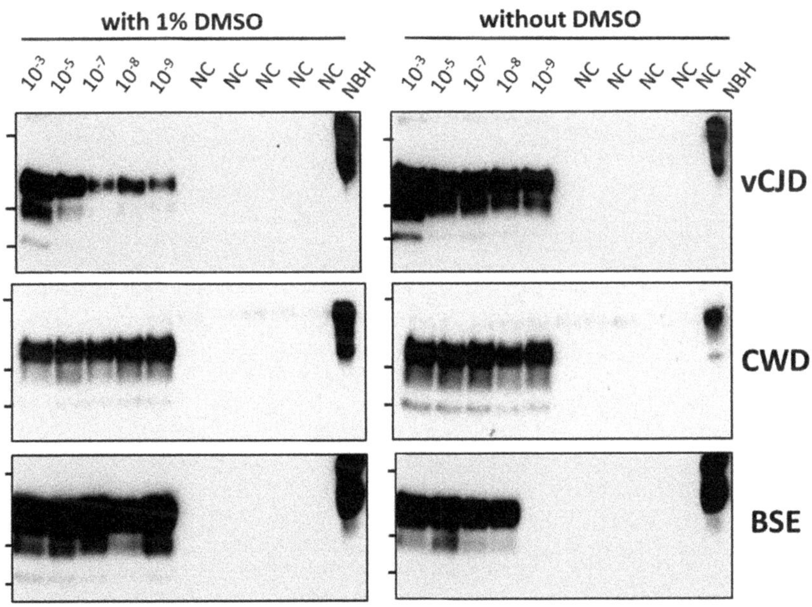

Figure 1. Effect of DMSO on the efficiency of prion replication. Since molecules to be tested are generally dissolved in DMSO, we tested whether addition of 1% DMSO (final concentration in the tube) produced any effect on prion replication. For this purpose, we incubated a series of dilutions of brain homogenate infected with vCJD, CWD and BSE with 1% DMSO and proceed to perform PMCA amplification in the presence of the respective substrate. After one round of 48 PMCA cycles (24 h), samples were analyzed by western blot after proteinase K digestion. Negative control (NC) consists of samples containing all materials except for PrPSc seeds. NBH correspond to the normal brain homogenate used for each amplification (i.e., wild type brain for rodent prions, transgenic mice expressing human PrP for vCJD prions, etc.). This is used as a migration control. Lanes on the left of each blot represent molecular weight standards (34 KDa, 26 KDa and 17 KDa). Please see the original Western blot image in the Supplementary Material.

Each compound was tested at a concentration of 100 µM using one round of PMCA and utilizing prions from 4 different species, including mouse RML, white-tailed deer CWD, cattle classical BSE, and human vCJD. Different dilutions of infected brain homogenate (from 10^{-3} to 10^{-7}) were used to seed prion replication. As positive controls, we used the same dilutions in the absence of any compound but included 1% DMSO (the vehicle used to dissolve the compounds), which did not change in any way prion replication in any of the species studied (Figures 1 and 2, left panels). The results show that only methylene blue (MB) was able to completely block prion replication in all species at this concentration (Figure 2). Interestingly, some molecules were able to inhibit prion replication in some species, but not others. For example, Congo red, tannic acid, quinacrine and curcumin partially inhibited mouse RML prion replication, but did not have any detectable effect with non-experimental prions in relevant species (Figure 2). The ability to study anti-prion compounds in various species at the same time represents one of the great advantages of using PMCA for screening.

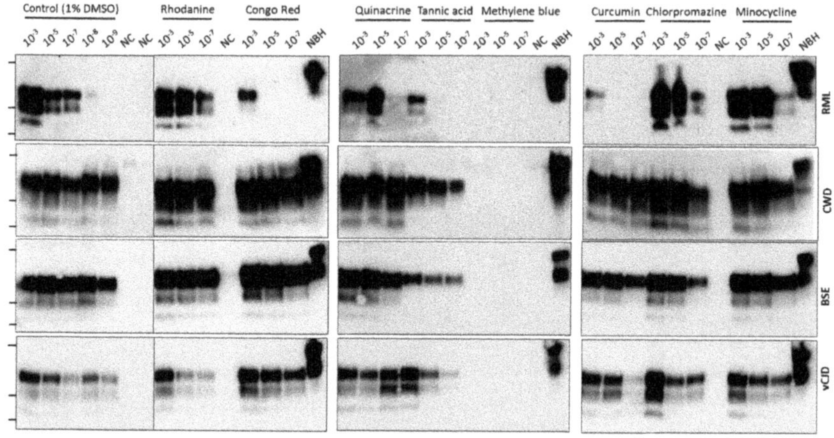

Figure 2. Using PMCA to evaluate anti-prion activity. Eight molecules previously reported as able to prevent prion replication in various models were tested at 100 μM concentration for their activity to prevent PrP^C to PrP^{Sc} conversion by PMCA of RML, CWD, BSE and vCJD prions. For testing inhibition we added 3 different quantities of PrP^{Sc} equivalent to a 10^{-3}, 10^{-5} and 10^{-7} dilution of infected brain homogenate. Control consists on PMCA in the absence of any compound but the vehicle used to solubilize the molecules (1% DMSO). Negative control (NC) consists of samples without addition of PrP^{Sc}. Samples were subjected to one round of 48 PMCA cycles (24 h) and analyzed by western blot after proteinase K digestion. NBH correspond to transgenic mice normal brain homogenate, used as a migration control. Lanes on the left of each blot represent molecular weight standards (34 KDa, 26 KDa and 17 KDa). Please see the original Western blot image in the Supplementary Material.

3.2. *Estimation of IC50 for Compounds' Activity*

Using PMCA, we can also estimate the half-maximal inhibitory concentration (IC50). For this purpose, we tested the inhibitory activity in prion replication of different dilutions of vCJD brain homogenate in the presence of distinct concentrations of MB (Figure 3A). The data shows that high concentrations of MB (>25 μM) completely block prion replication even when using high amounts of vCJD PrP^{Sc} (low dilutions of brain homogenate), whereas low MB concentrations (<2 μM) produced no significant effect on prion replication (Figure 3A). To calculate IC50 we plotted the inhibitory activity (expressed as the last dilution in which the signal is observed) versus the logarithm of MB concentration (Figure 3B). For MB, we estimated an IC50 of 7.7 μM against vCJD prions.

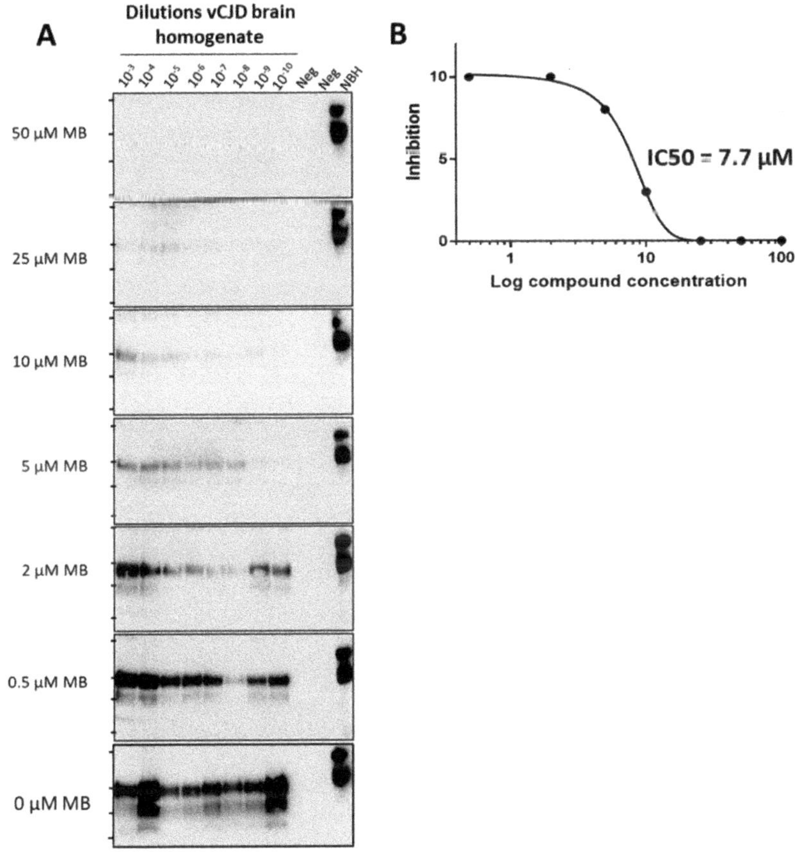

Figure 3. Estimation of IC50 for methylene blue anti-prion activity. (**A**), The effect of different concentrations of methylene blue (MB) on vCJD prion replication was evaluated at different dilutions of vCJD brain homogenate. Samples were incubated with the compound and 48 PMCA cycles (24 h) were done to asses prion replication by western blot. (**B**), The IC50 can be estimated from the experiment in panel A, by plotting the inhibitory activity (expressed as a last dilution in which signal is observed) versus the logarithm of the compound concentration. Negative control (Neg) consists of samples without addition of PrP^{Sc}. Samples were subjected to one round of 48 PMCA cycles (24 h) and analyzed by western blot after PK digestion. NBH correspond to transgenic mice normal brain homogenate, used as a migration control. Lanes on the left of each blot represent molecular weight standards. Please see the original Western blot image in the Supplementary Material.

3.3. Initial Structure-Activity Relationship Studies

To begin structure-activity relationship studies of the best inhibitor in this set, we searched for chemical derivatives of MB (3,7-bis(dimethylamino)-phenothiazin-5-ium chloride) and identified two compounds: azure A (N′,N′-dimethylphenothiazin-5-ium-3,7-diamine chloride) and thionine acetate (3,7-Diamino-5-phenothiazinium acetate) (Figure 4A) with similar chemical structure. We tested the effect of these derivatives in inhibiting prion replication of RML, CWD, BSE and vCJD PrP^{Sc}. The results show that while azure A retains activity, thionine acetate has a lower activity for CWD and vCJD (Figure 4B). Interestingly, all three compounds completely inhibited RML and BSE prions at a concentration of 100 μM.

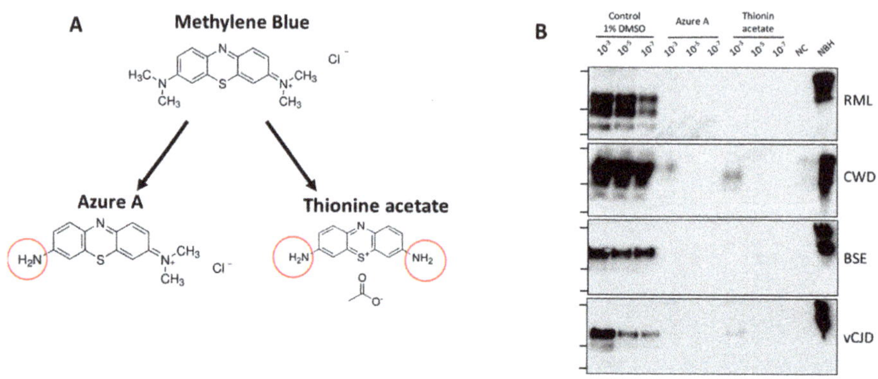

Figure 4. Preliminary structure-activity relationship studies. (**A**), Two close derivatives of MB, including azure A and thionine acetate were used to study the effect of chemical groups substitutions on MB activity. (**B**), Activity of the MB's derivatives was studied as in Figure 2. Negative control (NC) consists of samples without addition of PrP^{Sc}. Samples were subjected to one round of 48 PMCA cycles (24 h) and analyzed by western blot after PK digestion. NBH correspond to transgenic mice normal brain homogenate, used as a migration control. Lanes on the left of each blot represent molecular weight standards (34 KDa, 26 KDa and 17 KDa). Please see the original Western blot image in the Supplementary Material.

4. Discussion

PrDs remain 100% fatal and, hence, one of the largest unmet needs in the field is to identify drugs capable of stopping or slowing down the progression of these devastating diseases. The main obstacles to the development of effective anti-prion drugs are the rapidly progressing nature of the disease, the unorthodox nature of the prion infectious agent, the exponential increase in the quantity of prions during the disease because of self-propagating prion replication, the need for compounds to effectively cross the blood-brain barrier, and the rare and heterogeneous clinical presentation of the disease [6,7,43,44]. The key hallmark of PrDs is the misfolding and aggregation of the prion protein (PrP^{Sc}), which can self-propagate its misfolding at the expense of the cellular prion protein (PrP^{C}). Many studies have been conducted to identify and evaluate potential drug candidates [6,7,43,44]. A diversity of compounds and strategies have been studied, including various chemical and natural compounds targeting either PrP^{C},PrP^{Sc} or other putative players in the pathogenic mechanism [7,44–46]. Other proposed anti-prion treatments include passive and active immunization strategies, peptides, aptamers, and PrP^{C}-directed RNA interference techniques. A recent article by Zattoni and Legname [7], described a complete overview of the different reported strategies, including the list of compounds and the patents filed. Despite many publications and patents, only 6 clinical trials have been conducted so far to assess the therapeutic utility of diverse compounds, including flupirtine, quinacrine, doxycycline, pentosane polysulfate, the prion protein monoclonal antibody PRN100 and an anti-sense oligonucleotide [43]. Unfortunately, none of these trials produced significant therapeutic benefits.

In this study, we tested the effect of eight chemical compounds that have been shown previously to alter prion replication in different model systems. Prion replication was measured by PMCA, using simultaneously prions from four different species: mouse RML, human vCJD, cattle BSE and cervid CWD. Our results showed that five out of eight compounds tested (Congo red, tannic acid, curcumin, quinacrine and MB) produced some anti-prion activity (Figure 2). Interestingly, some of these molecules inhibited differentially some prions and not others. Of note, four of the five compounds only inhibited replication of experimental RML prions. These results suggest that searching for molecules using exclusively experimental prions in rodent models may not be translatable to human prions.

Out of the compounds tested, only MB produced a complete inhibition at the concentration tested in all prion species (Figure 2). However, since we measured prion replication at 24 h, it is not possible to rule out whether prion replication would still occur at longer times of the PMCA reaction. To study in more detail the inhibitory activity of MB, we used different concentrations of the molecule against vCJD prion replication (Figure 3). The results indicated that while concentrations of MB higher than 25 μM completely block prion replication, MB concentrations lower than 2 μM produced no significant effect on prion replication. We estimated the IC50 for MB as 7.7 μM (Figure 3). Finally, to begin attempting to understand the key chemical groups for activity, we tested the effect of two close derivatives of MB, azure A and thionine acetate (Figure 4). Thionine acetate only partially inhibited vCJD and CWD, while showing complete inhibition of RML and BSE prions (Figure 4). On the other hand, azure A showed an activity comparable to MB, except perhaps lower inhibitory capacity for CWD prions.

The results of this study suggest that PMCA is a powerful tool to identify and evaluate anti-prion compounds. The main advantage of PMCA is that it reproduces faithfully the process of prion replication while conserving the infectious and strain properties of prions. Furthermore, PMCA enables testing simultaneously various species and strains of prions, which is important since several molecules that were effective in murine prion models showed no activity in human studies. The main limitation of PMCA for drug screening is the low throughput of the assay. However, we believe this could be overcome by changing the format of the assay to ELISA plates, utilizing an easier readout (e.g., ELISA) and/or robotizing the assay. Another limitation is that PMCA can only test for inhibitors of prion replication and will miss molecules acting at other targets, such as prion neurotoxicity, clearance or expression.

Supplementary Materials: The following supporting information can be downloaded at: https://www.mdpi.com/article/10.3390/biom14091113/s1, The original Western blot image western blot image.

Author Contributions: Conceptualization, S.P., R.M. and C.S. Execution of experiments I.S. and A.T.-S. Data analysis, S.P. and C.S. Writing—original draft preparation, S.P.; writing—review and editing, R.M., and C.S. Funding acquisition, C.S. All authors have read and agreed to the published version of the manuscript.

Funding: This research was partially supported by grant P01AI077774 from the NIH to CS, and grants from the Creutzfeldt-Jakob Disease Foundation and NIH (1R01AI132695) to RM.

Institutional Review Board Statement: These studies were approved by the animal welfare committee, the biosafety committee and the institutional review board of the University of Texas Health Science Center at Houston (protocol number AWC-23-0020, approved on 28 April 2023).

Informed Consent Statement: Not applicable.

Data Availability Statement: All data and materials generated in this study will be made fully available to the scientific community by contacting the corresponding author.

Acknowledgments: We thank Glenn Telling for providing transgenic mice models of prion diseases to be used as PMCA substrates.

Conflicts of Interest: Claudio Soto is a Founder, Chief Scientific Officer and Member of the Board of Directors of Amprion Inc., a biotechnology company that focuses on the commercial use of PMCA and other seed amplification assays for high-sensitivity detection of misfolded protein aggregates involved in various neurodegenerative diseases. Sandra Pritzkow also has a conflict in relation to Amprion. The University of Texas Health Science Center at Houston has licensed patents and patent applications to Amprion. Rodrigo Morales is listed as an inventor in a patent associated with the PMCA technique. All other authors declare no conflicts.

References

1. Pritzkow, S.; Gorski, D.; Ramirez, F.; Soto, C. Prion Dissemination through the Environment and Medical Practices: Facts and Risks for Human Health. *Clin. Microbiol. Rev.* **2021**, *34*, e0005919. [CrossRef]

2. Prusiner, S.B. The prion diseases. *Brain Pathol.* **1998**, *8*, 499–513. [CrossRef] [PubMed]
3. Soto, C. Transmissible Proteins: Expanding the Prion Heresy. *Cell* **2012**, *149*, 968–977. [CrossRef] [PubMed]
4. Kraus, A.; Hoyt, F.; Schwartz, C.L.; Hansen, B.; Artikis, E.; Hughson, A.G.; Raymond, G.J.; Race, B.; Baron, G.S.; Caughey, B. High-resolution structure and strain comparison of infectious mammalian prions. *Mol. Cell* **2021**, *81*, 4540–4551. [CrossRef] [PubMed]
5. Moreno, J.A.; Telling, G.C. Insights into Mechanisms of Transmission and Pathogenesis from Transgenic Mouse Models of Prion Diseases. *Methods Mol. Biol.* **2017**, *1658*, 219–252. [CrossRef]
6. Zafar, S.; Noor, A.; Zerr, I. Therapies for prion diseases. *Handb. Clin. Neurol.* **2019**, *165*, 47–58. [CrossRef]
7. Zattoni, M.; Legname, G. Tackling prion diseases:A review of the patent landscape. *Expert. Opin. Ther. Pat.* **2021**, *31*, 1097–1115. [CrossRef]
8. Saborio, G.P.; Permanne, B.; Soto, C. Sensitive detection of pathological prion protein by cyclic amplification of protein misfolding. *Nature* **2001**, *411*, 810–813. [CrossRef]
9. Morales, R.; Duran-Aniotz, C.; Diaz-Espinoza, R.; Camacho, M.V.; Soto, C. Protein misfolding cyclic amplification of infectious prions. *Nat. Protoc.* **2012**, *7*, 1397–1409. [CrossRef]
10. Wang, F.; Pritzkow, S.; Soto, C. PMCA for ultrasensitive detection of prions and to study disease biology. *Cell Tissue Res.* **2023**, *392*, 307–321. [CrossRef]
11. Telling, G. Protein-based PCR for prion diseases? *Nat. Med.* **2001**, *7*, 778–779. [CrossRef] [PubMed]
12. Castilla, J.; Saá, P.; Hetz, C.; Soto, C. In vitro generation of infectious scrapie prions. *Cell* **2005**, *121*, 195–206. [CrossRef]
13. Concha-Marambio, L.; Pritzkow, S.; Moda, F.; Tagliavini, F.; Ironside, J.; Schulz, P.; Soto, C. Detection of prions in blood from patients with variant Creutzfeldt-Jakob disease. *Sci. Transl. Med.* **2016**, *8*, 370ra183. [CrossRef]
14. Castilla, J.; Saa, P.; Soto, C. Detection of prions in blood. *Nat. Med.* **2005**, *11*, 982–985. [CrossRef]
15. Castilla, J.; Morales, R.; Saa, P.; Barria, M.; Gambetti, P.; Soto, C. Cell-free propagation of prion strains. *EMBO J.* **2008**, *27*, 2557–2566. [CrossRef] [PubMed]
16. Castilla, J.; Gonzalez-Romero, D.; Saa, P.; Morales, R.; De, C.J.; Soto, C. Crossing the species barrier by PrP (Sc) replication in vitro generates unique infectious prions. *Cell* **2008**, *134*, 757–768. [CrossRef] [PubMed]
17. Barria, M.A.; Mukherjee, A.; Gonzalez-Romero, D.; Morales, R.; Soto, C. De novo generation of infectious prions in vitro produces a new disease phenotype. *PLoS Pathog.* **2009**, *5*, e1000421. [CrossRef]
18. do Carmo Ferreira, N.; Caughey, B. Cell-free prion protein conversion assays in screening for anti-prion drug candidates. *Curr. Opin. Pharmacol.* **2019**, *44*, 1–7. [CrossRef]
19. Manka, S.W.; Wenborn, A.; Collinge, J.; Wadsworth, J.D.F. Prion strains viewed through the lens of cryo-EM. *Cell Tissue Res.* **2023**, *392*, 167–178. [CrossRef]
20. Terry, C.; Harniman, R.L.; Sells, J.; Wenborn, A.; Joiner, S.; Saibil, H.R.; Miles, M.J.; Collinge, J.; Wadsworth, J.D.F. Structural features distinguishing infectious ex vivo mammalian prions from non-infectious fibrillar assemblies generated in vitro. *Sci. Rep.* **2019**, *9*, 376. [CrossRef]
21. Schmidt, C.; Fizet, J.; Properzi, F.; Batchelor, M.; Sandberg, M.K.; Edgeworth, J.A.; Afran, L.; Ho, S.; Badhan, A.; Klier, S.; et al. A systematic investigation of production of synthetic prions from recombinant prion protein. *Open Biol.* **2015**, *5*, 150165. [CrossRef] [PubMed]
22. Barret, A.; Tagliavini, F.; Forloni, G.; Bate, C.; Salmona, M.; Colombo, L.; De Luigi, A.; Limido, L.; Suardi, S.; Rossi, G.; et al. Evaluation of quinacrine treatment for prion diseases. *J. Virol.* **2003**, *77*, 8462–8469. [CrossRef]
23. Ghaemmaghami, S.; Ahn, M.; Lessard, P.; Giles, K.; Legname, G.; DeArmond, S.J.; Prusiner, S.B. Continuous quinacrine treatment results in the formation of drug-resistant prions. *PLoS Pathog.* **2009**, *5*, e1000673. [CrossRef]
24. Berry, D.B.; Lu, D.; Geva, M.; Watts, J.C.; Bhardwaj, S.; Oehler, A.; Renslo, A.R.; DeArmond, S.J.; Prusiner, S.B.; Giles, K. Drug resistance confounding prion therapeutics. *Proc. Natl. Acad. Sci. USA* **2013**, *110*, E4160–E4169. [CrossRef]
25. Caughey, B.; Race, R.E. Potent inhibition of scrapie-associated PrP accumulation by congo red. *J. Neurochem.* **1992**, *59*, 768–771. [CrossRef] [PubMed]
26. Poli, G.; Martino, P.A.; Villa, S.; Carcassola, G.; Giannino, M.L.; Dall'Ara, P.; Pollera, C.; Iussich, S.; Tranquillo, V.M.; Bareggi, S.; et al. Evaluation of anti-prion activity of congo red and its derivatives in experimentally infected hamsters. *Arzneimittelforschung* **2004**, *54*, 406–415. [CrossRef]
27. Rudyk, H.; Vasiljevic, S.; Hennion, R.M.; Birkett, C.R.; Hope, J.; Gilbert, I.H. Screening Congo Red and its analogues for their ability to prevent the formation of PrP-res in scrapie-infected cells. *J. Gen. Virol.* **2000**, *81*, 1155–1164. [CrossRef]
28. Demaimay, R.; Harper, J.; Gordon, H.; Weaver, D.; Chesebro, B.; Caughey, B. Structural aspects of Congo red as an inhibitor of protease-resistant prion protein formation. *J. Neurochem.* **1998**, *71*, 2534–2541. [CrossRef] [PubMed]
29. Caspi, S.; Halimi, M.; Yanai, A.; Sasson, S.B.; Taraboulos, A.; Gabizon, R. The anti-prion activity of Congo red. *Putative Mech. J. Biol. Chem.* **1998**, *273*, 3484–3489. [CrossRef]
30. Ryou, C.; Legname, G.; Peretz, D.; Craig, J.C.; Baldwin, M.A.; Prusiner, S.B. Differential inhibition of prion propagation by enantiomers of quinacrine. *Lab. Investig.* **2003**, *83*, 837–843. [CrossRef]
31. Korth, C.; May, B.C.; Cohen, F.E.; Prusiner, S.B. Acridine and phenothiazine derivatives as pharmacotherapeutics for prion disease. *Proc. Natl. Acad. Sci. USA* **2001**, *98*, 9836–9841. [CrossRef]

32. Collins, S.J.; Lewis, V.; Brazier, M.; Hill, A.F.; Fletcher, A.; Masters, C.L. Quinacrine does not prolong survival in a murine Creutzfeldt-Jakob disease model. *Ann. Neurol.* **2002**, *52*, 503–506. [CrossRef] [PubMed]
33. Geschwind, M.D.; Kuo, A.L.; Wong, K.S.; Haman, A.; Devereux, G.; Raudabaugh, B.J.; Johnson, D.Y.; Torres-Chae, C.C.; Finley, R.; Garcia, P.; et al. Quinacrine treatment trial for sporadic Creutzfeldt-Jakob disease. *Neurology* **2013**, *81*, 2015–2023. [CrossRef] [PubMed]
34. Collinge, J.; Gorham, M.; Hudson, F.; Kennedy, A.; Keogh, G.; Pal, S.; Rossor, M.; Rudge, P.; Siddique, D.; Spyer, M.; et al. Safety and efficacy of quinacrine in human prion disease (PRION-1 study):A patient-preference trial. *Lancet Neurol.* **2009**, *8*, 334–344. [CrossRef] [PubMed]
35. Caughey, B.; Raymond, L.D.; Raymond, G.J.; Maxson, L.; Silveira, J.; Baron, G.S. Inhibition of protease-resistant prion protein accumulation in vitro by curcumin. *J. Virol.* **2003**, *77*, 5499–5502. [CrossRef] [PubMed]
36. Kocisko, D.A.; Baron, G.S.; Rubenstein, R.; Chen, J.; Kuizon, S.; Caughey, B. New inhibitors of scrapie-associated prion protein formation in a library of 2000 drugs and natural products. *J. Virol.* **2003**, *77*, 10288–10294. [CrossRef]
37. Hyeon, J.W.; Kim, S.Y.; Lee, S.M.; Lee, J.; An, S.S.; Lee, M.K.; Lee, Y.S. Anti-Prion Screening for Acridine, Dextran, and Tannic Acid using Real Time-Quaking Induced Conversion: A Comparison with PrPSc-Infected Cell Screening. *PLoS ONE* **2017**, *12*, e0170266. [CrossRef]
38. Cavaliere, P.; Torrent, J.; Prigent, S.; Granata, V.; Pauwels, K.; Pastore, A.; Rezaei, H.; Zagari, A. Binding of methylene blue to a surface cleft inhibits the oligomerization and fibrillization of prion protein. *Biochim. Biophys. Acta (BBA)-Mol. Basis Dis.* **2013**, *1832*, 20–28. [CrossRef]
39. Villa, V.; Thellung, S.; Bajetto, A.; Gatta, E.; Robello, M.; Novelli, F.; Tasso, B.; Tonelli, M.; Florio, T. Novel celecoxib analogues inhibit glial production of prostaglandin E2, nitric oxide, and oxygen radicals reverting the neuroinflammatory responses induced by misfolded prion protein fragment 90-231 or lipopolysaccharide. *Pharmacol. Res.* **2016**, *113*, 500–514. [CrossRef]
40. Pickhardt, M.; Lawatscheck, C.; Borner, H.G.; Mandelkow, E. Inhibition of Tau Protein Aggregation by Rhodanine-based Compounds Solubilized Via Specific Formulation Additives to Improve Bioavailability and Cell Viability. *Curr. Alzheimer Res.* **2017**, *14*, 742–752. [CrossRef]
41. Stincardini, C.; Massignan, T.; Biggi, S.; Elezgarai, S.R.; Sangiovanni, V.; Vanni, I.; Pancher, M.; Adami, V.; Moreno, J.; Stravalaci, M.; et al. An antipsychotic drug exerts anti-prion effects by altering the localization of the cellular prion protein. *PLoS ONE* **2017**, *12*, e0182589. [CrossRef] [PubMed]
42. De, L.A.; Colombo, L.; Diomede, L.; Capobianco, R.; Mangieri, M.; Miccolo, C.; Limido, L.; Forloni, G.; Tagliavini, F.; Salmona, M. The efficacy of tetracyclines in peripheral and intracerebral prion infection. *PLoS ONE* **2008**, *3*, e1888.
43. Zerr, I. Investigating new treatments for Creutzfeldt-Jakob disease. *Lancet Neurol.* **2022**, *21*, 299–300. [CrossRef]
44. Panegyres, P.K.; Armari, E. Therapies for human prion diseases. *Am. J. Neurodegener. Dis.* **2013**, *2*, 176–186. [PubMed]
45. Charveriat, M.; Reboul, M.; Wang, Q.; Picoli, C.; Lenuzza, N.; Montagnac, A.; Nhiri, N.; Jacquet, E.; Gueritte, F.; Lallemand, J.Y.; et al. New inhibitors of prion replication that target the amyloid precursor. *J. Gen. Virol.* **2009**, *90*, 1294–1301. [CrossRef]
46. Bertsch, U.; Winklhofer, K.F.; Hirschberger, T.; Bieschke, J.; Weber, P.; Hartl, F.U.; Tavan, P.; Tatzelt, J.; Kretzschmar, H.A.; Giese, A. Systematic identification of antiprion drugs by high-throughput screening based on scanning for intensely fluorescent targets. *J. Virol.* **2005**, *79*, 7785–7791. [CrossRef] [PubMed]

biomolecules

MDPI

Article

A High-Throughput Screening of a Natural Products Library for Mitochondria Modulators

Emmanuel Makinde [1], Linlin Ma [1,2], George D. Mellick [1,2] and Yunjiang Feng [1,2,*]

[1] Griffith Institute for Drug Discovery, Griffith University, Brisbane, QLD 4111, Australia; emmanuel.makinde@griffithuni.edu.au (E.M.); linlin.ma@griffith.edu.au (L.M.); g.mellick@griffith.edu.au (G.D.M.)
[2] School of Environment and Science, Griffith University, Brisbane, QLD 4111, Australia
* Correspondence: y.feng@griffith.edu.au

Abstract: Mitochondria, the energy hubs of the cell, are progressively becoming attractive targets in the search for potent therapeutics against neurodegenerative diseases. The pivotal role of mitochondrial dysfunction in the pathogenesis of various diseases, including Parkinson's disease (PD), underscores the urgency of discovering novel therapeutic strategies. Given the limitations associated with available treatments for mitochondrial dysfunction-associated diseases, the search for new potent alternatives has become imperative. In this report, we embarked on an extensive screening of 4224 fractions from 384 Australian marine organisms and plant samples to identify natural products with protective effects on mitochondria. Our initial screening using PD patient-sourced olfactory neurosphere-derived (hONS) cells with rotenone as a mitochondria stressor resulted in 108 promising fractions from 11 different biota. To further assess the potency and efficacy of these hits, the 11 biotas were subjected to a subsequent round of screening on human neuroblastoma (SH-SY5Y) cells, using 6-hydroxydopamine to induce mitochondrial stress, complemented by a mitochondrial membrane potential assay. This rigorous process yielded 35 active fractions from eight biotas. Advanced analysis using an orbit trap mass spectrophotometer facilitated the identification of the molecular constituents of the most active fraction from each of the eight biotas. This meticulous approach led to the discovery of 57 unique compounds, among which 12 were previously recognized for their mitoprotective effects. Our findings highlight the vast potential of natural products derived from Australian marine organisms and plants in the quest for innovative treatments targeting mitochondrial dysfunction in neurodegenerative diseases.

Keywords: mitochondria modulators; high-throughput screening; natural products

1. Introduction

The mitochondrion, often referred to as the cell's "powerhouse" is pivotal for the functioning of eukaryotic cells, as it is responsible for most of the chemical energy supply needed to fuel cellular activities. This energy is mainly produced through oxidative phosphorylation (OXPHOS), a process that generates chemical energy stored as adenosine triphosphate (ATP), which is used to power complex biochemical processes in the cell [1,2]. In addition to their central role as the site of chemical energy generation, mitochondria are also crucial in regulating healthy cellular apoptosis, calcium homeostasis, biosynthesis of lipids and amino acids, and generating reactive oxygen species (ROS).

Neurons, with their high energy demands, contain hundreds of thousands of mitochondria, which are responsible for meeting most of their ATP needs for the functioning of the CNS [3–5]. The quality of these mitochondria is critical, as they must be highly functional to support the complex activities of neurons in the CNS [3]. Dysfunction in mitochondria, particularly in the OXPHOS system, has been linked to various diseases, notably neurodegenerative disorders like Parkinson's disease (PD), Huntington's disease, and Alzheimer's disease [2,3,6–9]. The mechanisms behind these links are multifaceted.

Firstly, diminished ATP production due to impaired mitochondrial function plays a significant role in the energy deficits observed in neurons affected by these diseases. Such deficits can compromise neuronal function and survival, potentially leading to cell death [3,5,10]. Secondly, about 90% of ROS are generated as by-products of the OXPHOS process [1,6,11,12]. Although ROS serve as signaling molecules under normal conditions, their overproduction or the failure of antioxidant defenses can induce oxidative stress, harming DNA, proteins, and lipids. This oxidative damage, a common feature of neurodegenerative diseases, likely drives further neuronal damage [3,5]. Thirdly, mitochondria play a key role in regulating intracellular calcium levels, which are crucial for various cellular processes, including neurotransmitter release, synaptic plasticity, and cell survival. Mitochondrial dysfunction can lead to dysregulated calcium homeostasis, exacerbating neuronal injury and death [1,6]. Fourthly, impaired mitophagy, a specific form of autophagy that removes damaged mitochondria from the cell, has been linked to neurodegenerative diseases such as PD [13,14]. Mutations in genes like PINK1 and Parkin, which are involved in mitophagy, can cause genetic PD, underscoring the importance of mitochondrial quality control in neurodegeneration [15,16]. Lastly, mitochondria play a crucial role in the intrinsic apoptosis pathway by releasing pro-apoptotic factors such as cytochrome C. Dysregulation of apoptotic signaling pathways due to mitochondrial dysfunction can trigger inappropriate neuronal cell death, contributing to neurodegeneration [12,17,18].

Recognizing the pivotal role of mitochondrial function in various diseases, extensive research has focused on identifying compounds that can modify mitochondrial functions. Compounds such as berberine, resveratrol, and epigallocatechin-3-gallate are known to trigger mitochondrial biogenesis and influence mitochondrial dynamics by promoting both fusion and fission [19–25]. Curcumin and its derivatives have been demonstrated to regulate mitochondrial dynamics to remedy dysfunction, and flavonoids like quercetin have shown potential in ameliorating memory impairment through mitochondrial regulation [26–31]. Additionally, compounds such as polydatin and acacetin have been found to induce mitophagy, enhancing mitochondrial function and offering protective effects in disease models [32,33]. Although these products show promise in enhancing mitochondrial function and offering disease model protection, the variability in the effectiveness of these compounds across different cell types and disease models highlights the need for more targeted approaches in their application and calls for novel, safe, and more potent mitochondrial modifiers.

Throughout history, natural products have served as the foundation of modern drug discovery. Notably, about half of FDA-approved drugs are either unmodified natural products or their synthetic derivatives [34,35]. Interest in natural product drug discovery seems to have waned over time; however, there has been a resurgence of interest in and commitment to natural product drug discovery in recent years [36,37]. This, in part, is due to technological advancement that has made it possible to screen thousands to hundreds of thousands of molecules against disease targets in a very short timeframe [36–38].

In our search for mitochondrial modulators with protective effects, we have established high-throughput screening assays and tested fractions sourced from a wide variety of Australian plants and marine sponges. These fractions have been obtained using a customized lead-like fractionation protocol developed in our laboratory [36]. To target lead-like natural products more efficiently, we diverged from the conventional approach of testing crude extracts. Instead, we directly examined 4224 fractions from 384 biota samples employing the CyQuant assay, a highly sensitive, quick, and robust cell viability assay [39]. Twenty fractions, representing 11 biota samples, were identified to protect the cells from rotenone, a mitochondrial complex I inhibitor [40]. Subsequently, the fractions from 11 active biota were further evaluated for their mitochondria modulatory activities using an MTT assay targeting mitochondrial nicotinamide adenine dinucleotide phosphate (NADPH)-dependent dehydrogenases [41,42], as well as a mitochondrial membrane potential (MMP) assay for protection against neurotoxins. Finally, the most active fractions were subjected to LC-Orbitrap MS analysis to identify natural products that can defend the mitochondria from toxic assaults (Figure 1).

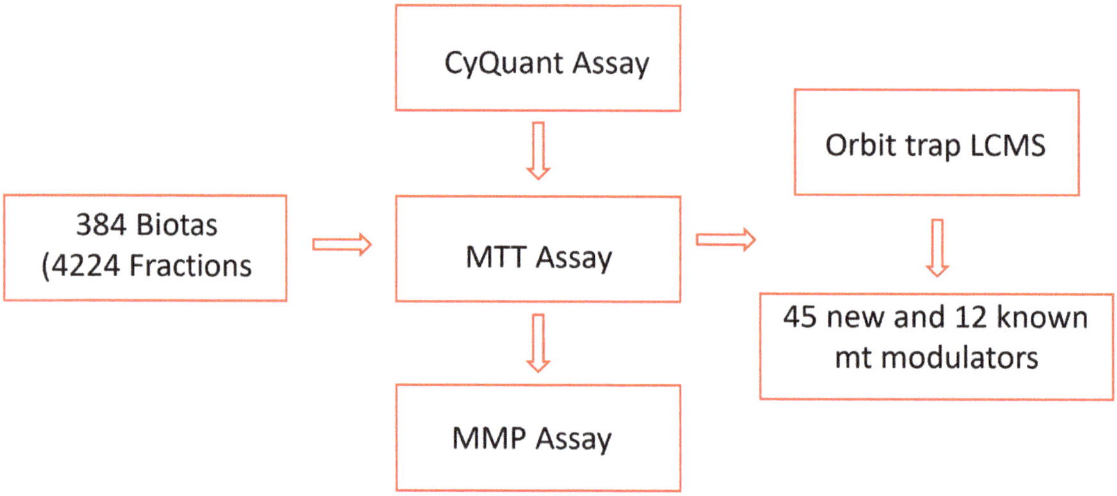

Figure 1. Flow chart of the study.

A common approach utilized to search for mitoprotective metabolites is to test compounds largely based on previously reported antioxidant or neuroprotective activity [43–45]. This work diverges from that by screening a large library of natural products to boost the chances of finding novel mitoprotective compounds. Overall, we have successfully established and implemented a robust process for the identification of mitoprotective compounds from natural products using a stepwise combination of three different assays, two cell lines, and two mitochondrial toxins. We have also described how each assay seamlessly dovetails with the next, leading to the identification of 57 metabolites, including 45 new mitoprotective compounds.

2. Materials and Methods

2.1. Ethics Statement

This work uses patient-derived cells which were collected under the ethical approval of the Griffith University ethics committee. Human olfactory neurosphere-derived cells were derived from nasal biopsies following approved ethical standards, as previously reported by Cook et al. [46]. The human neuroblastoma cell line SH-SY5Y was obtained from the ATCC (CRL-2266), Manassas, VA, USA.

2.2. Cell Culture

Human olfactory neurosphere-derived (hONS) cells were obtained from biopsies of the olfactory mucosa of PD patients and healthy controls [47,48]. The hONS cell line was cultured in DMEM/F-12 medium (Gibco, Invitrogen, Waltham, MA, USA) supplemented with 10% fetal bovine serum (Gibco, Invitrogen) and incubated at 37 °C with 5% CO_2.

The SH-SY5Y cells were cultured in DMEM/F-12 (Sigma Aldrich, St. Louis, MI, USA) medium supplemented with 10% fetal calf serum, 5% non-essential amino acids, and 5% glutamax (Gibco, Invitrogen) and incubated at 37 °C with 5% CO_2.

2.3. Plant Material

Biota samples were obtained from NatureBank, Vancouver, BC, Canada, a unique drug discovery platform that focuses on extracts and fractions of a diverse range of natural products sourced from Australian plants, fungi, and marine organisms [49]. The 384 biota samples screened in this project were Australian plants and marine sponges sourced from Queensland and were selected based on availability at the request time.

2.4. Extraction and Fractionation

The 384 biotas obtained from NatureBank were extracted using a modified lead-like extraction protocol previously established by NatureBank [36,50]. Briefly, plant material (600 mg) was washed in a solid-phase extraction cartridge with 4 mL of hexane to remove fatty components, followed by extraction with 4 mL of dichloromethane and methanol, successively. The hexane phase was discarded, while the dichloromethane and methanol extracts were combined and dried. The dried extracts were reconstituted in methanol and passed through polyamide gel to remove tannins, and the extracts were dried and stored. Marine sponges were extracted in a similar way with 4 mL dichloromethane/methanol (80:20 v/v) followed by 4 mL methanol.

Extracts were fractionated using the lead-like fractionation protocol developed by NatureBank [50]. Briefly, the lead-like extracts dissolved in DMSO were subjected to HPLC separations using a Phenomenex Onyx Monolithic C_{18} column (4.6 mm × 100 mm) with a gradient solvent system of MeOH:H_2O (0.1% TFA), as shown in Table 1. Fractions were collected every 60 s over 11 min. A total of 4224 fractions were collected for testing.

Table 1. HPLC solvent gradient.

Time	MeOH (%)	H_2O (%)	Flow Rate (mL)
0.01	10	90	4
3.00	50	50	4
3.01	50	50	3
6.50	100	0	3
7.00	100	0	3
7.00	100	0	4
8.00	100	0	4
9.00	10	90	4
11.00	10	90	4

2.5. Cell Viability Assays

CyQuant assay

A CyQUANT Cell Proliferation Assay Kit (Life Technologies, Carlsbad, CA, USA) was used to evaluate the cellular response of hONS cells to rotenone, as described by Murtaza et al. [47]. Cells were seeded at 625 cells per well in a 384-well plate and treated with 200 nm rotenone for 96 h. The fluorescence intensity of each sample was measured using the Synergy2 plate reader (Biotek Instruments, Winooski, VT, USA) set with an excitation of 485 nm and emission detection at 530 nm.

MTT assay

To enhance cell adhesion, all plates were pre-treated with 0.01% w/v poly-D-lysine (Sigma Aldrich, St. Louis, MI, USA) at least 3 h before the assay. Cells were seeded at an optimized density of 5×10^4 cells/well in a 96-well plate, leaving 3 empty wells as controls. The MTT assay was performed as previously described [51], the cells were incubated at 37 °C for 24 h, and then treated with 100 µg/mL of DMSO-solved fractions for 2 h at a final DMSO concentration of 0.5%. After pretreatment with fractions, cells were subjected to 6-OHDA challenge at the working concentration of 60 µM in FBS-free media, followed by incubation for another 24 h. To quantify cell viability based on mitochondrial function, the cells were incubated with 0.5 mg/mL MTT and incubated for 4 h at 37 °C. To dissolve insoluble formazan crystals produced from reducing MTT by the cells, 80 µL of 20% SDS was added, and plates were wrapped with foil and placed on an orbital shaker at 100 rpm for 4 h at room temperature. The absorbance of solubilized formazan products was then measured at 570 nm.

2.6. Mitochondrial Membrane Potential (MMP) Assay

A Codex Homogeneous Mitochondrial Membrane Potential Assay Kit (Codex BioSolutions, Inc., Gaithersburg, MD, USA) was used for the MMP assay; the assay was performed in line with the manufacturer's protocol and as described by Sakamuru et al. [52] with some modifications. To determine the appropriate toxin dose, cells were seeded at an optimized density of 5 × 10^4 cells/well in a 96-well plate. After 24 h, culture media were replaced with 100 µL media containing 6-OHDA concentrations ranging from 1.56 µM to 200 µM and incubated for 6 h at 37 °C. When fractions were tested, fractions solved in DMSO were added to the wells at 100 µg/mL 30 min after the addition of 6-OHDA at the optimized concentration. After incubation, the cells were loaded with the mitochondrial membrane potential indicator (m-MPI) and incubated at 37 °C for 30 min. The plate was subsequently washed once with 1X m-MPI assay buffer, after which 80 µL of 1X m-MPI assay buffer was added to each well for reading of the plate with a Spectramax iD5 Multi-Mode Microplate reader. The mitochondrial membrane potential was quantified by the ratio between the J-aggregate form of the mitochondrial MPI indicator (m-MPI) with green fluorescence (485 nm excitation and 535 nm emission) and the monomer form of the m-MPI with red fluorescence (540 nm excitation and 590 nm emission).

2.7. Untargeted Phytochemical Characterization of Active Fractions by HRMS

Acquisition of the phytochemical profile of fractions was carried out on a Vanquish™ Flex UHPLC system (Thermo Fisher Scientific, Waltham, MA, USA) connected to an Orbitrap Exploris 120 mass spectrometer (Thermo Scientific, Waltham, MA, USA). Separation of fractions was achieved on a Phenomenex Luna C18 column, (2.1 mm × 100 mm, 1.7 µm) using a mobile phase of A: 0.1% formic acid in water and B: methanol at a flow rate of 0.6 mL/min. The gradient program was as follows: 0 min, 10% B; 5 min, 10% B; 15 min, 100% B; 20 min, 10% B. The injection volume was 5 µL, and the column temperature was set at 35 °C.

Mass spectrometry data were recorded on an Orbitrap Exploris 140 mass spectrometer equipped with a heated ESI source and operated in the positive-ion mode with the following settings: ion spray voltage: 2.5 kV, sheath gas: 5.08 L min^{-1}, auxiliary gas: 9.37 L min^{-1}, ion transfer tube temperature: 320 °C, vaporizer temperature: 350 °C, scan range (m/z): 150–2000, and collision-energy voltage: 35 V. The full scan was operated at a mass resolution of 60,000 and MS^2 scan at 15,000. Data were acquired using Thermo Xcalibur software and analyzed with Compound Discoverer 3.3.

2.8. Data Analysis

Data analysis was performed using Graph Pad Prism version 10.1 for Microsoft windows (GraphPad Software, San Diego, CA, USA) using two-way analysis of variance, followed by Dunnett's multiple comparison test. Statistical significance was defined as * $p < 0.05$ and ** $p < 0.01$. All determinations were performed in triplicate (at least), and results are presented as means ± SDs.

3. Results and Discussion

3.1. Extraction and Fractionation

A total of 384 biota were randomly chosen from NatureBank, and the extracts and fractions were obtained using a modified lead-like extraction and fractionation protocol previously established by NatureBank [36,50]. This protocol has been optimized to prioritize molecules with drug-like physiochemical properties by frontloading extracts and subsequent fractions while excluding molecules that lack drug-like or lead-like characteristics [36,50]. Generally, the lead-like extraction and fractionation protocol retains components with a log P < 5 [36,50]. Extracts were then fractionated using HPLC, with fractions collected at 60 s intervals for 11 min (Figure 2), resulting in 4224 fractions for screening.

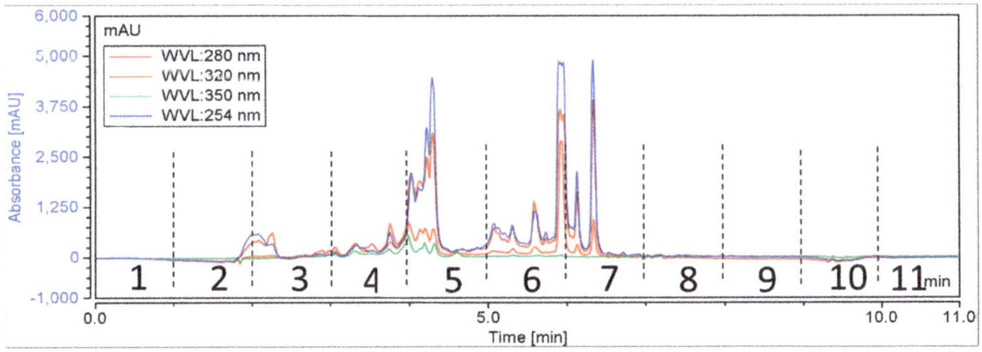

Figure 2. A schematic chromatogram illustrating the HPLC fractionation of a representative positive biota. Dotted lines indicate the time periods for each fraction.

3.2. Effect of Fractions on Cell Proliferation Using High-Throughput CyQuant Assay

The effect of fractions used in this study on cell proliferation was evaluated by exposing PD-sourced hONS cells to rotenone in the presence of test fractions. There is a well-established link between PD and mitochondrial dysfunction, as extensive research evidence has linked PD to features of mitochondrial dysfunction, such as oxidative stress, poor calcium homeostasis, disruption to the MMP, abnormal mitochondrial morphology, disruptions to the OXPHOS process as a result of reduced complex enzyme activity, imbalanced apoptosis, impaired mitophagy, and mitochondrial dynamics [3,14,53–56].

PD-patient derived hONS cells have been reported to show disease-specific alterations such as mitochondrial dysfunction and oxidative damage even in the absence of mitochondrial toxins [48]. Hence, it was possible to aggravate mitochondrial dysfunction in these cells and cause apoptosis using very low doses of neurotoxins, such as the 200 nm rotenone applied in this study. Rotenone is a well-known neurotoxin that induces mitochondrial dysfunction in cellular and animal disease models [40,47,57]. Its primary mechanism of action involves causing mitochondrial dysfunction by inhibiting mitochondrial complex I, an enzyme critical to the oxidative phosphorylation process [40]. Numerous studies have shown that the inhibition of mitochondrial complex I activity is accompanied by deleterious effects, such as impaired mitochondrial biogenesis and dynamics, oxidative stress, decreased ATP production, and apoptosis [40,57]. To identify modulators within the 4224 selected fractions affecting rotenone-induced toxicity in hONS cells, we first employed the CyQuant assay, a fluorescence-based method for quantifying and assessing cell proliferation and cytotoxicity [39]. We calculated Z-scores for all vehicle control-treated samples and set a threshold of a Z-score in the range of −2.5–2.5. The CyQuant values for all extract-treated samples were normalized to control-treated samples, and a relative Z-score was calculated based on the means and standard deviations of all control-treated samples. Figure 3 shows the relative Z-scores for all samples together with the threshold values as indicated by the two red lines. There were 108 fractions with relative Z-scores outside the threshold range; these constituted the initial hits. The 108 fractions were subsequently subjected to a confirmatory round (in triplicate) of the assay using the same parameters as the first. This resulted in 20 hit fractions with relative Z-scores ranging from −2.5 to 2.5. The 20 fractions were from 11 biotas (Table 2), which included eight Australian plants and three marine sponges (Figure 4).

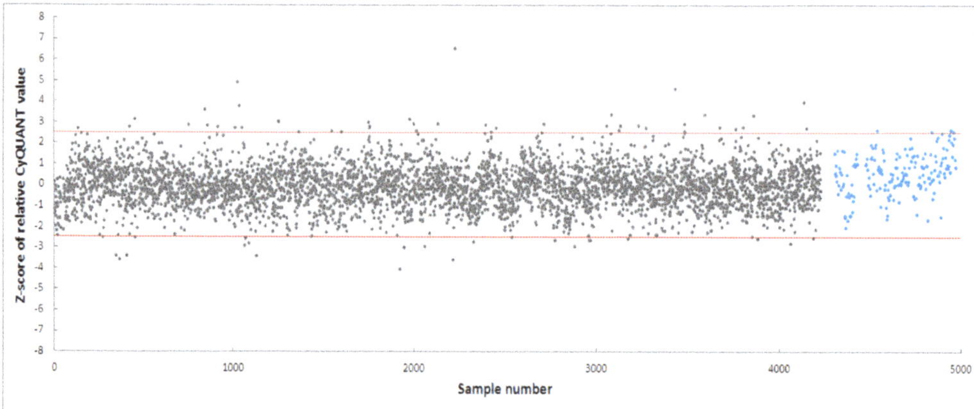

Figure 3. High-throughput screening of 4224 fractions obtained from NatureBank generating 108 hits. The two red lines indicate a relative Z-score of −2.5–2.5. Grey dots: fractions, blue dots: cells treated with vehicle only.

Table 2. List of eleven biotas from which hits were obtained during the second round of screening.

	Genus	Species	Source
1	*Balanops*	<None>	Plant
2	*Alnus*	*trabeculosa h-m.*	Plant
3	*Anredera*	<None>	Plant
4	*Furcraea*	<None>	Plant
5	*Cestrum*	<None>	Plant
6	*Ilex*	<None>	Plant
7	*Balanophora*	<None>	Plant
8	*Rhaphoxya*	3249	Marine
9	*Aaptos*	*aaptos*	Marine
10	*Dendrilla*	3106	Marine
11	*Ternstroemia*	<None>	Plant

3.3. Evaluating the Mitoprotective Effects of Hit Fractions by MTT Assay

To further validate the hits from the high-throughput CyQuant assay and assess their efficacy and potency, different cell models and assays were employed. Mitochondria exhibit tissue-specific properties and are known to show differences across tissues and cell lines [58]. Furthermore, PD patient-sourced hONS cells are primary cells, presenting diverse genetic and morphological characteristics among individual PD patients, which further complicates the validation process [48]. Given these considerations, we opted to confirm our findings from the CyQuant high-throughput screening using the human neuroblastoma cell line SH-SY5Y. SH-SY5Y cells resemble human dopaminergic neurons in many aspects and are widely used as model cells for studying mitochondrial dysfunction [59]. A recent review highlighted that approximately 41% of studies focusing on mitochondrial dysfunction used the SH-SY5Y cell line [1]. This is unsurprising, given that neuronal cells have high energy demands. Mitochondria play a crucial role in their health and functionality, catering to their high metabolic needs [1–3]. To establish a robust assay, we optimized conditions such as cell seeding density, the choice of neurotoxin and toxin concentration, and the DMSO tolerability of the cells. An optimized seeding density of 5.0×10^4 cells/well in a 96-well plate was established for SH-SY5Y cells, and three mitochondrial complex I inhibitory neurotoxins, namely, 6-hydroxydopamine (6-OHDA) [60,61], rotenone [40], and 1-methyl-4-phenylpyridinium (MPP^+), were tested [59]. DMSO was used as a carrier control.

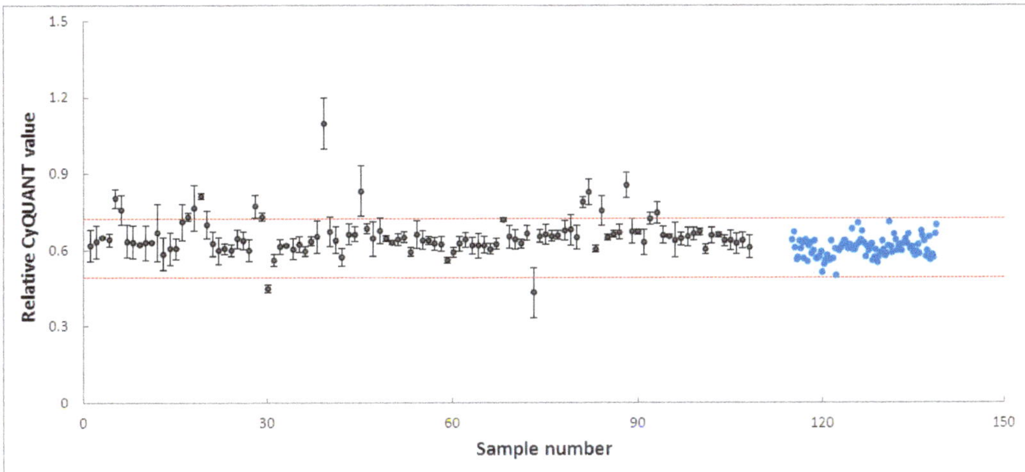

Figure 4. High-throughput screening of 108 fractions, leading to 20 fractions from 11 biotas. Grey dots: fractions, blue dots: cells treated with vehicle only. The two red lines indicate a relative Z-score of −2.5–2.5.

As shown in Figure 5, rotenone showed a relatively low apparent IC_{50} of 0.59 μM. However, it presented minimal efficacy of ~30% due to a significant solubility challenge, as it forms precipitates in culture media. MPP^+ has no solubility issues. However, it exhibited very low potency towards the SH-SY5Y cells. In contrast, 6-OHDA showed almost 100% efficacy and a moderate potency. Therefore, 6-OHDA was selected as the most suitable toxin to induce mitochondrial dysfunction, and a working concentration of 60 μM was used to maintain a cell viability range of 50–70%, allowing for scalability in the following large-scale experiments.

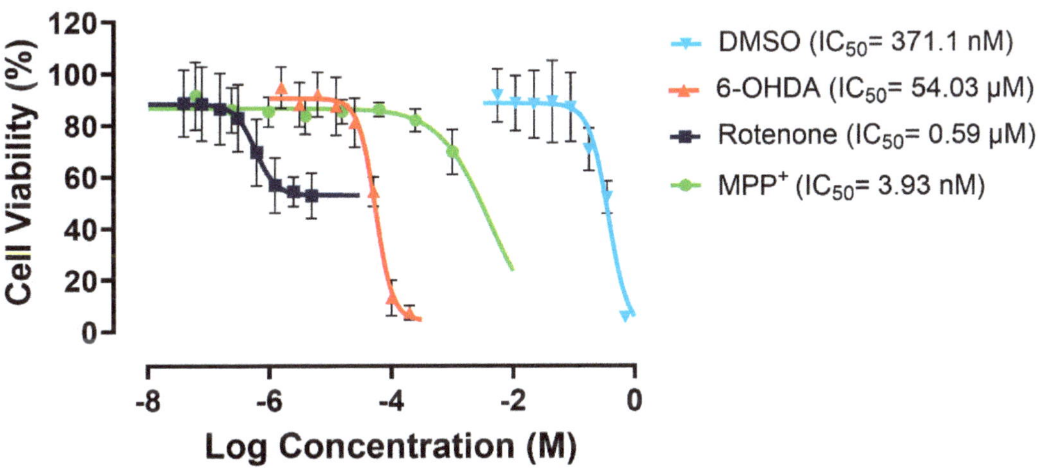

Figure 5. Dose–response curves for rotenone, MPP^+, 6-OHDA, and DMSO in SH-SY5Y cells. Assays were performed in three biological repeats, with two technical repeats for each biological repeat.

SH-SY5Y cells can tolerate a DMSO concentration of up to 1% v/v in culture media, maintaining at least 91% cell viability. To keep the DMSO concentration minimal without affecting the solubility of the fractions, the DMSO concentration was kept at 0.5% in the media containing tested fractions in the following studies.

For all 11 biotas, fractions 1–7 were tested for their protective effect on the mitochondria at concentrations of 100, 50, and 25 µg/mL, using the optimized MTT assay conditions (Figure 6). We excluded fractions 8–11 due to their minimal mass, which rendered them unsuitable for functional tests. The controls encompassed untreated cells, cells treated with DMSO, cells treated with 6-OHDA, and cells treated with 6-OHDA + DMSO. Treated groups were cells incubated with fractions in the presence of 6-OHDA and DMSO. Any fraction with statistically significant cell viability above that of cells treated with 6-OHDA and DMSO, which was 38.98%, was considered protective. Due to the number of data, only fractions showing significant protection against 6-OHDA-induced toxicity are shown in Figure 6. However, the full data sets for all 11 biotas, including inactive fractions, have been included in the Supplementary Materials. A total of 77 fractions were tested, and 40 (51.94%) fractions showed statistically significant protection ($p < 0.05$) against 6-OHDA-induced toxicity at a treatment concentration of 100 µg/mL (Figure 6A). Furthermore, 28 (36.36%) and 31 (40.26%) fractions showed statistically significant protection at lower concentrations of 50 and 25 µg/mL, respectively (Figure 6B,C).

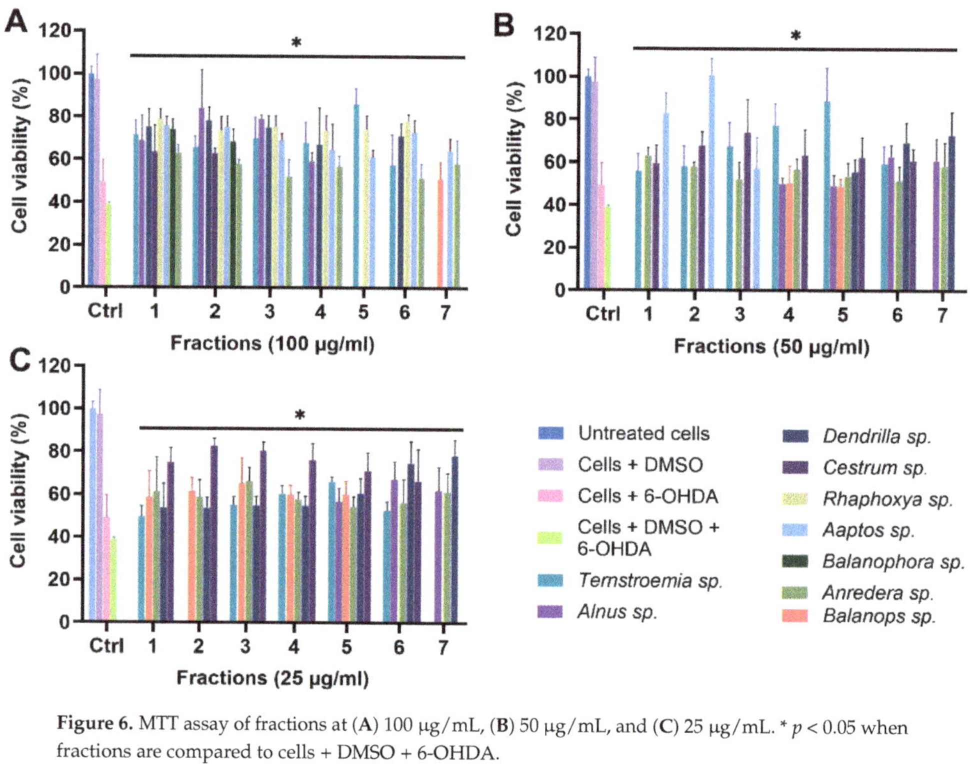

Figure 6. MTT assay of fractions at (**A**) 100 µg/mL, (**B**) 50 µg/mL, and (**C**) 25 µg/mL. * $p < 0.05$ when fractions are compared to cells + DMSO + 6-OHDA.

To a very large extent, the data from the MTT assay corroborate the results from the CyQuant assay, with 18 of the initial 20 hits from the CyQuant assay returning as positive despite the different cell model and assay. This, essentially, is a strong indicator that these fractions potentially contain protective mitochondrial modulators.

It is also noteworthy that fraction F7 of *Ternstroemia* sp. (Figure S12), F1–F7 of *Ilex* sp. (Figure S12), and F1–F7 of *Furcraea* sp. (Figure S13) showed significant toxicity to the mitochondria at both 50 and 25 μg/mL, resulting in lower cell viability when compared to the control group. These results suggest that these fractions could potentially contain mitochondrial modulators with toxic effects on the mitochondria. This is remarkable, and compounds responsible for this effect are potentially relevant in the treatment of diseases like cancer, with the mitochondrion strongly emerging as a target for cancer therapy, as there is increasing research evidence showing that inducing mitochondrial toxicity by targeting OXPHOS could be a potent way to destroy cancer cells [62–64].

3.4. Evaluating the Mitoprotective Effects of Fractions by Mitochondrial Membrane Potential (MMP) Assay

To further ascertain that the active fractions identified from the MTT assay were indeed acting on mitochondria, it was essential to use an assay with greater mitochondrial specificity. Hence, the active fractions were further tested using the mitochondrial membrane potential (MMP) assay.

ATP is generated in the mitochondria through the electron transport chain by creating an electrochemical gradient through a series of redox reactions [52,65]. This electrochemical gradient generates MMP, which is a direct measure of ATP production ability and a key metric for evaluating mitochondrial function and overall cellular health [65,66]. A decrease in MMP is one of the main features of mitochondrial dysfunction, and this decrease has been linked to apoptosis [52,66,67].

MMP was evaluated using the water-soluble mitochondrial membrane potential indicator (m-MPI), a dye that aggregates in healthy mitochondria as red fluorescent monomers emitting light at 590 nm [52]. When the MMP is depolarized, the m-MPI dye transitions to green fluorescent monomers with emission at 535 nm. The ratio between these two fluorescence values can then serve as a measure of MMP [52,67]. A total of 34 fractions with at least 60% efficacy at 100 μg/mL from the MTT assay were subjected to the MMP assay. We found that 6-OHDA induced mitochondrial dysfunction in SH-SY5Y cells by reducing MMP, with a mere 1.56 μM concentration decreasing MMP by more than 56% (Figure 7A). All the tested fractions, except for fraction F7 of *Dendrilla* sp. (11.42%), restored MMP to levels ranging between 36% and 134% when compared to cells treated with 6-OHDA and DMSO, where the level was 20.56% (Figure 7B). Notably, fraction F2 of *Balanophora* sp. fully restored MMP, followed by fractions F3 of *Dendrilla* sp. and F6 of *Aaptos* sp., which restored MMP to levels above 80%.

These findings corroborate the data from the MTT assay and further confirm that the selected fractions potentially contain components capable of rescuing mitochondria from toxic attacks.

3.5. Chemical Constituents of Selected Active Fractions

In Table 3, we present the top eight most active fractions, which exhibit MMP activities ranging from 60% to 133%. Additionally, we detail their CyQuant, MTT, and MMP activities, along with the corresponding mitoprotective compounds identified from these fractions.

These fractions were selected for orbitrap LCMS chemical profiling, and the data were analyzed using compound discoverer 3.3 with a native untargeted natural products identification workflow. Mass data were searched in the Chemspider database, while a spectral similarity search was performed in mZcloud for MS^2 fragmentation data of detected compounds. Compounds with at least 70% mZcloud match confidence and a mass error between −2 ppm and +2 ppm were selected, resulting in the identification of 57 compounds across eight fractions. Table 4 shows a list of identified compounds from all the positive biota fractions with their retention times, molecular formulae, and molecular weights. The extracted ion chromatogram and mass spectra of analyzed fractions are included in the Supplementary Materials (Figure S14–S20).

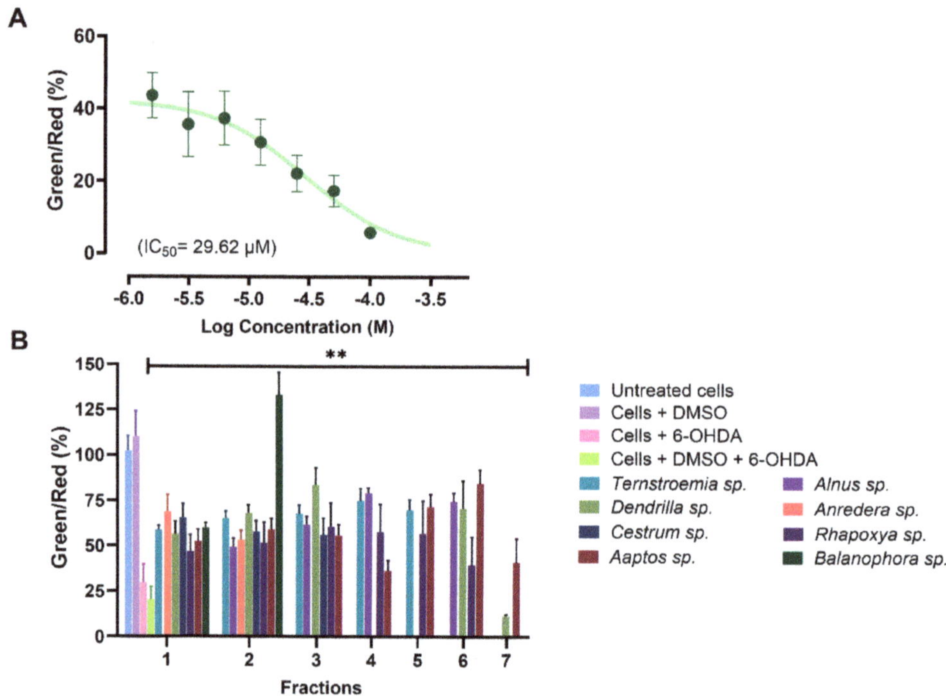

Figure 7. (**A**) Dose–response curve for SH-SY5Y cells exposed to 6-OHDA using the mitochondrial membrane potential (MMP) assay. (**B**) Analysis of selected active fractions identified in the MTT assay through the MMP assay. All the analyzed fractions, except for fraction 7 from *Dendrilla* sp., showed significant mitochondrial protective effects compared to the control group of cells treated with DMSO and 6-OHDA (** $p < 0.01$).

Table 3. Summary data for all eight fractions and identified mitoprotective compounds.

	CyQuant	MTT (%)	MMP (%)	Compound(s)
Cell + 6-OHDA + DMSO	×	38.98	20.58	-
***Ternstroemia* sp._F4**	√	67.70	75.04	1, 4, 6, 7, 9, 10, 12
***Cestrum* sp._F1**	√	63.39	65.69	4, 5, 8,
***Alnus* sp._F4**	√	59.24	79.15	1, 3, 10, 11
***Balanophora* sp._F2**	√	68.27	133.39	2
***Anredera* sp._F1**	√	65.82	68.76	10
***Aaptos* sp._F6**	√	72.83	84.52	-
***Dendrilla* sp._F3**	√	74.82	83.90	8, 10
***Rhaphoxya* sp._F3**	√	75.44	60.78	-

Table 4. Identified compounds from UHPLC-Orbitrap Ms.

	Name	RT (min)	MF	MW
	***Ternstroemia* sp._F4**			
1	NP-019811	0.56	$C_6H_7NO_2$	125.04765
2	5,7-Dihydroxy-2-(4-hydroxyphenyl)-6,8-bis [3,4,5-trihydroxy-6-(hydroxymethyl)tetrahydro-2H-pyran-2-yl]-4H-chromen-4-one	0.97	$C_{27}H_{30}O_{15}$	594.15819
3	(2S,3R,4S,5S,6R)-2-[4-(2-hydroxyethyl)phenoxy]-6-(hydroxymethyl)oxane-3,4,5-triol	1.06	$C_{14}H_{20}O_7$	300.12041

Table 4. *Cont.*

	Name	RT (min)	MF	MW
4	Esculin	1.07	$C_{15}H_{16}O_9$	340.07921
5	7-hydroxy-6-methoxy-2H-chromen-2-one	1.15	$C_{10}H_8O_4$	192.04201
6	Taxifolin	1.37	$C_{15}H_{12}O_7$	304.05798
7	Fraxetin	1.38	$C_{10}H_8O_5$	208.03696
8	3,4-Dihydroxybenzaldehyde	1.42	$C_7H_6O_3$	138.03163
9	5-(6-hydroxy-6-methyloctyl)-2,5-dihydrofuran-2-one	2.26	$C_{13}H_{22}O_3$	226.15672
10	Vanillin	2.45	$C_8H_8O_3$	152.04727
11	(2R,3S,4S,5R,6R)-2-(hydroxymethyl)-6-(2-phenylethoxy)oxane-3,4,5-triol	2.96	$C_{14}H_{20}O_6$	284.12577
12	Scopoletin	2.96	$C_{10}H_8O_4$	192.04216
13	4-Coumaric acid	3.00	$C_9H_8O_3$	164.04738
14	Isoliquiritigenin	3.30	$C_{15}H_{12}O_4$	256.07338
15	Quercetin	3.88	$C_{15}H_{10}O_7$	302.04265
16	5-hydroxy-3-(4-methoxyphenyl)-7-[(3,4,5-trihydroxy-6-{[(3,4,5-trihydroxy-6-methyloxan-2-yl)oxy]methyl}oxan-2-6-methyloxan-2-yl)oxy]methyl}oxan-2-yl)oxy]-4H-chromen-4-one	4.03	$C_{28}H_{32}O_{14}$	592.17931
17	3-amino-2-phenyl-2H-pyrazolo [4,3-c]pyridine-4,6-diol	5.48	$C_{12}H_{10}N_4O_2$	242.08023
18	(3aR,7aS,8S,9aR)-5,8-dimethyl-3-methylidene-2H,3H,3aH,4H,6H,7H,7aH,8H,9H,9aH-azuleno [6,5-b]furan-2,6-dione	7.16	$C_{15}H_{18}O_3$	246.12546
19	12-Aminododecanoic acid	8.90	$C_{12}H_{25}NO_2$	215.18846
	***Cestrum* sp._F1**			
1	Agmatine	0.45	$C_5H_{14}N_4$	130.12168
2	Nicotinic acid	0.83	$C_6H_5NO_2$	123.03203
3	N,N′-Diphenylguanidine	0.92	$C_{13}H_{13}N_3$	211.11072
4	Vanillin	2.44	$C_8H_8O_3$	152.04753
5	NP-007065	4.04	$C_8H_{10}O_3$	154.06290
6	4-Phenylbutyric acid	4.27	$C_{10}H_{12}O_2$	164.08365
7	Cantharidin	9.05	$C_{10}H_{12}O_4$	196.07384
8	NP-000925	9.87	$C_{17}H_{16}O_5$	300.09947
9	Pinocembrin	10.06	$C_{15}H_{12}O_4$	256.07314
	***Alnus* sp._F4**			
1	NP-019811	0.55	$C_6H_7NO_2$	125.04747
2	Nicotinic acid	0.87	$C_6H_5NO_2$	123.03188
3	2,3,4,9-Tetrahydro-1H-β-carboline-3-carboxylic acid	1.35	$C_{12}H_{12}N_2O_2$	216.08965
4	Fraxetin	2.15	$C_{10}H_8O_5$	208.03697
5	Syringic acid	3.13	$C_9H_{10}O_5$	198.05271
6	Isovanillic acid	3.33	$C_8H_8O_4$	168.04217
7	Isoferulic acid	3.33	$C_{10}H_{10}O_4$	194.05776
8	Naringenin	7.06	$C_{15}H_{12}O_5$	272.06819
9	Vanillin	8.01	$C_8H_8O_3$	152.04720
	***Balanophora* sp._F2**			
1	Nicotinamide	0.78	$C_6H_6N_2O$	122.04791
2	3′-Adenosine monophosphate (3′-AMP)	0.98	$C_{10}H_{14}N_5O_7P$	347.06261
3	3-(2-methylpropyl)-octahydropyrrolo [1,2-a]pyrazine-1,4-dione	3.03	$C_{11}H_{18}N_2O_2$	210.13658
4	NP-007065	4.00	$C_8H_{10}O_3$	154.06297

Table 4. *Cont.*

	Name	RT (min)	MF	MW
	***Anredera* sp._F1**			
1	L-Aspartic acid	0.66	$C_4H_7NO_4$	133.03739
2	Nicotinic acid	0.86	$C_6H_5NO_2$	123.03193
3	L-Phenylalanine	0.88	$C_9H_{11}NO_2$	165.07886
4	Vanillin	2.43	$C_8H_8O_3$	152.04729
5	4-Phenylbutyric acid	3.13	$C_{10}H_{12}O_2$	164.08362
	***Aaptos* sp._F6**			
1	Pulegone	7.95	$C_{10}H_{16}O$	152.12012
2	NP-022512	8.67	$C_{13}H_{19}NO$	205.14656
3	NP-019636	15.25	$C_9H_8O_4$	180.04219
4	Palmitoleic acid	16.69	$C_{16}H_{30}O_2$	254.22444
5	4-Methoxycinnamic acid	17.15	$C_{10}H_{10}O_3$	178.06291
6	8Z,11Z,14Z-Eicosatrienoic acid	17.66	$C_{20}H_{34}O_2$	306.25477
	***Dendrilla* sp._F3**			
1	Agmatine	0.49	$C_5H_{14}N_4$	130.12166
2	Trigonelline	0.82	$C_7H_7NO_2$	137.04752
3	Nicotinic acid	0.82	$C_6H_5NO_2$	123.03189
4	Vanillin	2.41	$C_8H_8O_3$	152.04744
5	NP-011220	2.77	$C_{11}H_{18}N_2O_2$	210.13654
6	4-Phenylbutyric acid	4.24	$C_{10}H_{12}O_2$	164.08368
7	Cantharidin	9.04	$C_{10}H_{12}O_4$	196.07358
8	(-)-Caryophyllene oxide	11.31	$C_{15}H_{24}O$	220.18263
9	4-Phenylbutyric acid	11.76	$C_{10}H_{12}O_2$	164.08377
	***Rhaphoxya* sp._F3**			
1	Nicotinic acid	0.82	$C_6H_5NO_2$	123.03196
2	3-[(4-hydroxyphenyl)methyl]-octahydropyrrolo [1,2-a]pyrazine-1,4-dione	1.24	$C_{14}H_{16}N_2O_3$	260.11604
3	NP-016455	2.39	$C_{11}H_{18}N_2O_4$	242.12662
4	NP-011220	2.77	$C_{11}H_{18}N_2O_2$	210.13680
5	Cyclo(leucylprolyl)	3.01	$C_{11}H_{18}N_2O_2$	210.13684
6	DL-2-(acetylamino)-3-phenylpropanoic acid	3.37	$C_{11}H_{13}NO_3$	207.08954
7	4-Phenylbutyric acid	3.38	$C_{10}H_{12}O_2$	164.08382
8	Cyclo(phenylalanyl-prolyl)	3.70	$C_{14}H_{16}N_2O_2$	244.12128
9	4-Methylumbelliferone hydrate	4.48	$C_{10}H_8O_3$	176.04732
10	2-Hydroxybenzothiazole	5.03	C_7H_5NOS	151.00932
11	4-Methoxycinnamaldehyde	7.03	$C_{10}H_{10}O_2$	162.06818
12	Bis(2-ethylhexyl) amine	8.66	$C_{16}H_{35}N$	241.27669
13	(1R,4aS)-7-(2-Hydroxypropan-2-yl)-1,4a-dimethyl-9-oxo-3,4,10,10a-carboxylic acid	10.63	$C_{20}H_{26}O_4$	330.18295
14	(-)-Caryophyllene oxide	12.54	$C_{15}H_{24}O$	220.18265

It is noteworthy that 12 compounds identified by LC-MS have been reported to protect against mitochondrial dysfunction, underscoring the robustness of our screening methodology (Figure 8). Research has shown that the majority of these compounds protect the mitochondria by reducing oxidative stress and modulating antioxidant defense systems, oxidative stress, and apoptotic markers [45,68,69]. In the SH-SY5Y model, fraxetin (**1**) protected against rotenone-induced apoptosis via the induction of the chaperone HSP70, crucial for maintaining mitochondrial functions [70]. Fraxetin also protects mitochondria by reducing ROS and apoptotic proteins, such as cytochrome c, Bax, and caspase-3 and -9, while upregulating the expression of antioxidant defense enzymes, such as SOD, GPX, and catalase [69,70]. Nicotinamide (**2**), a precursor to nicotinamide adenine dinucleotide (NAD), has been reported to prevent neurodegeneration in glaucoma by defending against mitochondrial dysfunction, boosting OXPHOS, and increasing mitochondrial size [68]. In addition to reducing ROS, regulating oxidative enzymes, and increasing the expression of complexes I and V, naringenin (**3**) upregulates ATP production, activates the PI3K/Akt/GSK-3β pathway, improves nuclear E2-related factor 2 (Nrf2) expression, and stabilizes MMP [45,71–73]. Isoliquiritigenin (**4**) inhibits apoptosis by promoting increased phosphorylation of glycogen synthase kinase-3β (GSK3β) and also enhances mitochondrial biogenesis by activation of AMP-activated protein kinase (AMPK) [74,75]. Reports have shown that pinocembrin (**5**) protects brain mitochondria structure and function by decreasing ROS, restoring MMP, and improving mitochondrial morphology [76]. This compound also exhibits antiapoptotic effects, restores the electron transport chain, and upregulates ATP and Nrf2 [77,78]. Other compounds like quercetin (**6**) [30,31,79], taxifolin (**7**) [80–82], agmatine (**8**) [83], esculin (**9**) [84], vanillin (**10**) [85], syringic acid (**11**) [86–88], and 4-coumaric acid (**12**) [43,89] have also been documented for their protective effects against mitochondrial dysfunction.

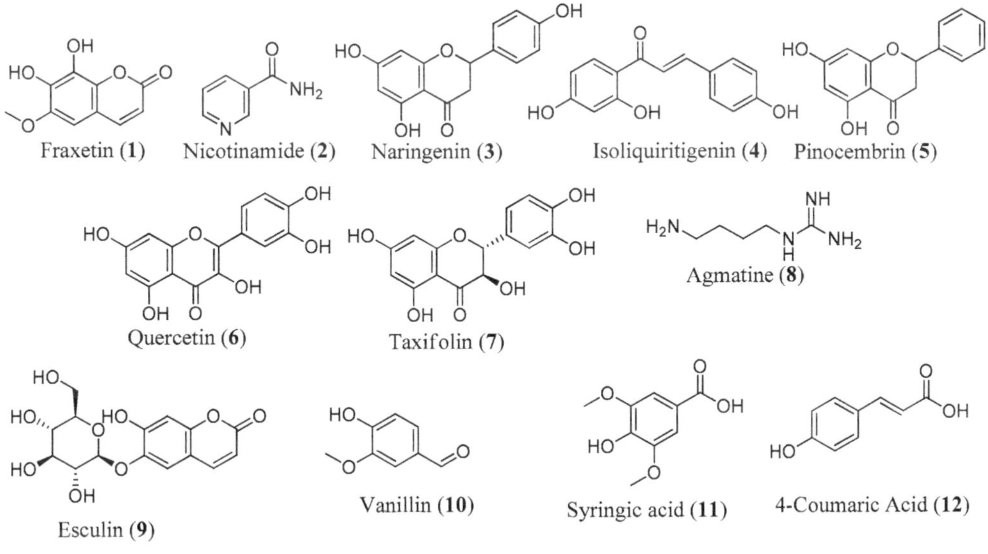

Figure 8. Structure of compounds with known mitoprotective activity.

We also identified 45 new mitoprotective compounds spanning diverse structural classes, such as alkaloids, coumarins, carboxylic acids, cinnamic acids, lipopeptides, terpenes, benzothiazoles, amines, amino acids, and fatty acids. While there are no reports in the literature on the mitoprotective activity of these metabolites, some of them have been reported to show activity in mitochondrial dysfunction-linked diseases. For example, trigonelline, an alkaloid previously isolated from plants such as fenugreek, Japanese radish, and coffee beans stands out for its potential to attenuate oxidative stress and show activity in mitochondrial dysfunction-linked conditions, such as diabetes, PD, AD, stroke, dementia, and depression [90]. Trigonelline improved memory function in AD and showed neuroprotective and antiapoptotic effects in a 6-OHDA-induced model of PD in rats [91–93]. Our work demonstrated that these beneficial effects may at least partially be attributed to its mitoprotective functions. Isoferulic acid and scopoletin exert their neuroprotective properties by decreasing ROS, activating the Nrf2 pathway, and suppressing apoptosis [94–97]. In this study, they were found to be promising candidates for the mitoprotective effects of *Alnus* sp. and *Ternstroemia* sp., respectively (Table 4).

4. Conclusions, Limitations, and Future Directions

In this study, we screened 4224 fractions derived from 384 biotas belonging to Australia's diverse flora and fauna for their protective effects against mitochondrial dysfunction. We were able to identify 20 hit fractions from 11 biota in the initial round of screening using assays based on rotenone-induced mitochondrial dysfunction in PD patient-derived hONS cell models. These initial findings were successfully validated by MTT and MMP assays in a 6-OHDA/SH-SY5Y model. Additionally, we identified 57 metabolites that are potentially responsible for the activity of the eight most active hits using HRMS. Twelve metabolites (**1**–**12**) have been previously reported to show protective effects against mitochondrial dysfunction.

These findings strongly indicate that the described methodologies provide a robust, effective, and rather quick approach to the screening of mitochondrial modulators, especially on a large scale. With the use of robotics, the screening approach described in this work can be easily scaled to screen tens of thousands of compounds. Also, this method can be readily adapted to screen for toxic mitochondrial modulators as a strategy for cancer therapy, rather than for mitoprotective compounds, which was the focus of this work.

However, a significant limitation of this work is the reliance on databases for identifying compounds, the results being inherently dependent on the contents of these databases. Consequently, there is a considerable risk that both known and new natural products have been overlooked. This challenge underscores the importance of further isolation and characterization of active compounds from these biotas to precisely determine the metabolites responsible for their mitoprotective properties and further testing of these metabolites to understand their mechanisms of actions.

Supplementary Materials: The following supporting information can be downloaded at: https://www.mdpi.com/article/10.3390/biom14040440/s1, Figure S1: HPLC Chromatogram of *Ternstroemia* sp., Figure S2: HPLC Chromatogram of *Alnus* sp., Figure S3: HPLC Chromatogram of *Balanops* sp., Figure S4: HPLC Chromatogram of *Anredera* sp., Figure S5: HPLC Chromatogram of *Cestrum* sp., Figure S6: HPLC Chromatogram of *Ilex* sp., Figure S7: HPLC Chromatogram of *Dendrilla* sp., Figure S8: HPLC Chromatogram of *Balanophora* sp., Figure S9: HPLC Chromatogram of *Aaptos* sp., Figure S10: HPLC Chromatogram of *Rhaphoxya* sp., Figure S11: HPLC Chromatogram of *Fucraea* sp., Figure S12: MTT assay of fractions at 100, 50, and 25 µg/mL, Figure S13: Base peak ion chromatogram and full mass spectra of *Ternstroemia* sp., fraction 4, Figure S14: Base peak ion chromatogram and full mass spectra of *Alnus* sp., fraction 4, Figure S15: Base peak ion chromatogram and full mass spectra of *Anredera* sp., fraction 1, Figure S16: Base peak ion chromatogram and full mass spectra of *Cestrum* sp., fraction 1, Figure S17: Base peak ion chromatogram and full mass spectra of *Dendrilla* sp., fraction, Figure S18: Base peak ion chromatogram and full mass spectra of *Balanophora* sp., fraction 2, Figure S19: Base peak ion chromatogram and full mass spectra of *Aaptos* sp., fraction 6, Figure S20: Base peak ion chromatogram and full mass spectra of *Rhaphoxya* sp., fraction 3.

Author Contributions: E.M. performed the experiment and wrote the manuscript. L.M., G.D.M. and Y.F. supervised the project and the writing of the manuscript; they also contributed to the development of its contents. All authors have read and agreed to the published version of the manuscript.

Funding: This research received no external funding.

Institutional Review Board Statement: Not applicable.

Informed Consent Statement: Not applicable.

Data Availability Statement: Data are available in the manuscript text or in the Supplementary Materials.

Acknowledgments: We acknowledge Stephen Wood and Jiangu Shan, who assisted with conducting and supervising the initial screening experiment. We also acknowledge Compounds Australia and NatureBank for supplying the biotas used in this work. Emmanuel Makinde would like to acknowledge Griffith University for providing a Griffith University Postgraduate Research Scholarship and a Griffith University International Postgraduate Research Scholarship.

Conflicts of Interest: The authors declare no conflicts of interest.

References

1. Makinde, E.; Ma, L.; Mellick, G.D.; Feng, Y. Mitochondrial Modulators: The Defender. *Biomolecules* **2023**, *13*, 226. [CrossRef] [PubMed]
2. Annesley, S.J.; Fisher, P.R. Mitochondria in Health and Disease. *Cells* **2019**, *8*, 680. [CrossRef] [PubMed]
3. Zambrano, K.; Barba, D.; Castillo, K.; Noboa, L.; Argueta-Zamora, D.; Robayo, P.; Arizaga, E.; Caicedo, A.; Cavilanes, A.W.D. Fighting Parkinson's Disease: The Return of the Mitochondria. *Mitochondrion* **2022**, *64*, 34–44. [CrossRef] [PubMed]
4. Rouaud, T.; Corbillé, A.-G.; Leclair-Visonneau, L.; de Guilhem de Lataillade, A.; Lionnet, A.; Preterre, C.; Damier, P.; Derkinderen, P. Pathophysiology of Parkinson's Disease: Mitochondria, Alpha-Synuclein and Much More.... *Rev. Neurol.* **2021**, *177*, 260–271. [CrossRef] [PubMed]
5. Grünewald, A.; Kumar, K.R.; Sue, C.M. New Insights into the Complex Role of Mitochondria in Parkinson's Disease. *Progress. Neurobiol.* **2019**, *177*, 73–93. [CrossRef] [PubMed]
6. Harrington, J.S.; Ryter, S.W.; Plataki, M.; Price, D.R.; Choi, A.M.K. Mitochondria in Health, Disease, and Aging. *Physiol. Rev.* **2023**, *103*, 2349–2422. [CrossRef] [PubMed]
7. Fernández-Moriano, C.; González-Burgos, E.; Gómez-Serranillos, M.P. Mitochondria-Targeted Protective Compounds in Parkinson's and Alzheimer's Diseases. *Oxid. Med. Cell Longev.* **2015**, *2015*, 408927. [CrossRef]
8. Arauna, D.; Furrianca, M.; Espinosa-Parrilla, Y.; Fuentes, E.; Alarcón, M.; Palomo, I. Natural Bioactive Compounds As Protectors Of Mitochondrial Dysfunction In Cardiovascular Diseases And Aging. *Molecules* **2019**, *24*, 4259. [CrossRef] [PubMed]
9. Wada, J.; Nakatsuka, A. Mitochondrial Dynamics and Mitochondrial Dysfunction in Diabetes. *Acta Med. Okayama* **2016**, *70*, 151–158. [CrossRef]
10. Rangaraju, V.; Lewis, T.L.; Hirabayashi, Y.; Bergami, M.; Motori, E.; Cartoni, R.; Kwon, S.-K.; Courchet, J. Pleiotropic Mitochondria: The Influence of Mitochondria on Neuronal Development and Disease. *J. Neurosci.* **2019**, *39*, 8200–8208. [CrossRef]
11. Vringer, E.; Tait, S.W.G. Mitochondria and Cell Death-Associated Inflammation. *Cell Death Differ.* **2023**, *30*, 304–312. [CrossRef] [PubMed]
12. Abate, M.; Festa, A.; Falco, M.; Lombardi, A.; Luce, A.; Grimaldi, A.; Zappavigna, S.; Sperlongano, P.; Irace, C.; Caraglia, M.; et al. Mitochondria as Playmakers of Apoptosis, Autophagy and Senescence. *Semin. Cell Dev. Biol.* **2020**, *98*, 139–153. [CrossRef] [PubMed]
13. Bai, Q.; Wang, Z.; Piao, Y.; Zhou, X.; Piao, Q.; Jiang, J.; Liu, H.; Piao, H.; Li, L.; Song, Y.; et al. Sesamin Alleviates Asthma Airway Inflammation by Regulating Mitophagy and Mitochondrial Apoptosis. *J. Agric. Food Chem.* **2022**, *70*, 4921–4933. [CrossRef] [PubMed]
14. Wang, H.; Dou, S.; Zhu, J.; Shao, Z.; Wang, C.; Xu, X.; Cheng, B. Ghrelin Protects against Rotenone-Induced Cytotoxicity: Involvement of Mitophagy and the AMPK/SIRT1/PGC1α Pathway. *Neuropeptides* **2021**, *87*, 102134. [CrossRef] [PubMed]
15. Barazzuol, L.; Giamogante, F.; Brini, M.; Calì, T. PINK1/Parkin Mediated Mitophagy, Ca2+ Signalling, and ER-Mitochondria Contacts in Parkinson's Disease. *Int. J. Mol. Sci.* **2020**, *21*, 1772. [CrossRef]
16. Salazar, C.; Ruiz-Hincapie, P.; Ruiz, L.M. The Interplay among PINK1/PARKIN/Dj-1 Network during Mitochondrial Quality Control in Cancer Biology: Protein Interaction Analysis. *Cells* **2018**, *7*, 154. [CrossRef] [PubMed]
17. Antunes, M.S.; Ladd, F.V.L.; Ladd, A.A.B.L.; Moreira, A.L.; Boeira, S.P.; Cattelan Souza, L. Hesperidin Protects against Behavioral Alterations and Loss of Dopaminergic Neurons in 6-OHDA-Lesioned Mice: The Role of Mitochondrial Dysfunction and Apoptosis. *Metab. Brain Dis.* **2021**, *36*, 153–167. [CrossRef] [PubMed]
18. Ben-Hail, D.; Begas-Shvartz, R.; Shalev, M.; Shteinfer-Kuzmine, A.; Gruzman, A.; Reina, S.; Pinto, V.D.; Shoshan-Barmatz, V. Novel Compounds Targeting the Mitochondrial Protein VDAC1 Inhibit Apoptosis and Protect against Mitochondrial Dysfunction *. *J. Biol. Chem.* **2016**, *291*, 24986–25003. [CrossRef]
19. Abudureyimu, M.; Yang, M.; Wang, X.; Luo, X.; Ge, J.; Peng, H.; Zhang, Y.; Ren, J. Berberine Alleviates Myocardial Diastolic Dysfunction by Modulating Drp1-Mediated Mitochondrial Fission and Ca^{2+} Homeostasis in a Murine Model of HFpEF. *Front. Med.* **2023**, *17*, 1219–1235. [CrossRef]
20. Qin, X.; Zhao, Y.; Gong, J.; Huang, W.; Su, H.; Yuan, F.; Fang, K.; Wang, D.; Li, J.; Zou, X.; et al. Berberine Protects Glomerular Podocytes via Inhibiting Drp1-Mediated Mitochondrial Fission and Dysfunction. *Theranostics* **2019**, *9*, 1698–1713. [CrossRef]
21. Hu, M.; Wang, R.; Chen, X.; Zheng, M.; Zheng, P.; Boz, Z.; Tang, R.; Zheng, K.; Yu, Y.; Huang, X.-F. Resveratrol Prevents Haloperidol-Induced Mitochondria Dysfunction through the Induction of Autophagy in SH-SY5Y Cells. *NeuroToxicology* **2021**, *87*, 231–242. [CrossRef] [PubMed]
22. Chen, J.; Liu, Q.; Wang, Y.; Guo, Y.; Xu, X.; Huang, P.; Lian, B.; Zhang, R.; Chen, Y.; Ha, Y. Protective Effects of Resveratrol Liposomes on Mitochondria in Substantia Nigra Cells of Parkinsonized Rats. *Ann. Palliat. Med.* **2021**, *10*, 2458468. [CrossRef] [PubMed]
23. Zhou, D.; Sun, M.-H.; Jiang, W.-J.; Li, X.-H.; Lee, S.-H.; Heo, G.; Niu, Y.-J.; Ock, S.A.; Cui, X.-S. Epigallocatechin-3-Gallate Protects Porcine Oocytes against Post-Ovulatory Aging through Inhibition of Oxidative Stress. *Aging* **2022**, *14*, 8633–8644. [CrossRef] [PubMed]

24. Zhu, T.-T.; Zhang, W.-F.; Luo, P.; He, F.; Ge, X.-Y.; Zhang, Z.; Hu, C.-P. Epigallocatechin-3-Gallate Ameliorates Hypoxia-Induced Pulmonary Vascular Remodeling by Promoting Mitofusin-2-Mediated Mitochondrial Fusion. *Eur. J. Pharmacol.* **2017**, *809*, 42–51. [CrossRef] [PubMed]
25. Chen, Y.; Chen, J.; Sun, X.; Shi, X.; Wang, L.; Huang, L.; Zhou, W. Evaluation of the Neuroprotective Effect of EGCG: A Potential Mechanism of Mitochondrial Dysfunction and Mitochondrial Dynamics after Subarachnoid Hemorrhage. *Food Funct.* **2018**, *9*, 6349–6359. [CrossRef]
26. Bagheri, H.; Ghasemi, F.; Barreto, G.E.; Rafiee, R.; Sathyapalan, T.; Sahebkar, A. Effects of Curcumin on Mitochondria in Neurodegenerative Diseases. *BioFactors* **2020**, *46*, 5–20. [CrossRef] [PubMed]
27. Jayaraj, R.L.; Tamilselvam, K.; Manivasagam, T.; Elangovan, N. Neuroprotective Effect of CNB-001, a Novel Pyrazole Derivative of Curcumin on Biochemical and Apoptotic Markers Against Rotenone-Induced SK-N-SH Cellular Model of Parkinson's Disease. *J. Mol. Neurosci.* **2013**, *51*, 863–870. [CrossRef]
28. Naserzadeh, P.; Mehr, S.N.; Sadabadi, Z.; Seydi, E.; Salimi, A.; Pourahmad, J. Curcumin Protects Mitochondria and Cardiomyocytes from Oxidative Damage and Apoptosis Induced by Hemiscorpius Lepturus Venom. *Drug Res.* **2018**, *68*, 113–120. [CrossRef] [PubMed]
29. Uğuz, A.C.; Öz, A.; Nazıroğlu, M. Curcumin Inhibits Apoptosis by Regulating Intracellular Calcium Release, Reactive Oxygen Species and Mitochondrial Depolarization Levels in SH-SY5Y Neuronal Cells. *J. Recept. Signal Transduct.* **2016**, *36*, 395–401. [CrossRef]
30. Wang, W.-W.; Han, R.; He, H.-J.; Li, J.; Chen, S.-Y.; Gu, Y.; Xie, C. Administration of Quercetin Improves Mitochondria Quality Control and Protects the Neurons in 6-OHDA-Lesioned Parkinson's Disease Models. *Aging* **2021**, *13*, 11738–11751. [CrossRef]
31. Vanani, A.R.; Mahdavinia, M.; Shirani, M.; Alizadeh, S.; Dehghani, M.A. Protective Effects of Quercetin against Oxidative Stress Induced by Bisphenol-A in Rat Cardiac Mitochondria. *Environ. Sci. Pollut. Res.* **2020**, *27*, 15093–15102. [CrossRef] [PubMed]
32. Li, T.; Liu, Y.; Xu, W.; Dai, X.; Liu, R.; Gao, Y.; Chen, Z.; Li, Y. Polydatin Mediates Parkin-Dependent Mitophagy and Protects against Mitochondria-Dependent Apoptosis in Acute Respiratory Distress Syndrome. *Lab. Investig.* **2019**, *99*, 819–829. [CrossRef] [PubMed]
33. Li, L.; Tan, H.-P.; Liu, C.-Y.; Yu, L.-T.; Wei, D.-N.; Zhang, Z.-C.; Lu, K.; Zhao, K.-S.; Maegele, M.; Cai, D.-Z.; et al. Polydatin Prevents the Induction of Secondary Brain Injury after Traumatic Brain Injury by Protecting Neuronal Mitochondria. *Neural Regen. Res.* **2019**, *14*, 1573. [CrossRef] [PubMed]
34. Gu, J.; Gui, Y.; Chen, L.; Yuan, G.; Lu, H.-Z.; Xu, X. Use of Natural Products as Chemical Library for Drug Discovery and Network Pharmacology. *PLoS ONE* **2013**, *8*, e62839. [CrossRef] [PubMed]
35. Newman, D.J.; Cragg, G.M. Natural Products as Sources of New Drugs over the Nearly Four Decades from 01/1981 to 09/2019. *J. Nat. Prod.* **2020**, *83*, 770–803. [CrossRef] [PubMed]
36. Camp, D.; Campitelli, M.; Carroll, A.R.; Davis, R.A.; Quinn, R.J. Front-Loading Natural-Product-Screening Libraries for Log P: Background, Development, and Implementation. *Chem. Biodivers.* **2013**, *10*, 524–537. [CrossRef]
37. Harvey, A.L.; Edrada-Ebel, R.; Quinn, R.J. The Re-Emergence of Natural Products for Drug Discovery in the Genomics Era. *Nat. Rev. Drug Discov.* **2015**, *14*, 111–129. [CrossRef] [PubMed]
38. Atanasov, A.G.; Zotchev, S.B.; Dirsch, V.M.; Supuran, C.T. Natural Products in Drug Discovery: Advances and Opportunities. *Nat. Rev. Drug Discov.* **2021**, *20*, 200–216. [CrossRef]
39. Jones, L.J.; Gray, M.; Yue, S.T.; Haugland, R.P.; Singer, V.L. Sensitive Determination of Cell Number Using the CyQUANT® Cell Proliferation Assay. *J. Immunol. Methods* **2001**, *254*, 85–98. [CrossRef]
40. Yarmohammadi, F.; Wallace Hayes, A.; Najafi, N.; Karimi, G. The Protective Effect of Natural Compounds against Rotenone-Induced Neurotoxicity. *J. Biochem. Mol. Toxicol.* **2020**, *34*, e22605. [CrossRef]
41. Ghasemi, M.; Turnbull, T.; Sebastian, S.; Kempson, I. The MTT Assay: Utility, Limitations, Pitfalls, and Interpretation in Bulk and Single-Cell Analysis. *Int. J. Mol. Sci.* **2021**, *22*, 12827. [CrossRef] [PubMed]
42. Stockert, J.C.; Blázquez-Castro, A.; Cañete, M.; Horobin, R.W.; Villanueva, Á. MTT Assay for Cell Viability: Intracellular Localization of the Formazan Product Is in Lipid Droplets. *Acta Histochem.* **2012**, *114*, 785–796. [CrossRef]
43. Park, S.; Kim, Y.A.; Lee, J.; Seo, H.; Nam, S.-J.; Jo, D.-G.; Hyun, D.-H. 4-Hydroxycinnamic Acid Attenuates Neuronal Cell Death by Inducing Expression of Plasma Membrane Redox Enzymes and Improving Mitochondrial Functions. *Food Sci. Human. Wellness* **2023**, *12*, 1287–1299. [CrossRef]
44. Mao, Z.; Tian, L.; Liu, J.; Wu, Q.; Wang, N.; Wang, G.; Wang, Y.; Seto, S. Ligustilide Ameliorates Hippocampal Neuronal Injury after Cerebral Ischemia Reperfusion through Activating PINK1/Parkin-Dependent Mitophagy. *Phytomedicine* **2022**, *101*, 154111. [CrossRef] [PubMed]
45. Kesh, S.; Kannan, R.R.; Balakrishnan, A. Naringenin Alleviates 6-Hydroxydopamine Induced Parkinsonism in SHSY5Y Cells and Zebrafish Model. *Comp. Biochem. Physiol. Part C Toxicol. Pharmacol.* **2021**, *239*, 108893. [CrossRef] [PubMed]
46. Cook, A.L.; Vitale, A.M.; Ravishankar, S.; Matigian, N.; Sutherland, G.T.; Shan, J.; Sutharsan, R.; Perry, C.; Silburn, P.A.; Mellick, G.D.; et al. NRF2 Activation Restores Disease Related Metabolic Deficiencies in Olfactory Neurosphere-Derived Cells from Patients with Sporadic Parkinson's Disease. *PLoS ONE* **2011**, *6*, e21907. [CrossRef] [PubMed]
47. Murtaza, M.; Shan, J.; Matigian, N.; Todorovic, M.; Cook, A.L.; Ravishankar, S.; Dong, L.F.; Neuzil, J.; Silburn, P.; Mackay-Sim, A.; et al. Rotenone Susceptibility Phenotype in Olfactory Derived Patient Cells as a Model of Idiopathic Parkinson's Disease. *PLoS ONE* **2016**, *11*, e0154544. [CrossRef] [PubMed]

48. Matigian, N.; Abrahamsen, G.; Sutharsan, R.; Cook, A.L.; Vitale, A.M.; Nouwens, A.; Bellette, B.; An, J.; Anderson, M.; Beckhouse, A.G.; et al. Disease-Specific, Neurosphere-Derived Cells as Models for Brain Disorders. *Dis. Models Mech.* **2010**, *3*, 785–798. [CrossRef]
49. NatureBank. Available online: https://www.griffith.edu.au/institute-drug-discovery/unique-resources/naturebank (accessed on 26 October 2023).
50. Camp, D.; Davis, R.A.; Campitelli, M.; Ebdon, J.; Quinn, R.J. Drug-like Properties: Guiding Principles for the Design of Natural Product Libraries. *J. Nat. Prod.* **2012**, *75*, 72–81. [CrossRef]
51. Vahabzadeh, G.; Soltani, H.; Barati, M.; Golab, F.; Jafari-Sabet, M.; Safari, S.; Moazam, A.; Mohamadrezaei, H. Noscapine Protects the H9c2 Cardiomyocytes of Rats against Oxygen–Glucose Deprivation/Reperfusion Injury. *Mol. Biol. Rep.* **2020**, *47*, 5711–5719. [CrossRef]
52. Sakamuru, S.; Zhao, J.; Attene-Ramos, M.S.; Xia, M. Mitochondrial Membrane Potential Assay. *Methods Mol. Biol.* **2022**, *2474*, 11–19. [CrossRef]
53. Wang, Z.; Wang, S.; Shao, Q.; Li, P.; Sun, Y.; Luo, L.; Yan, X.; Fan, Z.; Hu, J.; Zhao, J.; et al. Brain-Derived Neurotrophic Factor Mimetic, 7,8-Dihydroxyflavone, Protects against Myocardial Ischemia by Rebalancing Optic Atrophy 1 Processing. *Free Radic. Biol. Med.* **2019**, *145*, 187–197. [CrossRef] [PubMed]
54. Lei, P.; Tian, S.; Teng, C.; Huang, L.; Liu, X.; Wang, J.; Zhang, Y.; Li, B.; Shan, Y. Sulforaphane Improves Lipid Metabolism by Enhancing Mitochondrial Function and Biogenesis In Vivo and In Vitro. *Mol. Nutr. Food Res.* **2019**, *63*, 1800795. [CrossRef] [PubMed]
55. Akhtar, A.; Dhaliwal, J.; Sah, S.P. 7,8-Dihydroxyflavone Improves Cognitive Functions in ICV-STZ Rat Model of Sporadic Alzheimer's Disease by Reversing Oxidative Stress, Mitochondrial Dysfunction, and Insulin Resistance. *Psychopharmacology* **2021**, *238*, 1991–2009. [CrossRef] [PubMed]
56. Tian, S.; Lei, P.; Zhang, J.; Sun, Y.; Li, B.; Shan, Y. Sulforaphane Balances Ca2+ Homeostasis Injured by Excessive Fat via Mitochondria-Associated Membrane (MAM). *Mol. Nutr. Food Res.* **2021**, *65*, 2001076. [CrossRef] [PubMed]
57. Balakrishnan, R.; Elangovan, N.; Mohankumar, T.; Nataraj, J.; Manivasagam, T.; Thenmozhi, A.J.; Essa, M.M.; Akbar, M.; Khan, M.A.S. Isolongifolene Attenuates Rotenone-Induced Mitochondrial Dysfunction, Oxidative Stress and Apoptosis. *Front. Biosci. -Sch.* **2018**, *10*, 248–261. [CrossRef]
58. Herbers, E.; Kekäläinen, N.J.; Hangas, A.; Pohjoismäki, J.L.; Goffart, S. Tissue Specific Differences in Mitochondrial DNA Maintenance and Expression. *Mitochondrion* **2019**, *44*, 85–92. [CrossRef] [PubMed]
59. Kim, H.-Y.; Jeon, H.; Kim, H.; Koo, S.; Kim, S. Sophora Flavescens Aiton Decreases MPP+-Induced Mitochondrial Dysfunction in SH-SY5Y Cells. *Front. Aging Neurosci.* **2018**, *10*, 119. [CrossRef] [PubMed]
60. Glinka, Y.Y.; Youdim, M.B.H. Inhibition of Mitochondrial Complexes I and IV by 6-Hydroxydopamine. *Eur. J. Pharmacol. Environ. Toxicol. Pharmacol.* **1995**, *292*, 329–332. [CrossRef]
61. Guimarães, R.P.; Ribeiro, D.L.; Santos, K.B. dos; Godoy, L.D.; Corrêa, M.R.; Padovan-Neto, F.E. The 6-Hydroxydopamine Rat Model of Parkinson's Disease. *JoVE (J. Vis. Exp.)* **2021**, *176*, e62923. [CrossRef]
62. Baran, N.; Lodi, A.; Dhungana, Y.; Herbrich, S.; Collins, M.; Sweeney, S.; Pandey, R.; Skwarska, A.; Patel, S.; Tremblay, M.; et al. Inhibition of Mitochondrial Complex I Reverses NOTCH1-Driven Metabolic Reprogramming in T-Cell Acute Lymphoblastic Leukemia. *Nat. Commun.* **2022**, *13*, 2801. [CrossRef] [PubMed]
63. Basit, F.; Van Oppen, L.M.; Schöckel, L.; Bossenbroek, H.M.; Van Emst-De Vries, S.E.; Hermeling, J.C.; Grefte, S.; Kopitz, C.; Heroult, M.; Hgm Willems, P.; et al. Mitochondrial Complex I Inhibition Triggers a Mitophagy-Dependent ROS Increase Leading to Necroptosis and Ferroptosis in Melanoma Cells. *Cell Death Dis.* **2017**, *8*, e2716. [CrossRef] [PubMed]
64. Yap, T.A.; Daver, N.; Mahendra, M.; Zhang, J.; Kamiya-Matsuoka, C.; Meric-Bernstam, F.; Kantarjian, H.M.; Ravandi, F.; Collins, M.E.; Francesco, M.E.D.; et al. Complex I Inhibitor of Oxidative Phosphorylation in Advanced Solid Tumors and Acute Myeloid Leukemia: Phase I Trials. *Nat. Med.* **2023**, *29*, 115–126. [CrossRef] [PubMed]
65. Haider, S.Z.; Mohanraj, N.; Markandeya, Y.S.; Joshi, P.G.; Mehta, B. Picture Perfect: Imaging Mitochondrial Membrane Potential Changes in Retina Slices with Minimal Stray Fluorescence. *Exp. Eye Res.* **2021**, *202*, 108318. [CrossRef] [PubMed]
66. Zaib, S.; Hayyat, A.; Ali, N.; Gul, A.; Naveed, M.; Khan, I. Role of Mitochondrial Membrane Potential and Lactate Dehydrogenase A in Apoptosis. *Anti-Cancer Agents Med. Chem.* **2022**, *22*, 2048–2062. [CrossRef] [PubMed]
67. Sakamuru, S.; Attene-Ramos, M.S.; Xia, M. Mitochondrial Membrane Potential Assay. *Methods Mol. Biol.* **2016**, *1473*, 17–22. [CrossRef] [PubMed]
68. Tribble, J.R.; Otmani, A.; Sun, S.; Ellis, S.A.; Cimaglia, G.; Vohra, R.; Jöe, M.; Lardner, E.; Venkataraman, A.P.; Domínguez-Vicent, A.; et al. Nicotinamide Provides Neuroprotection in Glaucoma by Protecting against Mitochondrial and Metabolic Dysfunction. *Redox Biol.* **2021**, *43*, 101988. [CrossRef] [PubMed]
69. Sánchez-Reus, M.I.; Peinado, I.I.; Molina-Jiménez, M.F.; Benedí, J. Fraxetin Prevents Rotenone-Induced Apoptosis by Induction of Endogenous Glutathione in Human Neuroblastoma Cells. *Neurosci. Res.* **2005**, *53*, 48–56. [CrossRef] [PubMed]
70. Molina-Jiménez, M.F.; Sánchez-Reus, M.I.; Cascales, M.; Andrés, D.; Benedí, J. Effect of Fraxetin on Antioxidant Defense and Stress Proteins in Human Neuroblastoma Cell Model of Rotenone Neurotoxicity. Comparative Study with Myricetin and N-Acetylcysteine. *Toxicol. Appl. Pharmacol.* **2005**, *209*, 214–225. [CrossRef]
71. de Oliveira, M.R.; Brasil, F.B.; Andrade, C.M.B. Naringenin Attenuates H2O2-Induced Mitochondrial Dysfunction by an Nrf2-Dependent Mechanism in SH-SY5Y Cells. *Neurochem. Res.* **2017**, *42*, 3341–3350. [CrossRef]

72. de Oliveira, M.R.; Custódio de Souza, I.C.; Fürstenau, C.R. Promotion of Mitochondrial Protection by Naringenin in Methylglyoxal-Treated SH-SY5Y Cells: Involvement of the Nrf2/GSH Axis. *Chem.-Biol. Interact.* **2019**, *310*, 108728. [CrossRef] [PubMed]
73. Jin, Y.; Wang, H. Naringenin Inhibit the Hydrogen Peroxide-Induced SH-SY5Y Cells Injury Through Nrf2/HO-1 Pathway. *Neurotox. Res.* **2019**, *36*, 796–805. [CrossRef] [PubMed]
74. Choi, S.H.; Kim, Y.W.; Kim, S.G. AMPK-Mediated GSK3β Inhibition by Isoliquiritigenin Contributes to Protecting Mitochondria against Iron-Catalyzed Oxidative Stress. *Biochem. Pharmacol.* **2010**, *79*, 1352–1362. [CrossRef] [PubMed]
75. Yang, L.; Wang, D.; Zhang, Z.; Jiang, Y.; Liu, Y. Isoliquiritigenin Alleviates Diabetic Symptoms via Activating AMPK and Inhibiting mTORC1 Signaling in Diet-Induced Diabetic Mice. *Phytomedicine* **2022**, *98*, 153950. [CrossRef] [PubMed]
76. Guang, H.-M.; Du, G.-H. Protections of Pinocembrin on Brain Mitochondria Contribute to Cognitive Improvement in Chronic Cerebral Hypoperfused Rats. *Eur. J. Pharmacol.* **2006**, *542*, 77–83. [CrossRef] [PubMed]
77. Brasil, F.B.; de Almeida, F.J.S.; Luckachaki, M.D.; Dall'Oglio, E.L.; de Oliveira, M.R. Pinocembrin Pretreatment Counteracts the Chlorpyrifos-Induced HO-1 Downregulation, Mitochondrial Dysfunction, and Inflammation in the SH-SY5Y Cells. *Metab. Brain Dis.* **2021**, *36*, 2377–2391. [CrossRef] [PubMed]
78. Wang, H.; Liu, X.; Yang, H.; Jing, X.; Wang, W.; Liu, X.; Zhang, B.; Liu, X.; Shao, Y.; Cui, X. Activation of the Nrf-2 Pathway by Pinocembrin Safeguards Vertebral Endplate Chondrocytes against Apoptosis and Degeneration Caused by Oxidative Stress. *Life Sci.* **2023**, *333*, 122162. [CrossRef] [PubMed]
79. Cao, H.; Jia, Q.; Yan, L.; Chen, C.; Xing, S.; Shen, D. Quercetin Suppresses the Progression of Atherosclerosis by Regulating MST1-Mediated Autophagy in Ox-LDL-Induced RAW264.7 Macrophage Foam Cells. *Int. J. Mol. Sci.* **2019**, *20*, 6093. [CrossRef] [PubMed]
80. Cai, J.; Shi, G.; Zhang, Y.; Zheng, Y.; Yang, J.; Liu, Q.; Gong, Y.; Yu, D.; Zhang, Z. Taxifolin Ameliorates DEHP-Induced Cardiomyocyte Hypertrophy via Attenuating Mitochondrial Dysfunction and Glycometabolism Disorder in Chicken. *Environ. Pollut.* **2019**, *255*, 113155. [CrossRef]
81. Sun, X.; Chen, R.; Yang, Z.; Sun, G.; Wang, M.; Ma, X.; Yang, L.; Sun, X. Taxifolin Prevents Diabetic Cardiomyopathy in Vivo and in Vitro by Inhibition of Oxidative Stress and Cell Apoptosis. *Food Chem. Toxicol.* **2014**, *63*, 221–232. [CrossRef]
82. Turovskaya, M.V.; Gaidin, S.G.; Mal'tseva, V.N.; Zinchenko, V.P.; Turovsky, E.A. Taxifolin Protects Neurons against Ischemic Injury in Vitro via the Activation of Antioxidant Systems and Signal Transduction Pathways of GABAergic Neurons. *Mol. Cell. Neurosci.* **2019**, *96*, 10–24. [CrossRef] [PubMed]
83. Zhang, D.; Li, J.; Li, T. Agmatine Mitigates Palmitate (PA)-Induced Mitochondrial and Metabolic Dysfunction in Microvascular Endothelial Cells. *Hum. Exp. Toxicol.* **2022**, *41*, 9603271221110857. [CrossRef] [PubMed]
84. Serralha, R.S.; Rodrigues, I.F.; Bertolini, A.; Lima, D.Y.; Nascimento, M.; Mouro, M.G.; Punaro, G.R.; Visoná, I.; Rodrigues, A.M.; Higa, E.M.S. Esculin Reduces P2X7 and Reverses Mitochondrial Dysfunction in the Renal Cortex of Diabetic Rats. *Life Sci.* **2020**, *254*, 117787. [CrossRef] [PubMed]
85. Kamat, J.P.; Ghosh, A.; Devasagayam, T.P.A. Vanillin as an Antioxidant in Rat Liver Mitochondria: Inhibition of Protein Oxidation and Lipid Peroxidation Induced by Photosensitization. *Mol. Cell Biochem.* **2000**, *209*, 47–53. [CrossRef] [PubMed]
86. Liu, G.; Zhang, B.; Hu, Q.; Liu, X.; Chen, J. Syringic Acid Mitigates Myocardial Ischemia Reperfusion Injury by Activating the PI3K/Akt/GSK-3β Signaling Pathway. *Biochem. Biophys. Res. Commun.* **2020**, *531*, 242–249. [CrossRef] [PubMed]
87. Rashedinia, M.; Khoshnoud, M.J.; Fahlyan, B.K.; Hashemi, S.-S.; Alimohammadi, M.; Sabahi, Z. Syringic Acid: A Potential Natural Compound for the Management of Renal Oxidative Stress and Mitochondrial Biogenesis in Diabetic Rats. *Curr. Drug Discov. Technol.* **2021**, *18*, 405–413. [CrossRef] [PubMed]
88. Rashedinia, M.; Alimohammadi, M.; Shalfroushan, N.; Khoshnoud, M.J.; Mansourian, M.; Azarpira, N.; Sabahi, Z. Neuroprotective Effect of Syringic Acid by Modulation of Oxidative Stress and Mitochondrial Mass in Diabetic Rats. *BioMed Res. Int.* **2020**, *2020*, e8297984. [CrossRef] [PubMed]
89. My Tien Truong, T.; Jang, H.-J.; Ghosh, M.; Son, Y.-O.; Kang, I. P-Coumaric Acid Alleviates Skeletal Muscle Atrophy by Improving Muscular Inflammation and Mitochondrial Dysfunction in High-Fat and High-Sucrose Diet-Fed C57BL/6 Male Mice. *J. Funct. Foods* **2024**, *112*, 105979. [CrossRef]
90. Liang, Y.; Dai, X.; Cao, Y.; Wang, X.; Lu, J.; Xie, L.; Liu, K.; Li, X. The Neuroprotective and Antidiabetic Effects of Trigonelline: A Review of Signaling Pathways and Molecular Mechanisms. *Biochimie* **2023**, *206*, 93–104. [CrossRef]
91. Farid, M.M.; Yang, X.; Kuboyama, T.; Tohda, C. Trigonelline Recovers Memory Function in Alzheimer's Disease Model Mice: Evidence of Brain Penetration and Target Molecule. *Sci. Rep.* **2020**, *10*, 16424. [CrossRef]
92. Aktar, S.; Ferdousi, F.; Kondo, S.; Kagawa, T.; Isoda, H. Transcriptomics and Biochemical Evidence of Trigonelline Ameliorating Learning and Memory Decline in the Senescence-Accelerated Mouse Prone 8 (SAMP8) Model by Suppressing Proinflammatory Cytokines and Elevating Neurotransmitter Release. *GeroScience* **2024**, *46*, 1671–1691. [CrossRef] [PubMed]
93. Mirzaie, M.; Khalili, M.; Kiasalari, Z.; Roghani, M. Neuroprotective and Antiapoptotic Potential of Trigonelline in a Striatal 6-Hydroxydopamine Rat Model of Parkinson's Disease. *Neurophysiology* **2016**, *48*, 176–183. [CrossRef]
94. Gay, N.H.; Suwanjang, W.; Ruankham, W.; Songtawee, N.; Wongchitrat, P.; Prachayasittikul, V.; Prachayasittikul, S.; Phopin, K. Butein, Isoliquiritigenin, and Scopoletin Attenuate Neurodegeneration via Antioxidant Enzymes and SIRT1/ADAM10 Signaling Pathway. *RSC Adv.* **2020**, *10*, 16593–16606. [CrossRef] [PubMed]

95. Narasimhan, K.K.S.; Jayakumar, D.; Velusamy, P.; Srinivasan, A.; Mohan, T.; Ravi, D.B.; Uthamaraman, S.; Sathyamoorthy, Y.K.; Rajasekaran, N.S.; Periandavan, K. Morinda Citrifolia and Its Active Principle Scopoletin Mitigate Protein Aggregation and Neuronal Apoptosis through Augmenting the DJ-1/Nrf2/ARE Signaling Pathway. *Oxidative Med. Cell. Longev.* **2019**, *2019*, e2761041. [CrossRef] [PubMed]
96. Dilshara, M.G.; Lee, K.-T.; Jayasooriya, R.G.P.T.; Kang, C.-H.; Park, S.R.; Choi, Y.H.; Choi, I.-W.; Hyun, J.-W.; Chang, W.-Y.; Kim, Y.-S.; et al. Downregulation of NO and PGE2 in LPS-Stimulated BV2 Microglial Cells by Trans-Isoferulic Acid via Suppression of PI3K/Akt-Dependent NF-κB and Activation of Nrf2-Mediated HO-1. *Int. Immunopharmacol.* **2014**, *18*, 203–211. [CrossRef]
97. Meeprom, A.; Chan, C.B.; Sompong, W.; Adisakwattana, S. Isoferulic Acid Attenuates Methylglyoxal-Induced Apoptosis in INS-1 Rat Pancreatic β-Cell through Mitochondrial Survival Pathways and Increasing Glyoxalase-1 Activity. *Biomed. Pharmacother.* **2018**, *101*, 777–785. [CrossRef]

biomolecules

MDPI

Article

Neuroprotective Effects of the Nutraceutical Dehydrozingerone and Its C_2-Symmetric Dimer in a Drosophila Model of Parkinson's Disease

Maria Dolores Setzu [1,†], Ignazia Mocci [2,†], Davide Fabbri [3], Paola Carta [3], Patrizia Muroni [1], Andrea Diana [1], Maria Antonietta Dettori [3,*] and Maria Antonietta Casu [2,*]

1 Department of Biomedical Sciences, University of Cagliari, Monserrato, 09042 Cagliari, Italy; mdsetzu@unica.it (M.D.S.); pmuroni@unica.it (P.M.); diana@unica.it (A.D.)
2 Unit of Cagliari, CNR-Institute of Translational Pharmacology, Pula, 09050 Cagliari, Italy; ignazia.mocci@ift.cnr.it
3 Unit of Sassari, CNR-Institute of Biomolecular Chemistry, 07100 Sassari, Italy; davidegaetano.fabbri@cnr.it (D.F.); paola.carta@cnr.it (P.C.)
* Correspondence: mariaantonietta.dettori@cnr.it (M.A.D.); mariaantonietta.casu@ift.cnr.it (M.A.C.); Tel.: +39-079-2841224 (M.A.D.); +39-070-92435502 (M.A.C.)
† These authors contributed equally to this work.

Abstract: Parkinson's disease (PD) is a neurodegenerative disorder characterized by the loss of dopaminergic neurons responsible for unintended or uncontrollable movements. Mutations in the leucine-rich repeat kinase 2 locus contribute to genetic forms of PD. The fruit fly *Drosophila melanogaster* carrying this mutation (LRRK2-Dm) is an in vivo model of PD that develops motor impairment and stands for an eligible non-mammalian paradigm to test novel therapeutic approaches. Dehydrozingerone (DHZ) is a natural phenolic compound isolated from ginger and presents anti-inflammatory, antioxidant and neuroprotective properties, making it a potential therapeutic target for PD. We administered DHZ and its C_2-symmetric dimer (DHZ-DIM) at 0.5 and 1 mM for 14 and 21 days in the LRRK2-Dm, with the aim of assessing changes in rescuing motor behavior, brain dopaminergic neurons, mitochondria and synapses (T-bars). The shorter treatment with both molecules revealed efficacy at the higher dose, improving climbing behavior with a prevention of dopaminergic neuronal demise. After 21 days, a recovery of the motor disability, dopaminergic neuron loss, mitochondrial damage and T-bars failure was observed with the DHZ-DIM. Our data indicate that the DHZ-DIM exerts a more potent neuroprotective effect with respect to the monomer in LRRK2-Dm, prompting further investigation of these compounds in rodent models of PD.

Keywords: Parkinson's disease; nutraceutical compounds; dehydrozingerone; hydroxylated biphenyls; natural compounds; *Drosophila melanogaster*; LRRK2; neuroprotection

1. Introduction

Parkinson's disease (PD) is an age-related neurodegenerative disorder with typical manifestation of motor symptoms that includes bradykinesia, rigidity, postural instability, and tremor associated with several non-motor symptoms, namely, cognitive impairment, depression, sleep disturbance and olfactory deficits [1–5]. A pathological hallmark of the disease is the loss of dopaminergic neurons in the substantia nigra, though the processes underlying dopaminergic cell death remain unclear, as well as the exact etiology. However, some biochemical mechanisms are recognized as contributors to the neuropathology: first of all, an abnormal and toxic intracellular accumulation and aggregation of misfolded proteins, such as α-synuclein and parkin, which converge into the core of Lewy bodies. Moreover, neuroinflammation characterized by reactive microgliosis, oxidative stress caused by overproduction of reactive oxygen species, in combination with reactive metabolites of dopamine, and mitochondrial dysfunction are significantly present [6–8].

With regard to available therapies for PD, an effective preventive neuroprotective or disease-modifying cure is currently lacking, highlighting the urgent need for novel drugs or alternative strategies at least to halt the progression of the disease. In recent decades, many studies have adopted a more holistic approach based on metabolic amelioration achieved by specific dietetic programs, such as possible preventive therapy for neurodegenerative diseases. As a matter of fact, many nutraceuticals and food-derived bioactive compounds, by virtue of their intrinsic properties, could have a dramatic role in reducing the risk factors for the arising of chronic neurodegenerative diseases. Focused emphasis has been given to the use of polyphenols, present in most vegetables, e.g., flavonoids, phenolic acids or curcuminoids, as therapeutic natural compounds against inflammation, neurodegeneration, and oxidative stress [9–11].

Curcumin (Figure 1a), also known as diferuloylmethane, is the active component of the *Curcuma longa* (*Zingiberaceae* family) rhizome. This rhizome meets large appreciation as a spice in Indian curries and has garnered specific attention for its wide range of pharmacological activities [12–14]. Many studies in vitro and in vivo showed that curcumin possesses neuroprotection properties [15,16]. Moreover, it has been reported that curcumin is able to suppress PD-like phenotypes in flies [17–19].

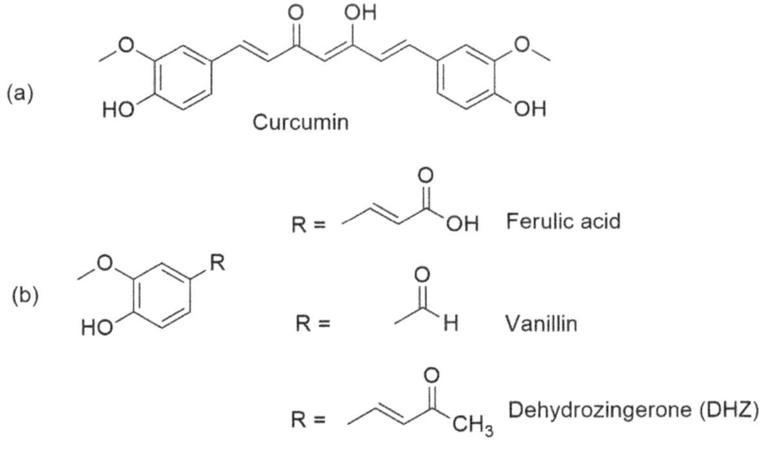

Figure 1. Chemical structures of (**a**) curcumin; (**b**) ferulic acid, vanillin and dehydrozingerone (DHZ).

Unfortunately, the pharmacological potential of curcumin is widely restricted because of its poor bioavailability, due to its chemical instability and rapid metabolic degradation into ferulic acid, vanillin and dehydrozingerone (DHZ) at physiological pH (Figure 1b) [17]. Therefore, it makes sense to explore the development of curcumin derivatives with enhanced bioavailability with consequent translational benefits to hinder PD by identification and production of more potent compounds in the context of phytotherapeutic options.

As mentioned above, DHZ, a structural half analogue of curcumin (CUR), is a natural phenolic compound extracted from ginger (*Zingiber officinale*) rhizome that exhibits enhanced solubility and stability in water compared to CUR. This property gives DHZ a tremendous advantage for biomedical applications where water solubility and stability are crucial factors.

It has been demonstrated that DHZ exhibits many biological activities and pharmacological properties, including anti-inflammatory, antioxidant, anti-obesity, anti-cancer, tyrosinase-inhibitory, neuroprotective, antidepressant and anti-fungal effects [18–28].

Often, hydroxylated biphenyls, have higher biological activities with respect to the corresponding monomer. Hydroxylated biphenyls are widely distributed in the plant kingdom and play an important role in biosystems due to their unique pharmacophore structure that is made up of two aromatic rings bridged by a single C–C bond. The high level of specificity of the hydroxylated biphenyls scaffolds with the catalytic domain of different proteins suggests that this moiety could be an effective scaffold for the discovery and design of new protein targets [29–31]. Consequently, the hydroxylated biphenyl framework offers an ideal molecular structure for structural modifications in the development of potential drug candidates [32].

In prior studies, we established that both DHZ and its symmetric dimer (DHZ-DIM) exhibit protective effects against lipid autoxidation [33], which is an important factor in the development of neurodegenerative disorders such as Alzheimer's and Parkinson's disease. Furthermore, DHZ-DIM exerted a potent anti-inflammatory, antioxidant, and antithrombotic activity on endothelial cells in combination with antiaggregating and cytoprotective properties, as demonstrated by its ability to partially inhibit the aggregation process of alpha-synuclein [34,35]. For these reasons, DHZ and DHZ-DIM could represent interesting candidates to reverse the symptoms of PD.

Although most PD cases seem sporadic, genetic factors may play a role in favoring the disease, and mutations in several specific genes have been related to familiar forms of PD. Among them, mutations in the leucine-rich repeat kinase 2 (LRRK2) gene have been correlated with late-onset autosomal dominant PD. Pathological mutations of LRRK2 have been found not only in 13% of familial forms but also in approximately 1–2% of idiopathic PD cases [36,37]. Furthermore, familial forms related to mutations in the LRRK2 gene show clinical symptoms indistinguishable from idiopathic forms of PD [38].

LRRK2 is a protein widely expressed within different brain areas, such as the cortex, striatum, hippocampus, cerebellum, and substantia nigra [39,40]. It is involved in GTPase and kinase activities associated with signal transduction cascades for synaptic vesicle trafficking, mitochondrial metabolism, and autophagy [41,42]. LRRK2 holds a double enzymatic core in the N-terminal and the C-terminal WD40 domain [43]. In particular, the coding variants G2019S, diffuse in Caucasian individuals [44], and G2385R, present in Asian population [45], in the WD40 domain, resulted in about 50% loss of kinase activity leading to a partial loss of function of LRRK2 [46]. These missense substitutions double the risk of developing PD [47–49].

Regarding the PD genetic approach, *Drosophila melanogaster* (Dm), commonly known as the fruit fly, is a powerful translational animal model for studying neurodegenerative diseases, as it carries nearly 75% homology with human disease genes [50]. Moreover, Dm has several advantages compared to mammalian models, from expandable population, quick life cycles, and easy genetic manipulation to low maintenance costs and less rigid ethical concerns. Notably, most of the genes implicated in familial forms of PD have an evolutionary counterpart in Dm [51,52]. Therefore, Dm carrying the LRRK2 loss-of-function mutation in the WD40 domain [53] develops the essential traits of the pathology, such as motor impairment, dopaminergic neuronal cell loss and mitochondrial abnormalities [54–56], providing precious information regarding PD pathophysiology mechanisms and a bona fide tool to firstly test novel therapeutic approaches to the disease.

Thus, the aim of this paper was to investigate the neuroprotective potential of DHZ and DHZ-DIM, prepared following new sustainable procedures, carrying out the reactions under microwave irradiation. These two compounds were tested on the Drosophila PD strain LRRK2 loss of function (LRRK) compared to wild type (w1118) in terms of physiological and brain morphological parameters that are severely compromised in mutant flies, such as longevity, motor activity, integrity of brain dopaminergic neurons, brain mitochondria and synapse abnormalities. After treatment with both molecules at doses of 0.5 and 1 mM and vehicle for 14 and 21 days after eclosion, there was a significant amelioration of motor performance, and prevention of dopaminergic neuron loss, mitochondrial damage, and synapse (T-bar) failure in LRRK mutants. Specifically, our data indicate that the DHZ-DIM exerts a more potent neuroprotective effect with respect to the monomer in this fruit fly PD model suggesting that these curcumin-related compounds could be promising medicaments for novel therapeutic scenarios toward LRRK2-linked PD.

2. Materials and Methods

2.1. General

Reagents were obtained from Sigma Aldrich (Munich, Germany) and were used without further purification. Microwave reactions were carried out on the MW instrument CEM-Discover SP MW (Matthews, NC, USA). 1H-NMR and 13C-NMR spectra were recorded in CDCl3 at 600 and 150 MHz, respectively, with a 600 MHz NMR spectrometer Bruker Avance III HD (Palo Alto, CA, USA). Chemical shifts are given in ppm (δ); multiplicities are indicated by s (singlet), d (doublet), or dd (doublet of doublets). Elemental analysis was performed using an elemental analyzer model 240 C Perkin Elmer (Waltham, MA, USA). Flash chromatography was carried out with silica gel 60 (230–400 mesh) VWR (Radnor, AF, USA) eluting with the appropriate solution in the stated *v:v* proportions. Reaction was monitored by analytical thin-layer chromatography (TLC) with 0.25 mm-thick silica gel plates 60 F 254 Sigma Aldrich (Munich, Germany). Melting point was determined on a 530 apparatus Büchi (Flawil, Switzerland) and is uncorrected. The purity of new compounds was judged to be >98% by 1H-NMR spectral determination.

2.2. Fly Stocks

Adult wild-type (WT; w1118) and LRRKWD40 (LRRK)-mutant (LRRKex1, #34750, from Bloomington Stock Center) Dm males were used. After emergence from pupae, WT or LRRK-mutant males were separated from females. WT and mutant flies were reared on a standard cornmeal–yeast–agar medium in controlled environmental conditions (24–25 °C; 60% relative humidity; light/dark = 12/12 h). For the treatment, groups of mutant and WT flies were reared on a standard medium supplemented with two concentrations of DHZ and DHZ-DIM (0.5 and 1 mM) for 14 and 21 days. The confirmation of DHZ and its dimer ingestion was assessed by the visualization of red (DHZ) and blue (DHZ-DIM) abdomen due to color-marked enriched medium (Figure S1).

2.3. Drugs

DHZ and DHZ-DIM were dissolved in DMSO (final concentration 0.5%) and added to the diet at the concentrations reported.

2.4. Climbing Assay

The climbing assay (negative geotaxis assay) was used to assess locomotor ability [57] in WT and LRRK mutants treated for 14 or 21 days with DHZ, DHZ-DIM or vehicle. Cohorts of at least 30 flies from each group, in three independent experiments, were subjected to the assay. Flies were placed individually in a vertically positioned plastic tube (length 10 cm; diameter 1.5 cm) and tapped to the bottom. Climbing time (s) was recorded upon crossing a line drawn at 6 cm from the bottom. The number of flies that could climb unto or above this line within 10 s was recorded and expressed as a percentage of the total flies tested. Data are expressed as the average ± standard error of the mean (SEM) from three experiment replications.

2.5. Survival Curves

In accordance with previous reports [58], WT and LRRK male flies were separated from females after emergence from pupae under CO_2 anesthesia. Cohorts of 60 flies from each group were collected in groups of 10–15 in vials containing a standard diet with drugs at 1 mM or VEH, monitored daily at 25 °C and changed with frequency throughout adult life. Data were collected from eclosion to death. For mortality analysis, Kaplan–Meier survival curves and statistical comparisons using the log-rank (Mantel–Cox) test and Gehan–Breslow–Wilcoxon test were utilized.

2.6. Immunohistochemistry

Six to ten flies from each experimental group were selected to perform free-floating fluorescent immunostaining for tyrosine hydroxylase (TH). Animals were anesthetized on ice, and brains were rapidly dissected and fixed in 4% paraformaldehyde in phosphate-buffered saline (PBS) for 2 h. Brains were then incubated with the TH rabbit primary antibody (1:100; AB 152 Millipore Corporation, Billerica, MA, USA) and 10% normal donkey serum in PBS + 0.3% Tween 20 (PBST), at 4 °C for 72 h. After rinsing, the brains were incubated with a donkey anti-rabbit Alexa Fluor 594 secondary antibody (1: 200 Jackson ImmunoResearch Europe, Newmarket, UK) and 10% normal donkey serum in PBST at 4 °C for another 72 h. Subsequently, the brains were mounted on glass slides, coverslipped with Vectashield and analyzed under a fluorescence-spinning disk confocal microscope (Crisel Instruments Rome, Italy). The brains were scanned through Z-stacks (63X objective, stack thickness 0.5 μm), and the number of TH-positive neurons of different clusters (PPL1, PPL2, PPM1/2, PPM3) in both hemispheres was counted via ImageJ software Fiji 1.46r (National Institutes of Health, Bethesda, MD, USA).

2.7. Electron Microscopy Analysis

The electron microscopy studies were performed in strict accordance with the general methodological procedures indicated by Casu et al. (2020) [56].

Briefly, flies from each experimental group (n = 5), were anesthetized on ice, and brains were rapidly dissected and fixed in a mixture of 1% paraformaldehyde and 1.25% glutaraldehyde in 0.15 M cacodylate buffer for 2 h. Brains were then post-fixed with 1% osmium tetroxide for 1 h and stained overnight with 0.5% uranyl acetate at 4 °C. After dehydration in a graded acetone series, brains were embedded in EPON resin. To identify the protocerebrum, where the dopaminergic posterior clusters reside, 1 μm semi-thin coronal sections of the whole brain were stained with toluidine blue. Ultrathin sections (90 nm) cut with a Reichert Supernova ultramicrotome were counterstained with uranyl acetate and lead citrate and observed under a JEOL JEM 1400 Plus electron microscope equipped with a CCD camera at an acceleration voltage of 80 kV.

For morphometric analysis, the mitochondria (total number), the percentage of mitochondria with swollen cristae (percentage of mitochondria displaying swollen cristae versus total mitochondria with discernible cristae) and the T-bar density were analyzed in the unitary area (25 μm^2) in the protocerebrum. Thirty to forty unitary fields were evaluated for each brain. In total, about 17,500 mitochondria and 4000 T-bars were randomly sampled on 792 non-overlapping micrographs at a magnification of 8000×. Swollen cristae were considered when the distance between two contiguous membranes of one crista doubled the average crista size. T-bars were unambiguously identified at presynaptic active zones by the presence of T-shaped electron-dense projections.

2.8. Statistics

Data are presented as means ± SEM. Two group comparisons have been analyzed by factorial two-way ANOVA with the strain and treatment as between-group factors. Before performing the analyses, datasets were checked for normal distribution using the Shapiro–Wilk test and for homogeneity of variances between the experimental groups with Bartlett's test. When the normal distribution of data and homogeneous variances across the experimental groups were obtained in all datasets, the factorial ANOVA was applied.

In all the other cases, when the transformation data did not reveal homogeneity of variances, non-parametric analysis by the Kruskal–Wallis comparisons test was performed. When parametric two-way ANOVA revealed statistically significant interactions, sources of significance were ascertained by pairwise post hoc analyses using the HSD Tukey's test. For mortality analysis, Kaplan–Meier survival and statistical comparisons with Gehan–Breslow–Wilcoxon tests were used. Statistical analyses were all carried out with PRISM, GraphPad 8.0.1 Software (2018), with the significance level set at $p < 0.05$.

3. Results

3.1. Chemistry

DHZ and DHZ-DIM were synthesized with comparable yields and purity, employing a method previously outlined by our group [34]. Notably, we followed more sustainable procedures, utilizing microwave irradiation to significantly reduce the reaction time from 12 h to just 30 min. DHZ was prepared by Claisen–Schmidt condensation reaction of vanillin and acetone in the presence of 1 N NaOH as base (90% yield). DHZ-DIM was prepared under the same conditions, starting from vanillin dimer [35] and acetone in the presence of 1 N LiOH as base (83% yield) (Scheme 1).

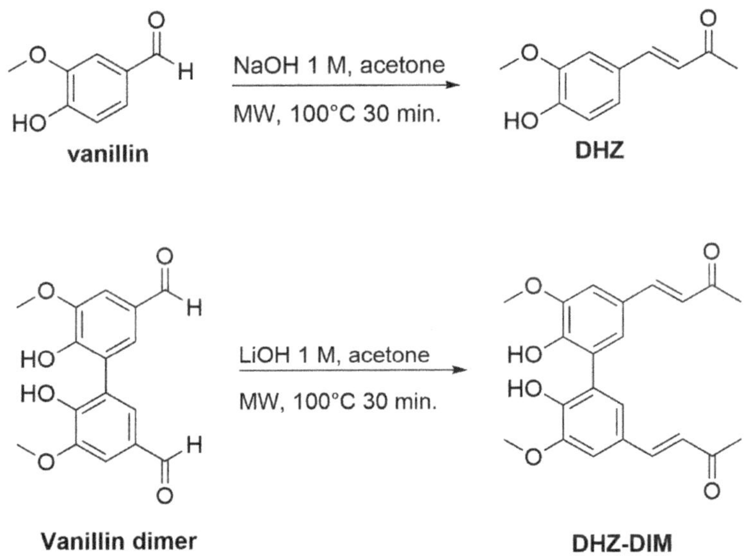

Scheme 1. Synthesis of dehydrozingerone (DHZ) and its dimer (DHZ-DIM).

3.1.1. Synthetic Procedures

[(E)-4-(4-Hydroxy-3-methoxyphenyl) but-3-en-2-one] (DHZ)

An aqueous solution of NaOH 1 N (5.2 mL, 5.2 mmol) was added to a vanillin solution (0.2 g, 1.3 mmol) in acetone (7 mL). The reaction mixture was stirred under MW irradiation at 100 °C for 30 min. The solvent was then roto-evaporated and 10% HCl was cautiously added. The obtained heterogeneous solution was extracted with ether, and the organic phase dried over Na_2SO_4 and evaporated to afford a brown solid compound. The crude material was finally purified by flash chromatography using CH_2Cl_2 as eluent to give DHZ as yellow solid (0.23 g, 90%): mp 126–127 °C; 1H NMR δ 2.39 (s, 3H), 3.85 (s, 3H), 5.99 (bs, 1H), 6.52 (d, J = 16.0 Hz, 1H), 6.90 (d, J = 8.0 Hz, Ar, 1H), 7.00 (d, J = 1.6 Hz, Ar, 1H), 7.06 (dd, J = 1.6, 8.0 Hz, Ar, 1H), 7.40 (d, J = 16.0 Hz, 1H); 13C NMR δ 27.29, 56.93, 109.30, 114.80, 123.94, 124.99, 126.95, 143.76, 146.84, 148.26, 198.46; Anal. Calcd for C11H12O3: C, 68.74; H, 6.29; Found: C, 68.93; H, 6.41.

[(3E,3′E)-4,4′-(6,6′-Dihydroxy-5,5′-dimethoxy-[1,1′-biphenyl]-3,3′-diyl) bis(but-3-en-2-one)] (DHZ-DIM)

An aqueous solution of LiOH 1 N (4.0 mL, 4.0 mmol) was added to vanillin dimer (0.2 g, 0.66 mmol) in acetone (5 mL). The reaction mixture was stirred under MW irradiation at 100 °C for 30 min. The solvent was then roto-evaporated and 10% HCl was cautiously added. The obtained precipitate was filtered, washed with water and dried to afford a DHZ-DIM as yellow solid compound (0.18 g, 83%): mp 242–243 °C; 1H NMR δ 2.35 (s, 6H), 3.99 (s, 6H), 5.28 (bs, 2H), 6.61 (d, J = 16.0 Hz, 2H), 7.1 (d, J = 2.0 Hz, Ar, 2H), 7.15 (d,

J = 2.0 Hz, Ar, 2H), 7.47 (d, J = 16.0 Hz, 2H); 13C NMR δ 27.35, 56.20, 108.67, 123.59, 125.28, 125.44, 126.60, 143.52, 145.45, 147.36, 198.28; Anal. Calcd for $C_{22}H_{22}O_6$: C, 69.10; H, 5.80; Found: C, 69.47; H, 5.70.

3.2. DHZ and DHZ-DIM Prevent Motor Impairment in LRRK: Climbing Test

Groups of LRRK and WT flies received both DHZ or DHZ-DIM at the two doses of 0.5 and 1 mM in their diet for 14 and 21 days to evaluate the effect of both molecules on locomotor ability. The climbing activity and the percentage of flies reaching the target within 10 s were measured. Remarkably, after 14 days of both molecules' treatment, a significant improvement in the climbing behavior of LRRK flies was observed only with the highest dose of 1 mM if compared to the LRRK vehicle ($p < 0.0001$ Figure 2a,b). On the other hand, at the same dose (1 mM), only DHZ-DIM was able to induce a recovery of the motor disability after 21 days of treatment ($p < 0.0001$). The administration of the same drugs at both concentrations did not significantly affect in the w1118 flies (Figure 2a–d).

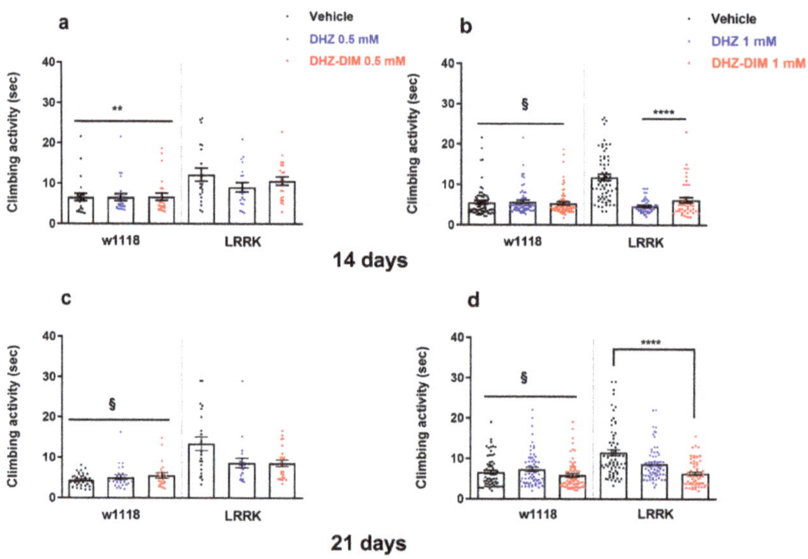

Figure 2. Effect of DHZ and DHZ-DIM at 0.5 mM and 1 mM on the climbing behavior parameters in both LRRK and w1118 WT flies and their respective control groups at 14 (**a**,**b**) and 21 days (**c**,**d**) of treatment. § and **** $p < 0.0001$ vs. LRRK vehicle; ** $p < 0.01$ vs. LRRK vehicle.

In the search for an additional parameter of motor ability, we counted the percentage of WT and LRRK flies completing the climbing test within 10 s (Figure 3). The resulting relative percentages confirmed the improvement in mutants treated for 14–21 days, regardless of DHZ and DHZ-DIM at the concentrations of 0.5 and 1 mM. In addition, the DHZ-DIM was more effective at the higher 1 mM dose on both treatment timelines.

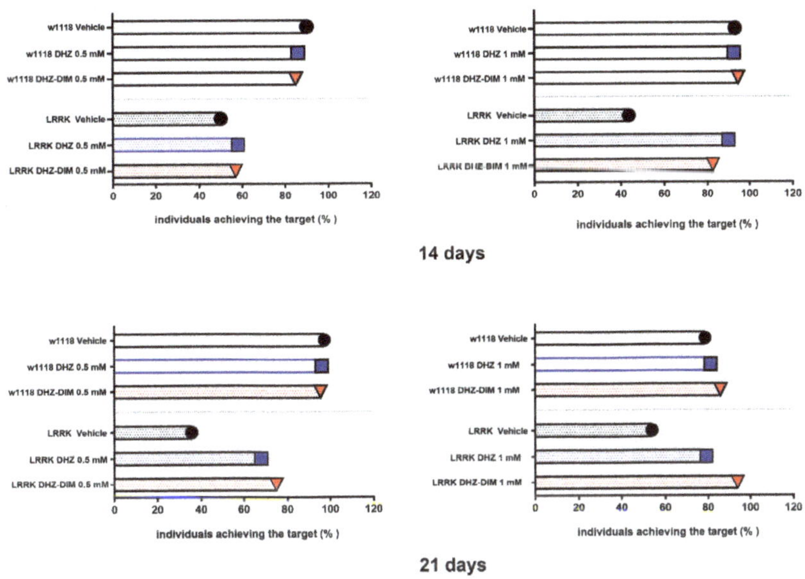

Figure 3. Percentage of w1118 and LRRK individuals reaching the target within 10 s.

3.3. *DHZ and DHZ-DIM Extended Longevity*

Assessment of fly life span using survival curves allowed monitoring of the effect of drugs on Dm survival throughout adulthood. LRRK displayed a shorter life span compared to WT w1118 ($p < 0.0001$, Figure 4a) flies, since mutant flies encountered a dramatic decay in their survival rate (about 50%) at 35–40 days after enclosure (with median survival at 47 days). In an attempt to prolong the LRRK life span, aimed at verifying to what extent the studied molecules were ad hoc effective, mutant flies were supplied with DHZ and DHZ-DIM at the higher concentration of 1 mM and compared with vehicle-treated flies. Both compounds significatively extended longevity of LRRK ($p < 0.0001$, Figure 4b) flies, although flies treated with DHZ-DIM showed a lower decay in the survival rate with the median survival about 70 days with respect to DHZ (around 54 days). Moreover, a small group of LRRK flies (about 20%) treated with the dimer achieved a survival rate at 80 days like w1118 flies.

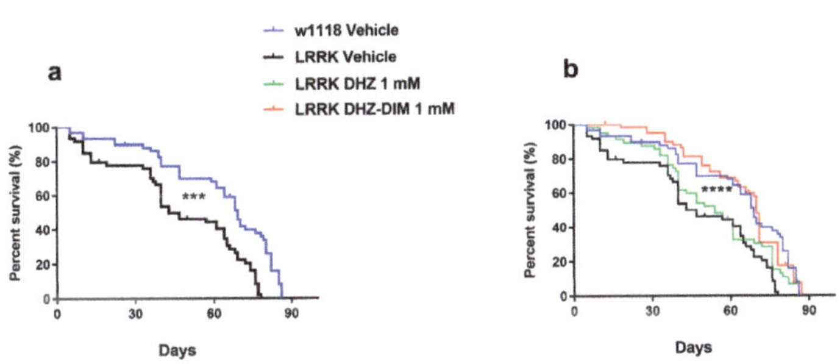

Figure 4. Life span. (**a**): reduced life span in LRRK compared to w1118 flies, both treated with vehicle; (**b**): survival rate observed in DHZ- and DHZ-DIM-treated LRRK flies at 1 mM compared to WT vehicle. Cumulative survival curve data are expressed as means ± SEM. *** $p < 0.001$ indicates significant difference between LRRK vehicle and WT vehicle. **** $p < 0.0001$ LRRK vehicle vs. LRRK treated with DHZ-DIM (Gehan–Breslow–Wilcoxon test).

3.4. DHZ and DHZ-DIM Prevent the Loss of Dopaminergic Neurons

After the climbing behavioral test, flies were euthanized to evaluate the effect of DHZ and DHZ-DIM treatment on TH-positive neurons in the brain posterior dopaminergic clusters (PPL1, PPL2, PPM1/2 and PPM3 (Figure 5)). Since both compounds displayed a maximal behavioral effect at 1 mM concentration, only brains from flies exposed for 14 and 21 days to the above concentration of DHZ and DHZ-DIM were processed for TH immunohistochemistry.

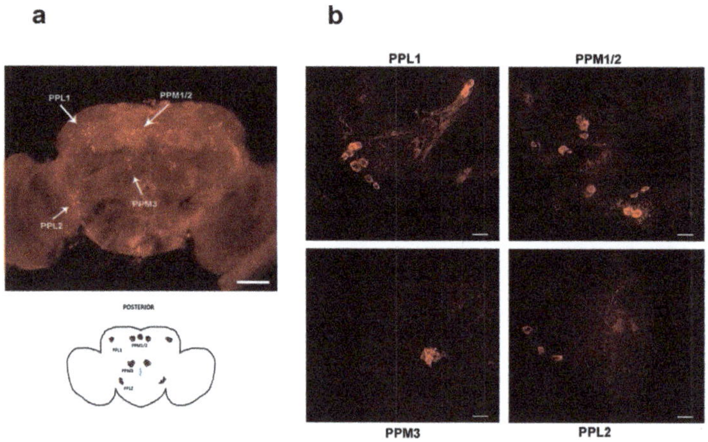

Figure 5. Representative image stacks showing the entire brain ((**a**): 10× scale bar: 100 µm) and the dopaminergic brain neurons in all posterior clusters ((**b**): 63×, scale bar: 10 µm).

The quantitative analysis of TH immunofluorescent neurons showed that both compounds at 1 mM significantly prevented the loss of dopaminergic neurons in all four posterior clusters after 14 days of treatment ($p < 0.0001$). However, at 21 days, a significant prevention of dopaminergic neuronal demise was surprisingly detected only with the DHZ-DIM treatment ($p < 0.0001$) (Figure 6).

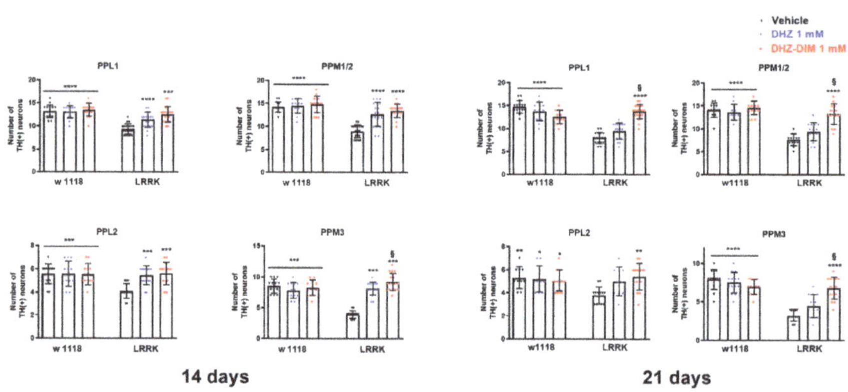

Figure 6. DHZ and DHZ-DIM prevent the loss of dopaminergic neurons. Both compounds at 1 mM significantly prevented the loss of dopaminergic neurons in all posterior clusters at 14 days of treatment, whereas at 21 days, a significant prevention of dopaminergic neuron loss was observed only with the DHZ-DIM treatment. * $p < 0.05$, ** $p < 0.01$, *** $p < 0.001$ and **** $p < 0.0001$ indicate significant differences between LRRK2 vehicle and LRRK2 treated with DHZ, DHZ-DIM or vehicle and treated WT groups. § $p < 0.0001$ LRRK treated with DHZ-DIM vs. LRRK treated with DHZ.

3.5. DHZ-DIM Is More Effective Than Monomer in Preventing the Mitochondrial Damage and the Loss of T-Bars in LRRK2 Drosophila

Transmission electron microscopy (TEM) analysis was conducted on LRRK and WT brain flies treated with vehicle, DHZ or DHZ-DIM at the dose of 1 mM for 21 days, since immunohistochemical results showed that both compounds elicited the maximal effects at 1 mM concentration, with particular reference to the latest time point treatment. As previously described, ultrastructural brain morphology showed axons and terminal boutons, mitochondria and T-bar presynaptic densities (Figure 7a–d). In particular, in vehicle-treated LRRK brain flies, several mitochondria displaying swollen cristae were present (Figure 7a).

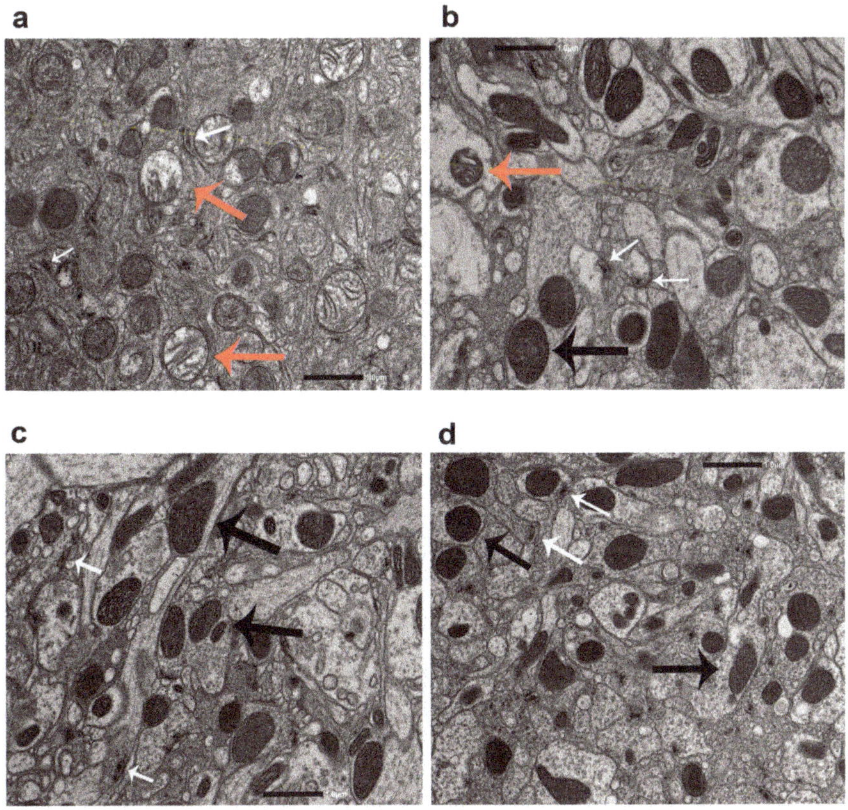

Figure 7. Representative images of mitochondria (8000X) acquired from the protocerebrum of LRRK and w1118 flies treated with vehicle (**a–d**), LRRK treated with DHZ (**b**) and DHZ-DIM (**c**) for 21 days. Red arrows: mitochondria with swollen cristae; black arrows: mitochondria with normal cristae; white arrows: T-bars.

The morphometric analysis did not reveal any difference in the number of total mitochondria between control and LRRK treated with DHZ or DHZ-DIM (Figure 8). The administration of both substances clearly decreased the occurrence of mitochondria with swollen cristae compared with vehicle administration, ($p < 0.0001$) and more relevantly, in accordance with other morphological and functional parameters, the treatment with DHZ-DIM was more worthwhile than DHZ in avoiding the formation of swollen mitochondria (Figure 8). Similarly, using both compounds, a significant increase in the number of T-bar in the presynaptic bouton active zones compared to vehicle treatment was observed ($p < 0.01$;

Figure 8), but the contribution of the dimer was definitely more impactful. By contrast, the exposure of WT flies to both monomer and dimer changed neither mitochondria number, morphology, nor the number of T-bars.

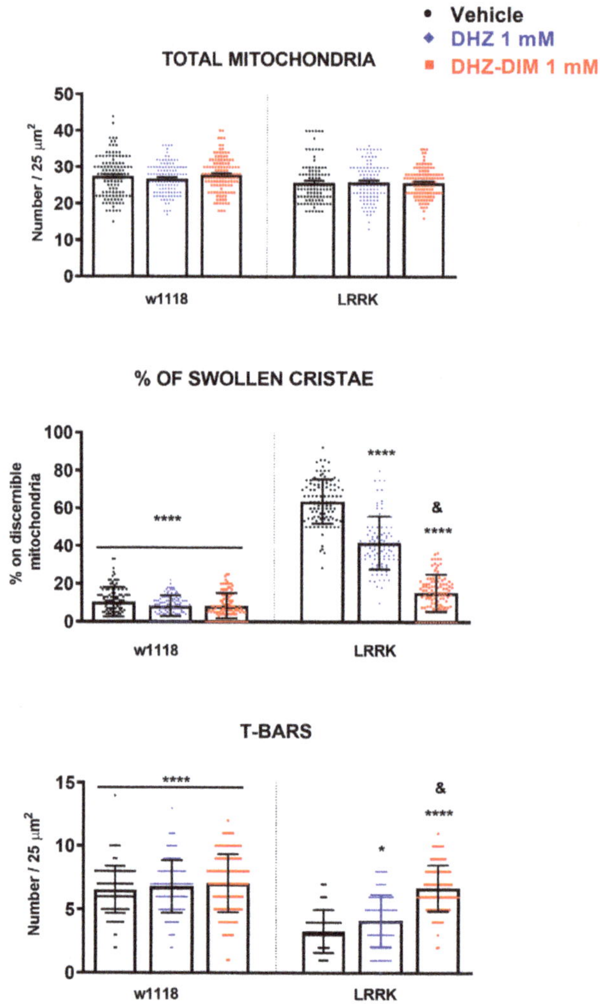

Figure 8. DHZ and DHZ-DIM effect on the number of total mitochondria, percentage of cristae and T-bars in the protocerebrum of LRRK after 21 days of treatment. * $p < 0.05$ and **** $p < 0.0001$ indicate significant differences between LRRK vehicle and LRRK treated with DHZ, DHZ-DIM or vehicle and treated WT groups. & $p < 0.0001$ LRRK treated with DHZ-DIM vs. LRRK treated with DHZ.

4. Discussion

In the present study, we investigated the neuroprotective activity of the two curcumin derivatives DHZ and its C_2 dimer in a transgenic Drosophila model of PD. We improved the process sustainability of DHZ and DHZ-DIM synthesis by carrying out the reactions under microwave (MW) irradiation. As matter of fact, MW technology presents numerous advantages in various chemical synthesis processes. This efficiency is attributed to the ability of MW in heating the reaction mixture both rapidly and uniformly, promoting faster and more complete chemical transformations compared to conventional methods.

Moreover, this technology minimizes the risk of hazardous reactions or the release of volatile substances, contributing to a safer working environment. The utilization of MW aligns with the broader goals of sustainability, promoting energy efficiency and reducing overall environmental impact. To our knowledge, this is the first study evaluating DHZ and DHZ-DIM as neuroprotective agents using an in vivo model of PD. Our data demonstrate that exposure to both compounds prevented motor deficits and protected against the progressive loss of dopaminergic neurons. The neuroprotective efficacy of the DHZ and DHZ-DIM could be related to their antioxidant action, since in previous studies, we found that both DHZ and its symmetric dimer (DHZ-DIM) exhibit protective effects against lipid autoxidation when used in combination with conventional antioxidants [33]. The antioxidant activity of DHZ-DIM is also associated with antiaggregating and cytoprotective properties, ascertained by its ability to partially inhibit the aggregation process of α-synuclein [35].

The methodological approach to the morphological study has taken into account that in the Drosophila brain, DA neurons are organized in distinct bilateral symmetric clusters with projections onto specific brain areas [59,60]. These dopaminergic neurons of posterior clusters analyzed in this study, such as PPL1, PPM1/2 and PPL2, innervate distinct regions of the mushroom bodies that are implicated with learning and memory [61]. Moreover, PPM3 neurons innervate the central complex, which is the area related to the control of motor activity.

Therefore, the neuronal rescue of the above anatomical circuits upon DHZ and DHZ-DIM directly correlates with the protective role for contrasting those brain areas dramatically affected by the progression of PD. In this context, DA neurons' degeneration in PD is further boosted by oxidative stress mechanisms also involving DA facing rapid oxidation. The dopamine autoxidation produced dopamine quinones and free radicals. Moreover, the cyclization of dopamine quinones forms aminochrome, which generates superoxide and down-regulates antioxidative nicotinamide adenine dinucleotide phosphate (NADPH) [59,60]. The susceptibility of the brain to oxidative stress is augmented by various factors, such as high oxygen demand, higher rates of oxidative metabolism and lower levels of protective antioxidants.

This critical scenario is further worsened in the PD genetic models, where LRRK2 mutation caused increased generation of ROS and cell toxicity. A proof of concept has been offered by a recent in vitro work showing that LRRK2 knockout provides resistance to oxidative stress and apoptosis, suggesting LRRK2 as a proapoptotic kinase [62]. Moreover, previous studies demonstrated that deletion of the WD40 domain prevents autophosphorylation [63,64], and on the other hand, the G2385R polymorphism in the WD40 domain, expressed in our drosophila PD model, increases the sensitivity of cells to hydrogen peroxide, suggesting a pro-apoptotic mechanism [65].

Moreover, since recent literature data presume the association of LRRK mutation with autophagy and lysosomal dysregulations [66], this appealing and convincing hypothesis deserves future deeper investigation, especially due to the neuroprotective role of DHZ and DHZ-DIM through modulation of these biological activities.

Considering that our compounds act as antioxidants, the observed neuroprotective ability on dopaminergic neurons matched well with the prevention of motor impairment in mutant flies.

It is important to note that after 14 days of treatment, DHZ and its dimer both effectively hindered the symptoms of Parkinson's diseases in LRRK flies, but only after a longer treatment of 21 days. The DHZ dimer is superior to the monomer in avoiding motor impairment and loss of dopaminergic neurons. These results are reasonably related to the aging progression that—by wide scientific consensus—is one of the major risk factors for developing PD. This is also suggested by the fact that there are many common features between PD and normal aging [67], including protein aggregation [47], increased oxidative stress [68], decreased mitochondrial function [69], dysfunction of the proteasome [70], and impairment of autophagy [71]. Therefore, in aged parkinsonian flies, the DHZ-DIM demonstrates greater efficacy compared to the monomer. This enhanced DHZ-DIM activity could be ascribed to the differences in the chemical structure and lipophilicity between the two molecules, and it makes sense to draw the hypothesis that the DHZ-DIM has a higher ability to cross the cell membrane and to interact with cell components more efficiently than the corresponding monomer. The superior ability of DHZ-DIM in protecting lipids from autoxidation has also been demonstrated, and additionally, its higher antioxidant properties and reactivity when compared to its corresponding monomer [33,72].

In our study, DHZ-DIM has been proven to be more decisive than monomer also in extending longevity of parkinsonian flies. This finding replicates the conclusion of several studies showing that dietary supplementation with compounds rich in polyphenols such as Avocado Persea americana, grape and grape seed extracts, gallic acid and very high doses of curcumin enhanced the life span of Drosophila models of Parkinson disease [73–75]. Polyphenols can delay oxidative reactions in cells by rapidly donating protons to radicals or by forming complexes with pro-oxidant metals. Furthermore, polyphenols can interact with receptors or enzymes in signal translation, promoting an antioxidant condition [10,76].

The neuroprotective effect on motor improvement and brain dopaminergic neurons of our compounds fits very well with the reduced mitochondrial damage in LRRK brain flies detected in our investigation, confirming previous findings related to the presence of damaged mitochondria in LRRK mutant flies, as indicated by multiple dilated cristae [56]. In that regard, oxidative phosphorylation within the mitochondria accounts for the majority of ATP production in neurons required for the transmission of nerve impulse. Various studies have suggested that mitochondrial dysfunction can contribute to increased levels of oxidative stress and can affect neuronal degeneration [77–79]. In postmortem studies, high oxidation of proteins and DNA stimulated parallel levels of lipid peroxidation such that reduction in glutathione was found in the substantia nigra in PD patients [80]. Moreover, mitochondrial complex I inhibition has been observed in PD patients, suggesting that the increased presence of ROS through complex I inhibition is one of the major contributors to DAergic neuronal cell death in PD, reinforcing the concept that this type of stress is dramatically involved in the pathology [1]. On top of such considerations, it cannot be overlooked that several genetic mutations, including LRRK2, are linked to mitochondrial dysfunction in PD pathogenesis. For example, LRRK2 interacts with the mitochondrial fusion proteins and mitochondrial outer membrane proteins [81,82]. LRRK2 mutations determined alterations in mitochondrial fusion and fission mechanisms, mitophagy and in mitochondrial DNA damage [6,83], all elements that increase ROS production, inhibition of peroxidase activity, and a consequent increment in oxidative stress [84].

The association of mitochondrial dysfunction and production of ROS represents a potential target for treating PD. Mitochondria-targeted antioxidants and flavonoids such as alpha-lipoic acid, hesperidin (flavanone rich in citrus), the flavonoid baicalein, the carotenoid lycopene and CUR have produced positive outcomes in in vitro and in vivo studies. Indeed, it has been demonstrated that these compounds can act on mitochondrial integrity, ATP production, mitochondrial membrane potentials, and GSH levels, halting increased ROS production and apoptosis and mitigating altered mitochondrial mechanisms [85–91].

In our study, after 21 days of treatment, DHZ-DIM proved to be more effective than monomer in reducing mitochondrial damage in LRRK brain flies, suggesting that DHZ-DIM can suppress mitochondrial dysfunction, as already demonstrated for the above antioxidants.

Although numerous studies in cell and animal models support the potential of antioxidants in treating PD, many of these results cannot be reproduced in humans.

As a matter of fact, clinical trials have not demonstrated any efficacy of creatine or coenzyme Q10 in patients with PD [92,93]. A possible explanation for these negative clinical results is that oxidative stress could be a downstream effect of mitochondrial dysfunction rather than a direct cause of PD neurodegeneration. Otherwise, of novel drug-delivery approaches may be required.

Finally, we noticed a recovery in the loss of T-bars in the mutant flies after the chronic treatment with DHZ, which was more significatively pronounced with DHZ-DIM. T-bars are the presynaptic active zones involved in neurotransmitter release in Drosophila [94]. LRRK2 binds synaptic vesicles through specific protein–protein interactions in the WD40 domain [95]. Remarkably, synaptic proteomic analysis showed that the G2385R variant impairs LRRK2 binding to key synaptic proteins, including synapsin, which may explain the loss of T-bars in the LRRK2 WD40 Drosophila model [49]. Moreover, evidence supports the role of mitochondria in synaptic plasticity by maintaining cytosolic calcium within physiological ranges [96]. Therefore, a reduced mitochondrial antioxidant function could also be linked to synaptic loss. The neuroprotective effect of DHZ-DIM could be related to its capacity to prevent mitochondrial damage and synaptic loss by its antioxidant activity.

5. Conclusions

In summary, the overall findings from this study indicated that DHZ-DIM, more than its monomer, possessed a strong neuroprotective effect through its ability to ameliorate the PD-like phenotypes in our Drosophila PD model. If validated in mammalian models of PD, it could be considered a promising compound for the design and development of novel nutraceutical agents with neuroprotective properties to be evaluated in humans. However, it should be also considered both that the additional demands to counteract pathological conditions and the practical inability to provide dietary intake with the necessary "overdose" of nutraceuticals present only in traces in available aliments is possibly decisive for a real impact against neurodegenerative diseases.

Although there is strong evidence that suggests to what extent dietary intake of natural compounds with antioxidant and anti-inflammatory properties may inhibit neurodegeneration in PD, more detailed studies are needed exploiting alternative neurodegenerative disease models in both mammals and Drosophila in combination with clinical trials. These further experiments will contribute to providing hope for future effective therapies against PD with natural compounds.

Supplementary Materials: The following supporting information can be downloaded at https://www.mdpi.com/article/10.3390/biom14030273/s1. Figure S1: Arrows indicate (a) red abdomen (DHZ) and (b) blue abdomen (DHZ-DIM) of LRRK Drosophila reared on medium food dye compared with an LRRK on standard food (c).

Author Contributions: Conceptualization, M.A.C., I.M., M.D.S. and M.A.D.; methodology, M.A.C., I.M., M.D.S., M.A.D., D.F. and P.C.; formal analysis, I.M., M.D.S. D.F. and P.C.; investigation, M.A.C., I.M., M.D.S., M.A.D., D.F., P.C. and P.M.; resources, M.A.C., and M.A.D.; data curation M.D.S., I.M., D.F. and P.C.; writing—original draft preparation, M.A.C.; writing—review and editing, M.A.C., I.M., M.D.S., M.A.D., D.F., P.C. and A.D. All authors have read and agreed to the published version of the manuscript.

Funding: This research was founded by the University of Cagliari, Italy, (FIR 2021) and by the CNR project FOE-2021 DBA.AD005.225 (Institute of Biomolecular Chemistry, Unit of Sassari, Italy).

Institutional Review Board Statement: Ethical review and approval were waived for this study due to the fact that the experimental use of Drosophila Melanogaster, not being vertebrates, but arthopodes, did not require ethical approval according to the European and Italian legal guidelines.

Informed Consent Statement: Not applicable.

Data Availability Statement: The data presented in this study are available on request from the corresponding author.

Acknowledgments: We acknowledge the CeSAR (Centro Servizi d'Ateneo per la Ricerca) of the University of Cagliari, Italy, for the fluorescent image stacks performed with a fluorescence spinning disk confocal microscope (Crisel Instruments) and for the electron microscopy analysis (JEOL JEM 1400 Plus electron microscope).

Conflicts of Interest: The authors declare no conflicts of interest.

References

1. Dauer, W.; Przedborski, S. Parkinson's Disease: Mechanisms and Models. *Neuron* **2003**, *39*, 889–909. [CrossRef] [PubMed]
2. Rodriguez-Oroz, M.C.; Jahanshahi, M.; Krack, P.; Litvan, I.; Macias, R.; Bezard, E.; Obeso, J.A. Initial Clinical Manifestations of Parkinson's Disease: Features and Pathophysiological Mechanisms. *Lancet Neurol.* **2009**, *8*, 1128–1139. [CrossRef] [PubMed]
3. Lees, A.J.; Hardy, J.; Revesz, T. Parkinson's Disease. *Lancet* **2009**, *373*, 2055–2066. [CrossRef] [PubMed]
4. Chaudhuri, K.R.; Odin, P. The Challenge of Non-Motor Symptoms in Parkinson's Disease. *Prog. Brain Res.* **2010**, *184*, 325–341. [CrossRef] [PubMed]
5. Erkkinen, M.G.; Kim, M.-O.; Geschwind, M.D. Clinical Neurology and Epidemiology of the Major Neurodegenerative Diseases. *Cold Spring Harb. Perspect. Biol.* **2018**, *10*, a033118. [CrossRef]
6. Park, J.-S.; Davis, R.L.; Sue, C.M. Mitochondrial Dysfunction in Parkinson's Disease: New Mechanistic Insights and Therapeutic Perspectives. *Curr. Neurol. Neurosci. Rep.* **2018**, *18*, 21. [CrossRef] [PubMed]
7. Blesa, J.; Trigo-Damas, I.; Quiroga-Varela, A.; Jackson-Lewis, V.R. Oxidative Stress and Parkinson's Disease. *Front. Neuroanat.* **2015**, *9*, 91. [CrossRef]
8. Hauser, D.N.; Hastings, T.G. Mitochondrial Dysfunction and Oxidative Stress in Parkinson's Disease and Monogenic Parkinsonism. *Neurobiol. Dis.* **2013**, *51*, 35–42. [CrossRef]
9. Scalbert, A.; Manach, C.; Morand, C.; Rémésy, C.; Jiménez, L. Dietary Polyphenols and the Prevention of Diseases. *Crit. Rev. Food Sci. Nutr.* **2005**, *45*, 287–306. [CrossRef]
10. Arias-Sánchez, R.A.; Torner, L.; Fenton Navarro, B. Polyphenols and Neurodegenerative Diseases: Potential Effects and Mechanisms of Neuroprotection. *Molecules* **2023**, *28*, 5415. [CrossRef]
11. Aryal, S.; Skinner, T.; Bridges, B.; Weber, J.T. The Pathology of Parkinson's Disease and Potential Benefit of Dietary Polyphenols. *Molecules* **2020**, *25*, 4382. [CrossRef]
12. Gupta, S.C.; Patchva, S.; Koh, W.; Aggarwal, B.B. Discovery of Curcumin, a Component of Golden Spice, and Its Miraculous Biological Activities. *Clin. Exp. Pharmacol. Physiol.* **2012**, *39*, 283–299. [CrossRef]
13. Maheshwari, R.K.; Singh, A.K.; Gaddipati, J.; Srimal, R.C. Multiple Biological Activities of Curcumin: A Short Review. *Life Sci.* **2006**, *78*, 2081–2087. [CrossRef]
14. Aggarwal, B.B.; Sung, B. Pharmacological Basis for the Role of Curcumin in Chronic Diseases: An Age-Old Spice with Modern Targets. *Trends Pharmacol. Sci.* **2009**, *30*, 85–94. [CrossRef]
15. Di Meo, F.; Margarucci, S.; Galderisi, U.; Crispi, S.; Peluso, G. Curcumin, Gut Microbiota, and Neuroprotection. *Nutrients* **2019**, *11*, 2426. [CrossRef]
16. Bhat, A.; Mahalakshmi, A.M.; Ray, B.; Tuladhar, S.; Hediyal, T.A.; Manthiannem, E.; Padamati, J.; Chandra, R.; Chidambaram, S.B.; Sakharkar, M.K. Benefits of Curcumin in Brain Disorders. *Biofactors* **2019**, *45*, 666–689. [CrossRef]
17. Wang, Y.J.; Pan, M.H.; Cheng, A.L.; Lin, L.I.; Ho, Y.S.; Hsieh, C.Y.; Lin, J.K. Stability of Curcumin in Buffer Solutions and Characterization of Its Degradation Products. *J. Pharm. Biomed. Anal.* **1997**, *15*, 1867–1876. [CrossRef]
18. Parihar, V.K.; Dhawan, J.; Kumar, S.; Manjula, S.N.; Subramanian, G.; Unnikrishnan, M.K.; Rao, C.M. Free Radical Scavenging and Radioprotective Activity of Dehydrozingerone against Whole Body Gamma Irradiation in Swiss Albino Mice. *Chem. Biol. Interact.* **2007**, *170*, 49–58. [CrossRef] [PubMed]
19. Kim, S.J.; Kim, H.M.; Lee, E.S.; Kim, N.; Lee, J.O.; Lee, H.J.; Park, N.Y.; Jo, J.Y.; Ham, B.Y.; Han, S.H.; et al. Dehydrozingerone Exerts Beneficial Metabolic Effects in High-Fat Diet-Induced Obese Mice via AMPK Activation in Skeletal Muscle. *J. Cell. Mol. Med.* **2015**, *19*, 620–629. [CrossRef] [PubMed]
20. Dettori, M.A.; Pisano, M.; Rozzo, C.; Delogu, G.; Fabbri, D. Synthesis of Hydroxylated Biphenyl Derivatives Bearing an α,β-Unsaturated Ketone as a Lead Structure for the Development of Drug Candidates against Malignant Melanoma. *ChemMedChem* **2021**, *16*, 1022–1033. [CrossRef] [PubMed]
21. Hampannavar, G.A.; Karpoormath, R.; Palkar, M.B.; Shaikh, M.S. An Appraisal on Recent Medicinal Perspective of Curcumin Degradant: Dehydrozingerone (DZG). *Bioorganic Med. Chem.* **2016**, *24*, 501–520. [CrossRef]
22. Moorkoth, S.; Prathyusha, N.S.; Manandhar, S.; Xue, Y.; Sankhe, R.; Pai, K.S.R.; Kumar, N. Antidepressant-like Effect of Dehydrozingerone from *Zingiber officinale* by Elevating Monoamines in Brain: In Silico and in Vivo Studies. *Pharmacol. Rep.* **2021**, *73*, 1273–1286. [CrossRef]
23. Lee, E.S.; Kang, J.S.; Kim, H.M.; Kim, S.J.; Kim, N.; Lee, J.O.; Kim, H.S.; Lee, E.Y.; Chung, C.H. Dehydrozingerone Inhibits Renal Lipotoxicity in High-fat Diet–Induced Obese Mice. *J. Cell. Mol. Med.* **2021**, *25*, 8725–8733. [CrossRef]
24. Pathak, N.; Cheruku, S.P.; Rao, V.; Vibhavari, R.J.A.; Sumalatha, S.; Gourishetti, K.; Rao, C.M.; Kumar, N. Dehydrozingerone Protects Temozolomide-Induced Cognitive Impairment in Normal and C6 Glioma Rats besides Enhancing Its Anticancer Potential. *3 Biotech* **2020**, *10*, 438. [CrossRef]
25. Tirunavalli, S.K.; Gourishetti, K.; Kotipalli, R.S.S.; Kuncha, M.; Kathirvel, M.; Kaur, R.; Jerald, M.K.; Sistla, R.; Andugulapati, S.B. Dehydrozingerone Ameliorates Lipopolysaccharide Induced Acute Respiratory Distress Syndrome by Inhibiting Cytokine Storm, Oxidative Stress via Modulating the MAPK/NF-κB Pathway. *Phytomedicine* **2021**, *92*, 153729. [CrossRef]
26. Liu, C.; Li, Y.; Wen, C.; Yan, Z.; Olatunji, O.J.; Yin, Z. Dehydrozingerone Alleviates Hyperalgesia, Oxidative Stress and Inflammatory Factors in Complete Freund's Adjuvant-Induced Arthritic Rats. *Drug Des. Dev. Ther.* **2022**, *16*, 3015–3022. [CrossRef]

27. Choi, J.G.; Kim, S.Y.; Jeong, M.; Oh, M.S. Pharmacotherapeutic Potential of Ginger and Its Compounds in Age-Related Neurological Disorders. *Pharmacol. Ther.* **2018**, *182*, 56–69. [CrossRef] [PubMed]
28. Begum, F.; Manandhar, S.; Kumar, G.; Keni, R.; Sankhe, R.; Gurram, P.C.; Beegum, F.; Teja, M.S.; Nandakumar, K.; Shenoy, R.R. Dehydrozingerone Promotes Healing of Diabetic Foot Ulcers: A Molecular Insight. *J. Cell Commun. Signal.* **2023**, *17*, 673–688. [CrossRef] [PubMed]
29. Hajduk, P.J.; Bures, M.; Praestgaard, J.; Fesik, S.W. Privileged Molecules for Protein Binding Identified from NMR-Based Screening. *J. Med. Chem.* **2000**, *43*, 3443–3447. [CrossRef] [PubMed]
30. Bringmann, G.; Gulder, T.; Gulder, T.A.M.; Breuning, M. Atroposelective Total Synthesis of Axially Chiral Biaryl Natural Products. *Chem. Rev.* **2011**, *111*, 563–639. [CrossRef]
31. Paquin, A.; Reyes-Moreno, C.; Bérubé, G. Recent Advances in the Use of the Dimerization Strategy as a Means to Increase the Biological Potential of Natural or Synthetic Molecules. *Molecules* **2021**, *26*, 2340. [CrossRef]
32. Pisano, M.; Pagnan, G.; Dettori, M.A.; Cossu, S.; Caffa, I.; Sassu, I.; Emionite, L.; Fabbri, D.; Cilli, M.; Pastorino, F.; et al. Enhanced Anti-Tumor Activity of a New Curcumin-Related Compound against Melanoma and Neuroblastoma Cells. *Mol. Cancer* **2010**, *9*, 137. [CrossRef]
33. Kancheva, V.; Slavova-Kazakova, A.; Fabbri, D.; Dettori, M.A.; Delogu, G.; Janiak, M.; Amarowicz, R. Protective Effects of Equimolar Mixtures of Monomer and Dimer of Dehydrozingerone with α-Tocopherol and/or Ascorbyl Palmitate during Bulk Lipid Autoxidation. *Food Chem.* **2014**, *157*, 263–274. [CrossRef]
34. Profumo, E.; Buttari, B.; D'Arcangelo, D.; Tinaburri, L.; Dettori, M.A.; Fabbri, D.; Delogu, G.; Riganò, R. The Nutraceutical Dehydrozingerone and Its Dimer Counteract Inflammation- and Oxidative Stress-Induced Dysfunction of In Vitro Cultured Human Endothelial Cells: A Novel Perspective for the Prevention and Therapy of Atherosclerosis. *Oxidative Med. Cell Longev.* **2016**, *2016*, 1246485. [CrossRef]
35. Marchiani, A.; Mammi, S.; Siligardi, G.; Hussain, R.; Tessari, I.; Bubacco, L.; Delogu, G.; Fabbri, D.; Dettori, M.A.; Sanna, D.; et al. Small Molecules Interacting with α-Synuclein: Antiaggregating and Cytoprotective Properties. *Amino Acids* **2013**, *45*, 327–338. [CrossRef]
36. Kumari, U.; Tan, E.K. LRRK2 in Parkinson's Disease: Genetic and Clinical Studies from Patients. *FEBS J.* **2009**, *276*, 6455–6463. [CrossRef]
37. Hernandez, D.G.; Reed, X.; Singleton, A.B. Genetics in Parkinson Disease: Mendelian versus Non-Mendelian Inheritance. *J. Neurochem.* **2016**, *139* (Suppl. 1), 59–74. [CrossRef]
38. Trinh, J.; Zeldenrust, F.M.J.; Huang, J.; Kasten, M.; Schaake, S.; Petkovic, S.; Madoev, H.; Grünewald, A.; Almuammar, S.; König, I.R.; et al. Genotype-Phenotype Relations for the Parkinson's Disease Genes SNCA, LRRK2, VPS35: MDSGene Systematic Review. *Mov. Disord.* **2018**, *33*, 1857–1870. [CrossRef] [PubMed]
39. Biskup, S.; Moore, D.J.; Celsi, F.; Higashi, S.; West, A.B.; Andrabi, S.A.; Kurkinen, K.; Yu, S.-W.; Savitt, J.M.; Waldvogel, H.J.; et al. Localization of LRRK2 to Membranous and Vesicular Structures in Mammalian Brain. *Ann. Neurol.* **2006**, *60*, 557–569. [CrossRef] [PubMed]
40. Taymans, J.-M.; Van den Haute, C.; Baekelandt, V. Distribution of PINK1 and LRRK2 in Rat and Mouse Brain. *J. Neurochem.* **2006**, *98*, 951–961. [CrossRef] [PubMed]
41. Wallings, R.; Manzoni, C.; Bandopadhyay, R. Cellular Processes Associated with LRRK2 Function and Dysfunction. *FEBS J.* **2015**, *282*, 2806–2826. [CrossRef] [PubMed]
42. Price, A.; Manzoni, C.; Cookson, M.R.; Lewis, P.A. The LRRK2 Signalling System. *Cell Tissue Res.* **2018**, *373*, 39–50. [CrossRef] [PubMed]
43. Mills, R.D.; Mulhern, T.D.; Cheng, H.-C.; Culvenor, J.G. Analysis of LRRK2 Accessory Repeat Domains: Prediction of Repeat Length, Number and Sites of Parkinson's Disease Mutations. *Biochem. Soc. Trans.* **2012**, *40*, 1086–1089. [CrossRef] [PubMed]
44. Berg, D.; Schweitzer, K.J.; Leitner, P.; Zimprich, A.; Lichtner, P.; Belcredi, P.; Brüssel, T.; Schulte, C.; Maass, S.; Nägele, T.; et al. Type and Frequency of Mutations in the LRRK2 Gene in Familial and Sporadic Parkinson's Disease. *Brain* **2005**, *128*, 3000–3011. [CrossRef] [PubMed]
45. Tan, E.K.; Schapira, A.H. Uniting Chinese across Asia: The LRRK2 Gly2385Arg Risk Variant. *Eur. J. Neurol.* **2008**, *15*, 203–204. [CrossRef] [PubMed]
46. Rudenko, I.N.; Kaganovich, A.; Hauser, D.N.; Beylina, A.; Chia, R.; Ding, J.; Maric, D.; Jaffe, H.; Cookson, M.R. The G2385R Variant of Leucine-Rich Repeat Kinase 2 Associated with Parkinson's Disease Is a Partial Loss-of-Function Mutation. *Biochem. J.* **2012**, *446*, 99–111. [CrossRef]
47. Tan, E.K.; Peng, R.; Wu, Y.R.; Wu, R.M.; Wu-Chou, Y.H.; Tan, L.C.; An, X.K.; Chen, C.M.; Fook-Chong, S.; Lu, C.S. LRRK2 G2385R Modulates Age at Onset in Parkinson's Disease: A Multi-Center Pooled Analysis. *Am. J. Med. Genet. Part B Neuropsychiatr. Genet.* **2009**, *150B*, 1022–1023. [CrossRef]
48. Ross, O.A.; Soto-Ortolaza, A.I.; Heckman, M.G.; Aasly, J.O.; Abahuni, N.; Annesi, G.; Bacon, J.A.; Bardien, S.; Bozi, M.; Brice, A.; et al. Association of LRRK2 Exonic Variants with Susceptibility to Parkinson's Disease: A Case-Control Study. *Lancet Neurol.* **2011**, *10*, 898–908. [CrossRef]
49. Carrion, M.D.P.; Marsicano, S.; Daniele, F.; Marte, A.; Pischedda, F.; Di Cairano, E.; Piovesana, E.; von Zweydorf, F.; Kremmer, E.; Gloeckner, C.J.; et al. The LRRK2 G2385R Variant Is a Partial Loss-of-Function Mutation That Affects Synaptic Vesicle Trafficking through Altered Protein Interactions. *Sci. Rep.* **2017**, *7*, 5377. [CrossRef] [PubMed]

50. Bilen, J.; Bonini, N.M. *Drosophila* as a Model for Human Neurodegenerative Disease. *Annu. Rev. Genet.* **2005**, *39*, 153–171. [CrossRef] [PubMed]
51. Reiter, L.T.; Potocki, L.; Chien, S.; Gribskov, M.; Bier, E. A Systematic Analysis of Human Disease-Associated Gene Sequences in *Drosophila melanogaster*. *Genome Res.* **2001**, *11*, 1114–1125. [CrossRef]
52. Aryal, B.; Lee, Y. Disease Model Organism for Parkinson Disease: *Drosophila melanogaster*. *BMB Rep.* **2019**, *52*, 250–258. [CrossRef]
53. Lee, S.; Imai, Y.; Gehrke, S.; Liu, S.; Lu, B. The Synaptic Function of LRRK2. *Biochem. Soc. Trans.* **2012**, *40*, 1047–1051. [CrossRef]
54. Hewitt, V.L.; Whitworth, A.J. Mechanisms of Parkinson's Disease: Lessons from *Drosophila*. *Curr. Top. Dev. Biol.* **2017**, *121*, 173–200. [CrossRef]
55. De Rose, F.; Marotta, R.; Poddighe, S.; Talani, G.; Catelani, T.; Setzu, M.D.; Solla, P.; Marrosu, F.; Sanna, E.; Kasture, S.; et al. Functional and Morphological Correlates in the *Drosophila* LRRK2 Loss-of-Function Model of Parkinson's Disease: Drug Effects of *Withania somnifera* (Dunal) Administration. *PLoS ONE* **2016**, *11*, e0146140. [CrossRef]
56. Casu, M.A.; Mocci, I.; Isola, R.; Pisanu, A.; Boi, L.; Mulas, G.; Greig, N.H.; Setzu, M.D.; Carta, A.R. Neuroprotection by the Immunomodulatory Drug Pomalidomide in the *Drosophila* LRRK2^{WD40} Genetic Model of Parkinson's Disease. *Front. Aging Neurosci.* **2020**, *12*, 31. [CrossRef] [PubMed]
57. Diana, A.; Collu, M.; Casu, M.A.; Mocci, I.; Aguilar-Santelises, M.; Setzu, M.D. Improvements of Motor Performances in the *Drosophila* LRRK2 Loss-of-Function Model of Parkinson's Disease: Effects of Dialyzed Leucocyte Extracts from Human Serum. *Brain Sci.* **2020**, *10*, 45. [CrossRef] [PubMed]
58. Mocci, I.; Casu, M.A.; Sogos, V.; Liscia, A.; Angius, R.; Cadeddu, F.; Fanti, M.; Muroni, P.; Talani, G.; Diana, A.; et al. Effects of Memantine on Mania-like Phenotypes Exhibited by *Drosophila Shaker* Mutants. *CNS Neurosci. Ther.* **2023**, *29*, 1750–1761. [CrossRef] [PubMed]
59. Monastirioti, M. Biogenic Amine Systems in the Fruit Fly *Drosophila melanogaster*. *Microsc. Res. Tech.* **1999**, *45*, 106–121. [CrossRef]
60. Lima, S.Q.; Miesenböck, G. Remote Control of Behavior through Genetically Targeted Photostimulation of Neurons. *Cell* **2005**, *121*, 141–152. [CrossRef] [PubMed]
61. Zars, T. Behavioral Functions of the Insect Mushroom Bodies. *Curr. Opin. Neurobiol.* **2000**, *10*, 790–795. [CrossRef]
62. Quintero-Espinosa, D.A.; Sanchez-Hernandez, S.; Velez-Pardo, C.; Martin, F.; Jimenez-Del-Rio, M. LRRK2 Knockout Confers Resistance in HEK-293 Cells to Rotenone-Induced Oxidative Stress, Mitochondrial Damage, and Apoptosis. *Int. J. Mol. Sci.* **2023**, *24*, 10474. [CrossRef] [PubMed]
63. Jaleel, M.; Nichols, R.J.; Deak, M.; Campbell, D.G.; Gillardon, F.; Knebel, A.; Alessi, D.R. LRRK2 Phosphorylates Moesin at Threonine-558: Characterization of How Parkinson's Disease Mutants Affect Kinase Activity. *Biochem. J.* **2007**, *405*, 307–317. [CrossRef] [PubMed]
64. Jorgensen, N.D.; Peng, Y.; Ho, C.C.-Y.; Rideout, H.J.; Petrey, D.; Liu, P.; Dauer, W.T. The WD40 Domain Is Required for LRRK2 Neurotoxicity. *PLoS ONE* **2009**, *4*, e8463. [CrossRef]
65. Tan, E.K.; Zhao, Y.; Skipper, L.; Tan, M.G.; Di Fonzo, A.; Sun, L.; Fook-Chong, S.; Tang, S.; Chua, E.; Yuen, Y.; et al. The LRRK2 Gly2385Arg Variant Is Associated with Parkinson's Disease: Genetic and Functional Evidence. *Hum. Genet.* **2007**, *120*, 857–863. [CrossRef] [PubMed]
66. Lo, C.H.; Zeng, J. Defective Lysosomal Acidification: A New Prognostic Marker and Therapeutic Target for Neurodegenerative Diseases. *Transl. Neurodegener.* **2023**, *12*, 29. [CrossRef] [PubMed]
67. Rodriguez, M.; Rodriguez-Sabate, C.; Morales, I.; Sanchez, A.; Sabate, M. Parkinson's Disease as a Result of Aging. *Aging Cell* **2015**, *14*, 293–308. [CrossRef] [PubMed]
68. Zhou, C.; Huang, Y.; Przedborski, S. Oxidative Stress in Parkinson's Disease: A Mechanism of Pathogenic and Therapeutic Significance. *Ann. N. Y. Acad. Sci.* **2008**, *1147*, 93–104. [CrossRef]
69. Henchcliffe, C.; Beal, M.F. Mitochondrial Biology and Oxidative Stress in Parkinson Disease Pathogenesis. *Nat. Clin. Pract. Neurol.* **2008**, *4*, 600–609. [CrossRef]
70. Cook, C.; Petrucelli, L. A Critical Evaluation of the Ubiquitin–Proteasome System in Parkinson's Disease. *Biochim. Et Biophys. Acta Mol. Basis Dis.* **2009**, *1792*, 664–675. [CrossRef]
71. Pan, T.; Kondo, S.; Le, W.; Jankovic, J. The Role of Autophagy-Lysosome Pathway in Neurodegeneration Associated with Parkinson's Disease. *Brain* **2008**, *131*, 1969–1978. [CrossRef]
72. Slavova-Kazakova, A.K.; Koleva, L.; Kancheva, V.D.; Delogu, G. Comparative Study of Antioxidant Potential of Curcumin and Its Degradation Products–Vanillin, Ferulic Acid and Dehydrozingerone. *Bulg. Chem. Comm.* **2018**, *50*, 158–163.
73. Ortega-Arellano, H.F.; Jimenez-Del-Rio, M.; Velez-Pardo, C. Dmp53, Basket and drICE Gene Knockdown and Polyphenol Gallic Acid Increase Life Span and Locomotor Activity in a *Drosophila* Parkinson's Disease Model. *Genet. Mol. Biol.* **2013**, *36*, 608–615. [CrossRef]
74. Ortega-Arellano, H.F.; Jimenez-Del-Rio, M.; Velez-Pardo, C. Neuroprotective Effects of Methanolic Extract of Avocado *Persea americana* (var. Colinred) Peel on Paraquat-Induced Locomotor Impairment, Lipid Peroxidation and Shortage of Life Span in Transgenic Knockdown Parkin *Drosophila melanogaster*. *Neurochem. Res.* **2019**, *44*, 1986–1998. [CrossRef] [PubMed]
75. Siddique, Y.H.; Naz, F.; Jyoti, S. Effect of Curcumin on Lifespan, Activity Pattern, Oxidative Stress, and Apoptosis in the Brains of Transgenic *Drosophila* Model of Parkinson's Disease. *Biomed Res. Int.* **2014**, *2014*, 606928. [CrossRef] [PubMed]
76. Mandel, S.; Youdim, M.B.H. Catechin Polyphenols: Neurodegeneration and Neuroprotection in Neurodegenerative Diseases. *Free Radic. Biol. Med.* **2004**, *37*, 304–317. [CrossRef] [PubMed]

77. Tsang, A.H.K.; Chung, K.K.K. Oxidative and Nitrosative Stress in Parkinson's Disease. *Biochim. Biophys. Acta* **2009**, *1792*, 643–650. [CrossRef] [PubMed]
78. Kausar, S.; Wang, F.; Cui, H. The Role of Mitochondria in Reactive Oxygen Species Generation and Its Implications for Neurodegenerative Diseases. *Cells* **2018**, *7*, 274. [CrossRef] [PubMed]
79. Yang, D.; Li, T.; Liu, Z.; Arbez, N.; Yan, J.; Moran, T.H.; Ross, C.A.; Smith, W.W. LRRK2 Kinase Activity Mediates Toxic Interactions between Genetic Mutation and Oxidative Stress in a *Drosophila* Model: Suppression by Curcumin. *Neurobiol. Dis.* **2012**, *47*, 385–392. [CrossRef]
80. Reale, M.; Pesce, M.; Priyadarshini, M.; Kamal, M.A.; Patruno, A. Mitochondria as an Easy Target to Oxidative Stress Events in Parkinson's Disease. *CNS Neurol. Disord. Drug Targets* **2012**, *11*, 430–438. [CrossRef]
81. Wang, X.; Yan, M.H.; Fujioka, H.; Liu, J.; Wilson-Delfosse, A.; Chen, S.G.; Perry, G.; Casadesus, G.; Zhu, X. LRRK2 Regulates Mitochondrial Dynamics and Function through Direct Interaction with DLP1. *Hum. Mol. Genet.* **2012**, *21*, 1931–1944. [CrossRef]
82. Hsieh, C.-H.; Shaltouki, A.; Gonzalez, A.E.; Bettencourt da Cruz, A.; Burbulla, L.F.; St Lawrence, E.; Schüle, B.; Krainc, D.; Palmer, T.D.; Wang, X. Functional Impairment in Miro Degradation and Mitophagy Is a Shared Feature in Familial and Sporadic Parkinson's Disease. *Cell Stem Cell* **2016**, *19*, 709–724. [CrossRef]
83. Mancini, A.; Mazzocchetti, P.; Sciaccaluga, M.; Megaro, A.; Bellingacci, L.; Beccano-Kelly, D.A.; Di Filippo, M.; Tozzi, A.; Calabresi, P. From Synaptic Dysfunction to Neuroprotective Strategies in Genetic Parkinson's Disease: Lessons From LRRK2. *Front. Cell. Neurosci.* **2020**, *14*, 158. [CrossRef]
84. Angeles, D.C.; Ho, P.; Chua, L.L.; Wang, C.; Yap, Y.W.; Ng, C.; Zhou, Z.D.; Lim, K.-L.; Wszolek, Z.K.; Wang, H.Y.; et al. Thiol Peroxidases Ameliorate LRRK2 Mutant-Induced Mitochondrial and Dopaminergic Neuronal Degeneration in *Drosophila*. *Hum. Mol. Genet.* **2014**, *23*, 3157–3165. [CrossRef]
85. Jin, H.; Kanthasamy, A.; Ghosh, A.; Anantharam, V.; Kalyanaraman, B.; Kanthasamy, A.G. Mitochondria-Targeted Antioxidants for Treatment of Parkinson's Disease: Preclinical and Clinical Outcomes. *Biochim. Biophys. Acta* **2014**, *1842*, 1282–1294. [CrossRef]
86. Zhang, H.; Jia, H.; Liu, J.; Ao, N.; Yan, B.; Shen, W.; Wang, X.; Li, X.; Luo, C.; Liu, J. Combined R-α–Lipoic Acid and Acetyl-L-Carnitine Exerts Efficient Preventative Effects in a Cellular Model of Parkinson's Disease. *J. Cell. Mol. Med.* **2010**, *14*, 215–225. [CrossRef]
87. Abdin, A.A.; Sarhan, N.I. Intervention of Mitochondrial Dysfunction-Oxidative Stress-Dependent Apoptosis as a Possible Neuroprotective Mechanism of α-Lipoic Acid against Rotenone-Induced Parkinsonism and L-Dopa Toxicity. *Neurosci. Res.* **2011**, *71*, 387–395. [CrossRef] [PubMed]
88. Tamilselvam, K.; Braidy, N.; Manivasagam, T.; Essa, M.M.; Prasad, N.R.; Karthikeyan, S.; Thenmozhi, A.J.; Selvaraju, S.; Guillemin, G.J. Neuroprotective Effects of Hesperidin, a Plant Flavanone, on Rotenone-Induced Oxidative Stress and Apoptosis in a Cellular Model for Parkinson's Disease. *Oxidative Med. Cell. Longev.* **2013**, *2013*, 102741. [CrossRef] [PubMed]
89. Wang, Y.-H.; Yu, H.-T.; Pu, X.-P.; Du, G.-H. Baicalein Prevents 6-Hydroxydopamine-Induced Mitochondrial Dysfunction in SH-SY5Y Cells via Inhibition of Mitochondrial Oxidation and up-Regulation of DJ-1 Protein Expression. *Molecules* **2013**, *18*, 14726–14738. [CrossRef] [PubMed]
90. Kaur, H.; Chauhan, S.; Sandhir, R. Protective Effect of Lycopene on Oxidative Stress and Cognitive Decline in Rotenone Induced Model of Parkinson's Disease. *Neurochem. Res.* **2011**, *36*, 1435–1443. [CrossRef] [PubMed]
91. Liu, Z.; Yu, Y.; Li, X.; Ross, C.A.; Smith, W.W. Curcumin Protects against A53T Alpha-Synuclein-Induced Toxicity in a PC12 Inducible Cell Model for Parkinsonism. *Pharmacol. Res.* **2011**, *63*, 439–444. [CrossRef] [PubMed]
92. Beal, M.F.; Oakes, D.; Shoulson, I.; Henchcliffe, C.; Galpern, W.R.; Haas, R.; Juncos, J.L.; Nutt, J.G.; Voss, T.S.; Ravina, B.; et al. A Randomized Clinical Trial of High-Dosage Coenzyme Q10 in Early Parkinson Disease: No Evidence of Benefit. *JAMA Neurol.* **2014**, *71*, 543–552. [CrossRef] [PubMed]
93. Kieburtz, K.; Tilley, B.C.; Elm, J.J.; Babcock, D.; Hauser, R.; Ross, G.W.; Augustine, A.H.; Augustine, E.U.; Aminoff, M.J.; Bodis-Wollner, I.G.; et al. Effect of Creatine Monohydrate on Clinical Progression in Patients with Parkinson Disease: A Randomized Clinical Trial. *JAMA* **2015**, *313*, 584–593. [CrossRef] [PubMed]
94. Fouquet, W.; Owald, D.; Wichmann, C.; Mertel, S.; Depner, H.; Dyba, M.; Hallermann, S.; Kittel, R.J.; Eimer, S.; Sigrist, S.J. Maturation of Active Zone Assembly by *Drosophila* Bruchpilot. *J. Cell Biol.* **2009**, *186*, 129–145. [CrossRef]
95. Piccoli, G.; Onofri, F.; Cirnaru, M.D.; Kaiser, C.J.O.; Jagtap, P.; Kastenmüller, A.; Pischedda, F.; Marte, A.; von Zweydorf, F.; Vogt, A.; et al. Leucine-Rich Repeat Kinase 2 Binds to Neuronal Vesicles through Protein Interactions Mediated by Its C-Terminal WD40 Domain. *Mol. Cell. Biol.* **2014**, *34*, 2147–2161. [CrossRef]
96. Zaichick, S.V.; McGrath, K.M.; Caraveo, G. The Role of Ca^{2+} Signaling in Parkinson's Disease. *Dis. Models Mech.* **2017**, *10*, 519–535. [CrossRef]

biomolecules

MDPI

Article

Ladostigil Reduces the Adenoside Triphosphate/ Lipopolysaccharide-Induced Secretion of Pro-Inflammatory Cytokines from Microglia and Modulate-Immune Regulators, TNFAIP3, and EGR1

Fanny Reichert [1], Keren Zohar [2], Elyad Lezmi [3], Tsiona Eliyahu [2], Shlomo Rotshenker [1], Michal Linial [2] and Marta Weinstock [4,*]

[1] Department of Medical Neurobiology, Institute for Medical Research Israel-Canada (IMRIC), Faculty of Medicine, Hebrew University of Jerusalem, Jerusalem 91120, Israel; funarei@gmail.com (F.R.); shlomor@ekmd.huji.ac.il (S.R.)

[2] Department of Biological Chemistry, Institute of Life Sciences, Hebrew University of Jerusalem, Jerusalem 91904, Israel; keren.zohar@mail.huji.ac.il (K.Z.); tsiona.e@mail.huji.ac.il (T.E.); michall@cc.huji.ac.il (M.L.)

[3] Department of Genetics, Institute of Life Sciences, Hebrew University of Jerusalem, Jerusalem 91904, Israel; elyad.lezmi@mail.huji.ac.il

[4] Institute of Drug Research, School of Pharmacy, Hebrew University of Jerusalem, Jerusalem 91120, Israel

* Correspondence: martar@ekmd.huji.ac.il; Tel.: +972-2-6758731

Abstract: Treatment of aging rats for 6 months with ladostigil (1 mg/kg/day) prevented a decline in recognition and spatial memory and suppressed the overexpression of gene-encoding pro-inflammatory cytokines, TNFα, IL1β, and IL6 in the brain and microglial cultures. Primary cultures of mouse microglia stimulated by lipopolysaccharides (LPS, 0.75 μg/mL) and benzoyl ATPs (BzATP) were used to determine the concentration of ladostigil that reduces the secretion of these cytokine proteins. Ladostigil (1×10^{-11} M), a concentration compatible with the blood of aging rats in, prevented memory decline and reduced secretion of IL1β and IL6 by ≈50%. RNA sequencing analysis showed that BzATP/LPS upregulated 25 genes, including early-growth response protein 1, (Egr1) which increased in the brain of subjects with neurodegenerative diseases. Ladostigil significantly decreased Egr1 gene expression and levels of the protein in the nucleus and increased TNF alpha-induced protein 3 (TNFaIP3), which suppresses cytokine release, in the microglial cytoplasm. Restoration of the aberrant signaling of these proteins in ATP/LPS-activated microglia in vivo might explain the prevention by ladostigil of the morphological and inflammatory changes in the brain of aging rats.

Keywords: aging rats; NFκB; NLRP3 inflammasome; P2x7 receptor; primary murine microglia; RNA-seq

1. Introduction

Microglia are the resident immune cells in the brain. In a healthy adult brain, microglia have a ramified morphology and long processes [1]. They are involved in the integration of new neurons into neuronal circuits, which is important for learning, memory, and cognition [2]. In response to injury, microglia contract their processes, assume an amoeboid shape, and proliferate and migrate toward the site of injury. ATP, released from injured neurons, stimulates purinergic receptors on the microglial membrane, triggering an efflux of K^+ that activates the nucleotide-binding oligomerization domain-(NOD)-LRR and pyrin domain inflammasome (NLRP3) and converts procaspase-1 to caspase-1 [3]. This enables the processing and secretion of IL1β and other pro-inflammatory cytokines [4].

Microglia have retracted processes in the aging brain [5] in which cell damage results from a decline in mitochondrial activity and antioxidant defense mechanisms [6,7]. They also have higher levels of pro-inflammatory cytokines and express cytokine receptors, which could contribute to neurodegeneration and memory impairment [8,9].

Basal levels of TNFα are necessary for regulating synaptic transmission and plasticity [10], while IL1β levels are needed to regulate long-term potentiation (LTP) that underlies learning and memory [11]. However, excess amounts of these cytokines can impair these cellular processes [12], as demonstrated by the direct injection of IL1β into the brain, which inhibits hippocampal LTP [13,14]. Also, greater levels of brain IL6 are linked to synapse loss and deficits in avoidance learning in mice [15].

Oxidative stress and cytokines activate signaling pathways, such as the mitogen-activated protein kinase (MAPK) family of proteins in immune cells [16]. MAPKs consist of three main families: extracellular signal-regulated kinase (ERK), c-Jun N-terminal kinase (JNK), and p38 [17]. IL1β activates MAPK p38 and JNK and increases the nuclear factor kappa-light-chain-enhancer of activated B cells (NFκB) [18], which is strongly associated with age in mice and humans [19]. NFκB is increased in the brains of subjects with neurodegenerative diseases [20]. This finding prompted a search for novel therapies to slow age-related memory impairment by inhibiting the nuclear translocation of NFκB [21].

For this purpose, we developed ladostigil (6-(N-ethyl, N-methyl carbamyloxy)-N propargyl-1(R)-aminoindan hemitartrate), which significantly reduced the mitochondrial potential in cells subjected to oxidative–nitrative stress and decreased malonaldehyde, a measure of oxidative stress, in the cerebral hemispheres of mice, induced by the injection of a lipopolysaccharide (LPS) [22] (Panarsky, 2012). In primary mouse microglial cultures activated by LPS, ladostigil reduced the nuclear translocation of NFκB and phosphorylation of ERK1/2 and p38 and downregulated the gene expression of TNFα, IL6, and IL1β [23]. In addition, when ladostigil was administered at a dose of 1 mg/kg/day for 6 months to 16-month-old rats, it prevented a decline in recognition and spatial memory [24]. It also suppressed the increase in mRNA of TNFα, IL6, and IL1β induced by aging [23] (Panarsky et al., 2012) and genes adversely affected by synaptic function in brain regions associated with learning and memory [25].

The aim of the current study was to obtain a better understanding of the mechanism through which ladostigil reduces cytokine release from microglia. The preparation of such primary microglial cultures causes the loss of some membrane receptors that are activated in the intact brain [26]. However, they still retain purinergic receptors for adenosine and ATP, released in response to neuronal injury [27]. The activation of the (P2x7R) subtype results in the processing and secretion of TNFα and IL1β in response to LPS [28]. The secretion of IL1β can be achieved by adding 2′-3′-O-(4-benzoyl benzoyl) adenosine 5′-triphosphate (BzATP), an agonist of P2x7R [29].

We first looked for the concentrations of ladostigil that would maximally inhibit the secretion of IL1β, IL6, and TNFα proteins from microglia activated by a combination of BzATP and LPS. Then, we measured the levels of ladostigil in the blood of the aging rats in which a dose of 1 mg/kg/day had prevented memory decline to ascertain whether there was enough of the drug in vivo to have affected cytokine release from microglia. Lastly, we sought additional information about the cellular processes involved in these actions of ladostigil, using RNA sequencing (RNA-seq) to perform a detailed analysis of its effect on gene expression in the microglia.

2. Materials and Methods

2.1. Animals

Male Balb/C mice and Wistar rats purchased from Harlan Laboratories (Jerusalem, Israel) were used in accordance with the National Research Council's guide for the care and use of laboratory animals. The Animal Care and Use Committee of the Hebrew University approval #MD-19-15710-4 was for the mice, #MD-08-11537-3 was for the rats.

2.2. Compounds and Reagents

Ladostigil was a gift from Spero Biopharma (Jerusalem, Israel). Dulbecco's Modified Eagle Medium/Nutrient Mixture F-12 (DMEM/F12) was used. Gentamycin sulfate and L-Glutamine were obtained from Biological Industries (Beit-Haemek, Israel) and BzATP, bovine serum albumin (BSA), and LPS were from Escherichia coli 055:B5, purified by trichloracetic acid extraction from Sigma-Aldrich Israel Ltd. (Rechovot, Israel).

2.3. Preparation of Microglia

Primary microglia were isolated from the brains of neonatal Balb/C mice, as previously described [30]. The brains were stripped of their meninges and enzymatically dissociated. Cells were plated on Poly-L-lysine-coated flasks for one week, re-plated for 1 to 2 h on bacteriological plates, and non-adherent cells were washed away. Microglia were propagated by incubation in 20% of the medium, and conditioned by the L-cell line that produces mouse-CSF. They were identified by morphology and positive immune reactivity to P2y12, F4/80, complement receptor-3, and Galectin-3/MAC-2 [30].

2.4. Measurement of Cytokines

Cytokines were measured as previously described [31] using ELISA Max deluxe sets (Biolegend, San Diego, CA, USA) for secreted TNFα and IL6 proteins and ELISA DuoSet (R&D Systems, Minneapolis, MN, USA) for secreted and cell-associated IL1β, according to the manufacturer's instructions. BSA (0.1%) was used to provide the necessary protein in place of the fetal calf serum used in our previous experiments [23], which was shown to contain substances that can inhibit cytokine release [32]. We also ascertained that the concentration of LPS (0.75 µg/mL) given together with BzATP did not affect cell viability after 3 and 24 h using the MTT assay described by Denizot and Lang [33].

The effect of ladostigil on the secretion of cytokine proteins was measured at concentrations of 1×10^{-13} M to 1×10^{-9} M by adding it with BSA to microglia for 2 h before LPS (0.75 µg/mL) and BzATP (400 µM). Other microglia were treated similarly with the steroid budesonide as a reference standard at concentrations of 1×10^{-13}–1×10^{-11} M. Measurements of cytokine secretion were made 8 h after the addition of LPS and BzATP. Since the levels of secreted IL1βTNF were still low, samples were concentrated 2- to 4-fold by an Amicon ultra-centrifugal filter device (Merck-Millipore, Tullagreen, Carrigtwohill, Co Cork, Ireland). ELISA was used to quantify the levels of cytokine proteins. Protein content in the microglia lysate was measured by Bradford, and levels of cytokines were calculated and presented as pg/µg of microglial protein. Each concentration of ladostigil was tested in 18–30 replicates for TNFα and IL6 and 13–18 replicates for IL1β.

2.5. Measurement of Ladostigil in Rat Plasma

Measurements were made in 6 male Wistar rats aged 22 months weighing 720–790 gm in which ladostigil (1 mg/kg/day) had been administered for 6 months in the drinking water, which prevented the development of learning and memory deficits [24]. The rats were lightly restrained while blood samples (0.2 µL) were taken from the tail vein between 09:00 and 12:00 into heparinized Eppendorf tubes. At the end of the experiment, blood samples were also taken by cardiac puncture from some of the rats after terminal anesthesia. They were centrifuged at 4 °C and 20,800 g for 10 min, and the plasma was stored at −80 °C until analysis by liquid chromatography–mass spectroscopy analysis. After precipitating plasma proteins with methanol, ladostigil was detected by an AB Sciex (Framingham, MA, USA) Triple Quad™ 5500 mass spectrometer in positive ion mode by electrospray ionization and a multiple reaction monitoring mode of acquisition using rivastigmine hemitartarate as an internal standard, as described in Moradov et al. [34].

2.6. RNA-Seq of Microglia

Ladostigil (1×10^{-10} M) was added to microglia for 2 h before BzATP/LPS, as described above. Cells were harvested before and 8 h after the addition of BzATP/LPS.

Total RNA was extracted using the RNeasy Plus Universal Mini Kit (QIAGEN), according to the manufacturer's protocol. Total RNA samples (1 µg RNA) were enriched for mRNAs by pull-down of poly (A). Libraries were prepared using a KAPA Stranded mRNA-Seq Kit, according to the manufacturer's protocol, and sequenced using Illumina NextSeq 500 to generate 85 bp single-end reads (a total of 25–30 million reads per sample).

2.6.1. Bioinformatic Analysis

Next-generation sequencing data underwent quality control using FastQC, version 0.11.9 (accessed on 15 March 2021). They were then preprocessed using Trimmomatic [35] and aligned to the reference genome GRCm38 with the STAR aligner [36] using default parameters. Genomic loci were annotated using GENCODE version M25 [37]. Genes with low expression were filtered out of the dataset by setting a threshold of a minimum of two counts per million in three samples.

2.6.2. Gene Module Classification

Pair-wise differential analyses were performed on all three BzATP/LPS time points, and genes with an FDR < 0.01 were considered. Only the genes with an absolute log fold-change of >0.5 across two consecutive time points were labeled up- or downregulated.

2.7. Immunocytochemistry

For immunocytochemistry, microglia cells were plated on 12 mm round glass coverslips in 24-well sterile plates (NUNC A/S, Roskilde, Denmark) in DMEM and low glu/10% FCS-HI. Non-adherent cells were washed out after 3–4 h. Adherent microglia were incubated overnight in 0.1% BSA/DMEM/F12 and then used in experiments identical to those carried out for testing cytokine secretion. To study the expression of TNF alpha-induced protein 3 (TNFaIP3, A20) protein, ladostigil (1×10^{-10} M) was added to the microglia for 2 h before BzATP/LPS, and measurements were made after 8 h. TNFaIP3 was visualized by immunofluorescence confocal microscopy (Zeiss Confocal LSM 980) using an antibody against TNFaIP3 (A20; Abcam # 92324). Microglia were fixed for 15 min in 4% methanol-free formaldehyde, permeabilized for 10 min in 0.1% Triton X100, and blocked for 1 h in 10% FCS in PBS. Anti-TNFaIP3 Ab (diluted 1/200 in PBS/FCS) was added to the microglia in wet chambers overnight at 4 °C. Cy3-labeled secondary Ab goat and anti-rabbit (in PBS/FCS) were applied for 1 h followed by Alexa Fluor 488phalloidin and Dapi staining. Randomly sampled microglia were scanned by a confocal microscope at one plane that ran through the middle of their nuclei. Immunofluorescence levels in the cytoplasm of Cy3-labeled TNFaIP3 were determined by IMARIS software, Version 10.1. Optical slices of cells, 1 µm thick, were scanned sequentially and used to produce the shown maximal intensity projection images (Zeiss Zen 3.3 software). We estimated the concentration of TNFaIP3 by determining the intensity/unit area to take into account any differences among the microglia in the volume of their cytoplasm. By sampling all the cells in the same plane that runs through the center of the cell nucleus, we neutralized any preferential localization of TNFaIP3 within the cell.

To study the expression of early growth response (EGR) 1 protein, the same protocol was used as TNFAIP3 protein with some modifications. Ladostigil (1×10^{-10} M) was added to microglia for 2 h before BzATP/LPS, and measurements were made after 3 h. EGR1 protein (red) was visualized by immunofluorescence microscopy using a monoclonal antibody against Egr1 (cell signaling, #4153). Randomly selected low-power fields that were scanned in the same plane that runs through the center of cell nuclei were used to determine the percentage of microglia that displayed positive EGR1 protein immunoreactivity in their nuclei.

2.8. Statistics

The cytokine quantification data were analyzed in samples of at least 24 replicates using one-way analysis of variance (ANOVA) by IBM SPSS Statistics Version 25 followed by Duncan's post hoc test. The assumption of the homogeneity of variances was verified using the Brown–Forsythe test for equality of group variances. Comparing two experimental groups, a two-sample *t*-test was performed. Results from experiments on cell viability, cytokine secretion from microglia, and TNFaIP3 in microglial cytoplasm are presented as mean ± SEM. Plasma levels of ladostigil are presented as mean ± SD. *p*-values of < 0.05 were considered statistically significant. Measures of TNFAIP3 in microglia were analyzed by a Kruskal–Wallis non-parametric test, and Egr1 in microglial nuclei was analyzed by Dunnett's multiple comparison test. Principal component analysis (PCA) was performed using the R-base function "prcomp". EdgeR was used to perform RNA read counts by the trimmed mean of the M-values normalization of RNA (TMM) and differential expression analysis [38]. Gene-set and KEGG pathway enrichment analyses were performed using the "goana" and "kegga" functions (respectively) in the "limma" R package [39]. Figures were generated using the ggplot2 R package.

3. Results

3.1. Ladostigil Concentration in the Plasma of Old Rats

The mean (±SD) plasma concentration of ladostigil in samples taken from six rats after they had been given ladostigil (1 mg/kg/day) in the drinking fluid for six months was 2.39 ± 1.08 ng/mL (8.75 ± 3.95 nM).

3.2. Effect of Ladostigil on Cytokine Release from Activated Microglia

The effect of BSA (0.1%) with LPS 0.75 μg/mL and BzATP (400 μM) on cell viability in arbitrary units after 3 h was 0.026 ± 0.002 and 0.029 ± 0.004 after 24 h. It did not differ from BSA, which was 0.025 ± 0.002 at both time points. The lowest concentration of ladostigil tested in microglia that significantly decreased cytokine secretion induced by BzATP/LPS was 1×10^{-13} M for TNFα and IL6, and 1×10^{-12} M for IL1β. Maximal reductions of ≈50% for IL6 and IL1β were obtained by ladostigil (1×10^{-11}–1×10^{-9} M). At all concentrations of ladostigil and budesonide tested, the reductions of IL6 were greater than TNFα, ($p < 0.001$). Reductions of IL1β by ladostigil (1×10^{-9} M) and budesonide (1×10^{-11} M) were also greater than TNFα (Figure 1). The greater effect of ladostigil on the release of IL6 than TNFα, which is also seen after budesonide, may be due to the differential regulation of these cytokines by EGR1 [40].

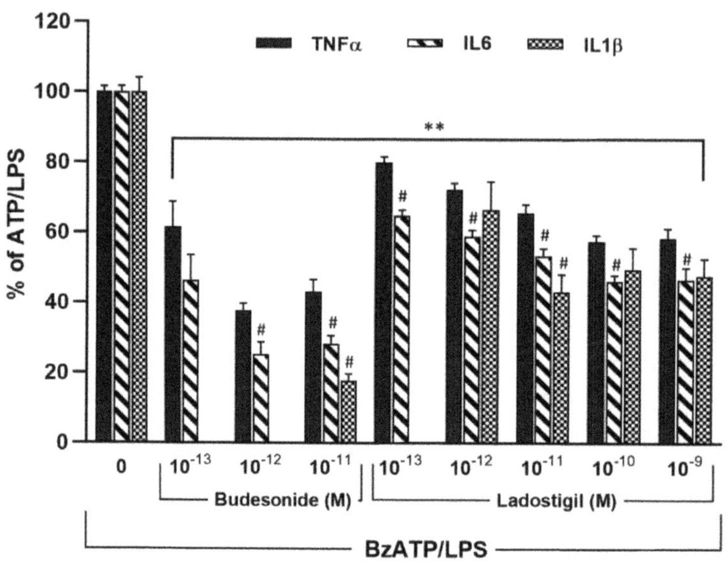

Figure 1. A dose–related reduction in the release of TNFα, IL6, and IL1β from microglia activated by BzATP/LPS. Ladostigil or budesonide was added 2 h before BzATP and LPS in the presence of 0.1%

BSA. ANOVA for TNFα; $F_{5,275} = 85.4$, $p < 0.0001$; IL6; $F_{5,282} = 141.7$, $p < 0.0001$; IL1β; $F_{4,73} = 18.14$, $p < 0.0001$. All concentrations of ladostigil and budesonide tested reduced the three cytokines; ** $p < 0.01$. This was significantly different from the value for TNFα; # $p < 0.05$.

3.3. Effect of Ladostigil on Genes in Microglia Assessed by RNA-seq

The reproducibility of the normalized RNA-seq read counts was assessed by performing PCA analysis for each biological sample. The biological replicates clustered tightly together, confirming the low variability within each experimental group. The sample variability (within and between groups) is illustrated by the first two principal components that comprise >83% of the variation. Only four DE genes were affected by ladostigil treatment in resting, unstimulated microglia (Figure 2A), but the expression of 25 genes was significantly altered 8 h after their activation by BzATP/LPS (Figure 2B) when ladostigil produced its inhibitory effect on cytokine secretion.

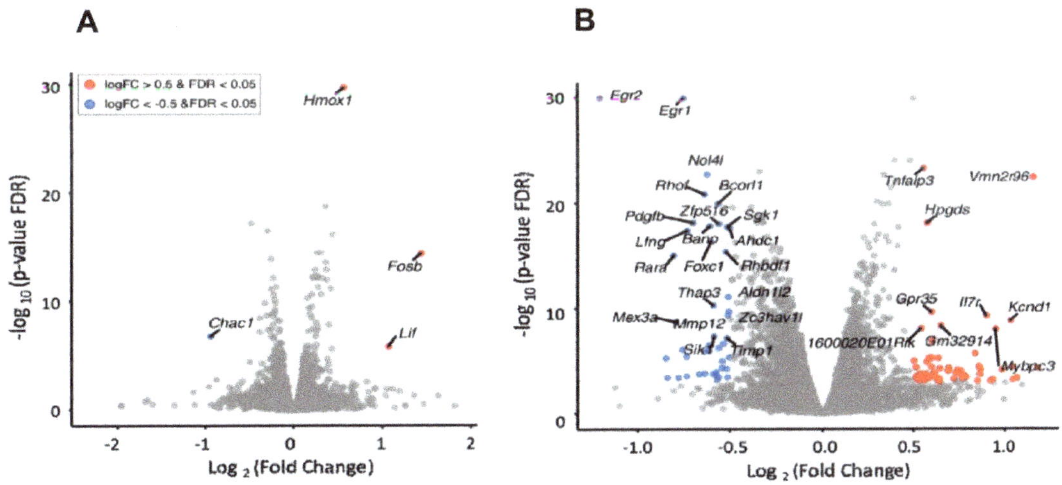

Figure 2. Ladostigil treatment alters gene expression in BzATP/LPS-activated microglia. (**A**) Volcano plots showing the log-fold change vs. −log10 (*p*-value), as calculated by edgeR, and differential expression analysis in ladostigil-treated microglia compared to untreated cells. (**B**) Volcano plots showing the log-fold change vs. −log10 (*p*-value) of ladostigil-treated microglia 8 h after the addition of BzATP/LPS. Blue dots indicate genes that are downregulated and red dots indicate genes that are upregulated by ladostigil. Cut-off value, ±0.5.

Among these were early-growth response proteins 1 and 2 (Egr1 and Egr2), matrix metalloproteinase (Mmp), Mmp 12, the tissue inhibitor of metalloprotease 1 (Timp1), and platelet-derived growth factor β (Pdgf-β) which were all downregulated by ladostigil. TNFaIP3 was upregulated (Figure 3A,B).

The connected genes with a STRING score of >0.6 are shown in Figure 4. The network connectivity is highly significant (*p*-value: 2.2×10^{-4}).

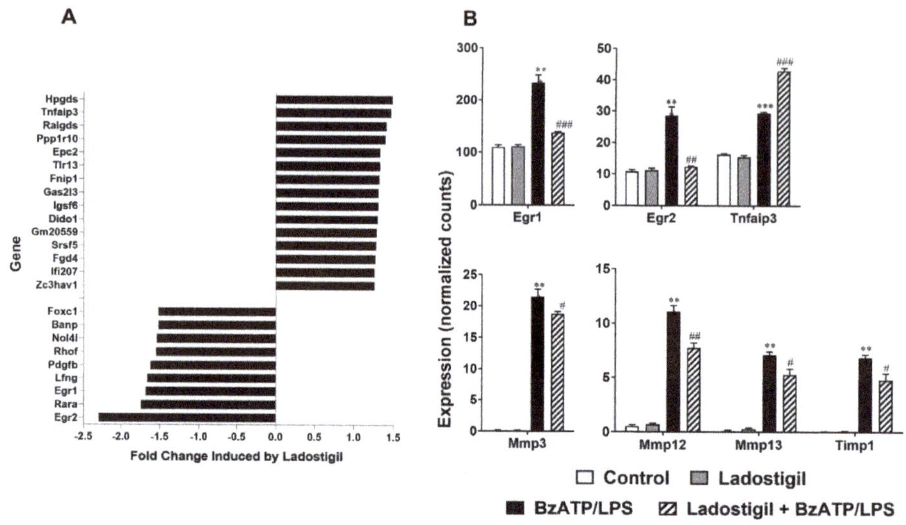

Figure 3. Genes altered by ladostigil treatment. (**A**) Bars show the change caused by ladostigil in gene expression in microglia after BzATP/LPS-induced activation. Bars on the right show genes that were upregulated and bars on the left show genes that were downregulated. Cut-off value, ±0.5. (**B**) Selected, differentially expressed genes after BzATP/LPS treatment, with or without ladostigil. Significantly different from the unstimulated control, ** $p < 0.01$, *** $p < 0.001$; significant effect of ladostigil, # $p < 0.05$; ## $p < 0.01$ ### $p < 0.001$.

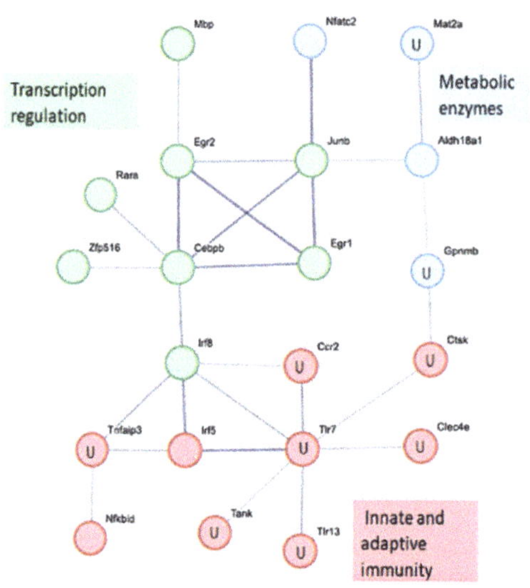

Figure 4. STRING protein–protein interaction network. Differentially expressed genes 8 h after BzATP/LPS activation with and without ladostigil. All clusters are labeled by their main cellular functions, and the larger connected network (red color) is partitioned into sub-clusters for functional annotation. The letter U in the node indicates upregulated genes. The other nodes are downregulated genes. STRING protein–protein interaction enrichment, $p = 2.2 \times 10^{-4}$.

3.4. Ladostigil Treatment Increases TNFaIP3 Protein in BzATP/LPS-Activated Microglia

The addition of BzATP/LPS to microglia caused a small but significant increase in TNFaIP3 protein in the cytoplasm ($p < 0.05$). This was increased further ($p < 0.001$) by the addition of ladostigil (1×10^{-10} M) (Figure 5 and Table 1).

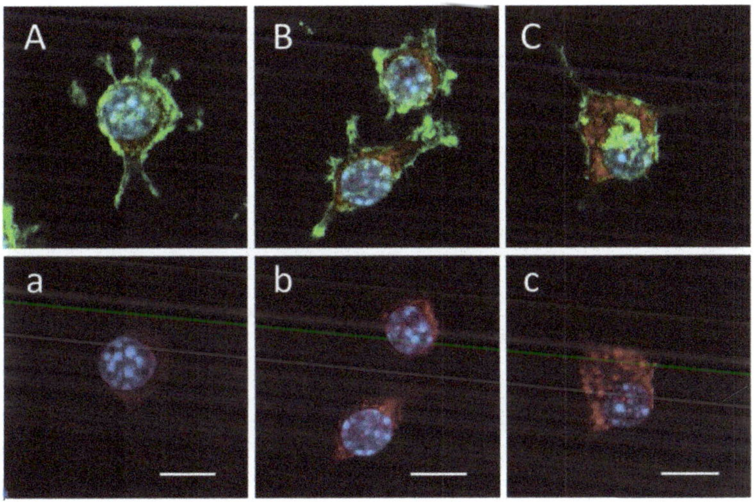

Figure 5. Ladostigil increases levels of TNFAIP3 protein immunoreactivity in microglial cytoplasm in the presence of BzATP/LPS. Three representative high-power immunofluorescence confocal microscopy images of single microglia are displayed: (**A**,**a**), medium with 0.1% BSA; (**B**,**b**) medium plus BzATP/LPS; and (**C**,**c**) medium plus ladostigil + BzATP/LPS. TNFAIP3 protein was visualized (red) by immunocytochemistry using an antibody against TNFAIP3 protein, F-actin (green) by Alexa 488-labeled phalloidin, and nuclei (blue) by DAPI staining. Optical slices of phagocytes, 1 µm thick, were scanned sequentially and used to produce the shown maximal intensity projection images (Zeiss Zen 3.3 software). TNFAIP3 protein immunoreactivity is detected in the cytoplasm but not the nuclei (**a**–**c**). In fields (**A**–**C**), TNFAIP3 protein-positive immunoreactivity, F-actin and nuclei are displayed. In fields (**a**–**c**), only TNFAIP3 protein-positive immunoreactivity and nuclei are displayed. Calibration bars: 10 µm.

Table 1. Quantification of TNFAIP3 protein in the microglial cytoplasm.

Treatment	N	Mean Fluorescence Intensity ± SEM (AU/µm^2)
Medium	32	19.8 ± 0.6
BzATP/LPS	53	23.3 ± 1.0 *
Ladostigil + BzATP/LPS	58	28.8 ± 0.9 ***###

Randomly selected fields (as those shown in Figure 5) were used. N = number of fields sampled. Significance of difference by ANOVA, $p < 0.0001$, and Bonferroni's multiple comparison test; medium vs. BzATP/LPS * $p < 0.05$ and medium vs. ladostigil + BzATP/LPS, *** $p < 0.001$; BzATP/LPS vs. ladostigil + BzATP/LPS, ### $p < 0.001$.

3.5. Ladostigil Treatment Decreases EGR1 Protein in the Nucleus of BzATP/LPS-Activated Microglia

The addition of BzATP/LPS to microglia significantly increased the number of microglia containing EGR1 protein in their nuclei ($p < 0.001$). This was decreased significantly ($p < 0.001$) by ladostigil (Figure 6 and Table 2).

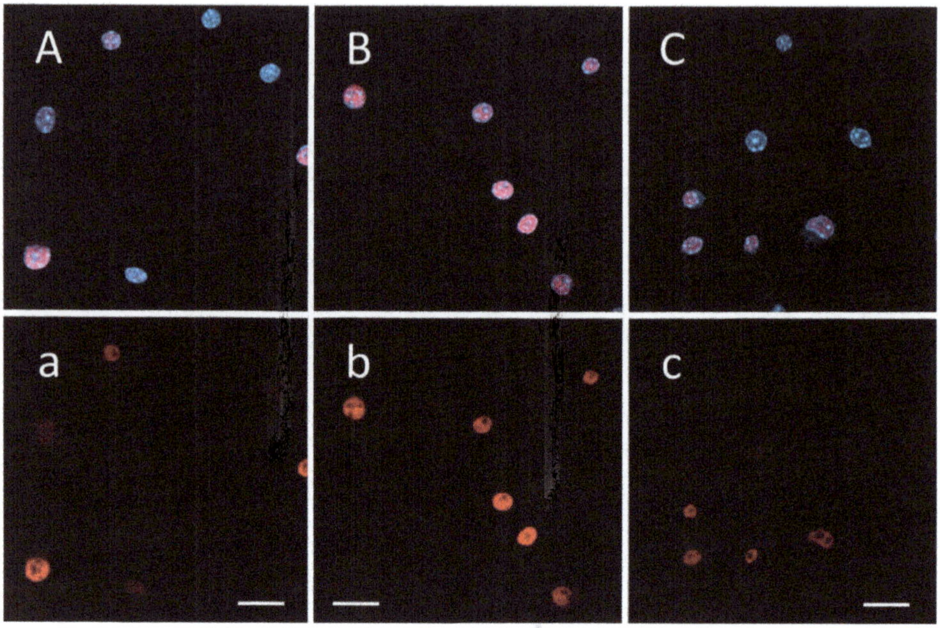

Figure 6. Ladostigil decreases the number of microglia displaying positive EGR1 protein immunoreactivity in their nuclei induced by BzATP/LPS treatment. Three representative low-power immunofluorescence confocal microscopy fields are displayed: (**A**,**a**) medium with 0.1% BSA; (**B**,**b**) medium plus BzATP/LPS; (**C**,**c**) ladostigil + BzATP/LPS. Egr1 was visualized using immunocytochemistry and a monoclonal antibody against EGR1; nuclei (blue) were visualized using DAPI staining, and EGR1 (red), overlaying nuclei appear pink. In fields (**A**–**C**), both EGR1 protein-positive immunoreactivity and nuclei are displayed. In fields (**a**–**c**), only EGR1 protein-positive immunoreactivity is displayed. Calibration bars: 20 μm.

Table 2. Quantification of EGR1 protein immunoreactivity in microglial nuclei.

Treatment	N	Mean Percent of Nuclei with Positive EGR1 Protein Immunoreactivity ± SEM
Medium	32	19.0 ± 4.2
BzATP/LPS	53	82.3 ± 3.7 ***
Ladostigil + BzATP/LPS	58	56.0 ± 3.9 ***###

Randomly selected fields (as those shown in Figure 5) were used. N = number of fields sampled. Significance of difference by ANOVA, $p < 0.0001$, and Bonferroni's multiple comparison test; medium vs. BzATP/LPS and medium vs. ladostigil + BzATP/LPS, *** $p < 0.001$; BzATP/LPS vs. ladostigil + BzATP/LPS, ### $p < 0.001$.

4. Discussion

The dysfunction of mitochondria and the generation of reactive oxygen species (ROS) occur in the aging brain and are early contributory events to neurodegeneration and Alzheimer's disease (AD) [6,7]. ROS releases ATP [41], which activates purinergic A_{2A} receptors ($A_{2A}R$) on the microglia membrane, causing them to retract their processes [42]. $A_{2A}R$ and P2x7 receptors (P2x7R) are upregulated in the brains of patients with AD [43] and in the hippocampus of aging rats with memory impairment. The stimulation of $A_{2A}R$ in the brain by ATP further increases neurodegeneration [44], while the activation of P2x7R on microglia releases several pro-inflammatory cytokines [28]. In the current study, the addition of BzATP to activate P2x7R in microglial cultures in addition to LPS enabled the measurement of IL1β protein secretion. Both IL1β and IL6 decreased by ≈50% at concentrations of ladostigil of 0.01–1 nM.

Chronic treatment with ladostigil (1 mg/kg/day) in aging rats prevented the upregulation of $A_{2A}R$ and memory decline [25]. The mean plasma concentration of ladostigil in these rats was found to be 2.39 ± 1.08 ng/mL or 8.75 ± 3.95 nM. However, ladostigil was not measured in the brain of the aging rats, and peak concentrations in the cerebral cortex in young adult rats after acute oral administration of 5 mg/kg were ≈25% of those in plasma (unpublished observations) Assuming a similar plasma-cortical ratio after the chronic administration of 1 mg/kg/day in aged rats, a concentration of ≈2.2 nM is obtained. After s.c. injection of 5 mg/kg in mice, the concentration of ladostigil in the brain was also 25–50% of that in plasma [34]. Since a dose of 1 mg/kg/day also decreased the gene expression of TNFα, IL6, and IL1β in the brain [23], it is reasonable to assume that there could have been enough ladostigil in the brain of the aging rats to have reduced cytokine release.

Cytokine secretion from microglia activated by LPS and BzATP was accompanied by a significant upregulation of transcription factors, Egr1, Egr2, and PDGF-β. Egr1 and Egr2 are downstream signaling targets of P2x7R [40] that increase in LPS-activated mixed astrocyte–microglial cultures [45]. Egr1 is rapidly and transiently induced in different cell types in response to a variety of stimuli, including oxidative stress, radiation injury, electrical stimulation, and neurotransmitter activity. It is activated by intracellular pathways, including MAPKs, ERK, and p38. Egr1/Krox24 gene expression in the hippocampus was shown to be related to the severity of AD in human subjects [46]. Egr1 also accelerates tau phosphorylation and the processing of amyloid precursor protein to β-amyloid in a mouse model of AD [47]. Much less is known about the role of Egr2 as a mediator of inflammation.

PDGF-β is released in the cerebral spinal fluid (CSF) from pericytes and is a specific marker for pericyte injury associated with a loss of integrity in the blood–brain barrier [48], which declines in normal aging and more rapidly in AD. A correlation was found between age and PDGF-β in CSF, with the highest levels found in subjects with mild cognitive impairment (MCI) and AD [49].

Ladostigil decreased the expression of Egr1 and Egr2 transcripts in BzATP/LPS-activated microglia and Mmps 12, all of which regulate cytokine release [50], and it also reduced the expression of PDGF-β. It significantly reduced the amount of EGR1 protein in the nucleus three h after it had been elevated by BzATP/LPS. On the other hand, ladostigil upregulated the gene expression of the ubiquitin-modifying enzyme, TNFaIP3, and increased the levels of this protein in the microglial cytoplasm. TNFaIP3 terminates the activation of NFκB in response to stimulation by LPS, IL1β, TNFα, IL6, or CD40 [51]. TNFaIP3 prevents the NFκB-dependent upregulation of NLRP3 and conversion of pro-IL1β to mature IL1β through the binding of its A20-like zinc finger domain to ubiquitin chains [52]. It also blocks IKKα/β activation by the upstream kinase, Tak1 [53]. Moreover, the brains of mice lacking TNFaIP3 have a larger number of microglia with shorter and fewer processes, resembling those after chronic infection or aging [54]. Together, these observations suggest that the elevation of TNFaIP3 could protect the organism against inflammatory conditions occurring in the aged brain [19].

An increase in cytosolic TNFaIP3 by ladostigil via the alteration of various feedback-controlling mechanisms [55] could be responsible for the reduction in the phosphorylation of ERK and/or p38 and the decrease in nuclear EGR1. This, in turn, explains how ladostigil reduced the formation and secretion of cytokines in BzATP/LPS-activated microglia in the current study. Restoring the aberrant signaling of these genes and their proteins by ladostigil to normal enables us to explain how they prevented the morphological and inflammatory changes in the brain regions of aging rats [25] and the attenuation of the decline in memory in the whole brain and hippocampal volumes in elderly subjects with MCI [56].

Author Contributions: F.R. isolated and cultured primary microglia, performed all experiments (cytokines and the extraction of total RNA for RNA-seq), immunocytochemistry, and confocal immunofluorescence microscopy; S.R. contributed to the design and analysis of the experiments on cytokine release from microglia and supervised confocal microscopy; K.Z. and E.L. performed the RNA sequential analyses; T.E. was responsible for the molecular preparation of RNA libraries; M.L. supervised the molecular RNA-seq analyses; M.W. designed the study, performed ladostigil measurements in rat plasma, and wrote the manuscript. All authors have read and agreed to the published version of the manuscript.

Funding: Research funds were given to M.W. by The Hebrew University of Jerusalem.

Institutional Review Board Statement: The study was conducted in accordance with the National Research Council's guide for the care and use of laboratory animals. The Animal Care and Use Committee of the Hebrew University approval #MD-19-15710-4 was for the mice, and #MD-08-11537-3 was for the rats. The dates of approval are 12 March 2019 for #MD-19-15710-4 and 15 August 2008 for MD-08-11537-3.

Informed Consent Statement: Not applicable.

Data Availability Statement: RNA-seq data files were deposited in ArrayExpress under accession E-MTAB-10450.

Conflicts of Interest: The rights for ladostigil are currently held by The Hebrew University of Jerusalem. A private Israeli company has the option to license ladostigil from the University. M.W. is a shareholder in this company.

References

1. Nimmerjahn, A.; Kirchhoff, F.; Helmchen, F. Resting microglial cells are highly dynamic surveillants of brain parenchyma in vivo. *Science* **2005**, *308*, 1314–1318. [CrossRef] [PubMed]
2. Wake, H.; Moorhouse, A.J.; Miyamoto, A.; Nabekura, J. Microglia: Actively surveying and shaping neuronal circuit structure and function. *Trends Neurosci.* **2013**, *36*, 209–217. [CrossRef]
3. Gustin, A.; Kirchmeyer, M.; Koncina, E.; Felten, P.; Losciuto, S.; Heurtaux, T.; Tardivel, A.; Heuschling, P.; Dostert, C. NLRP3 Inflammasome Is Expressed and Functional in Mouse Brain Microglia but Not in Astrocytes. *PLoS ONE* **2015**, *10*, e0130624. [CrossRef] [PubMed]
4. Katsumoto, A.; Takeuchi, H.; Takahashi, K.; Tanaka, F. Microglia in Alzheimer's Disease: Risk Factors and Inflammation. *Front. Neurol.* **2018**, *9*, 978. [CrossRef]
5. Koellhoffer, E.C.; McCullough, L.D.; Ritzel, R.M. Old Maids: Aging and Its Impact on Microglia Function. *Int. J. Mol. Sci.* **2017**, *18*, 769. [CrossRef] [PubMed]
6. Cai, Q.; Tammineni, P. Mitochondrial Aspects of Synaptic Dysfunction in Alzheimer's Disease. *J. Alzheimers Dis.* **2017**, *57*, 1087–1103. [CrossRef]
7. Muller, W.E.; Eckert, A.; Kurz, C.; Eckert, G.P.; Leuner, K. Mitochondrial dysfunction: Common final pathway in brain aging and Alzheimer's disease--therapeutic aspects. *Mol. Neurobiol.* **2010**, *41*, 159–171. [CrossRef]
8. Angelova, D.M.; Brown, D.R. Microglia and the aging brain: Are senescent microglia the key to neurodegeneration? *J. Neurochem.* **2019**, *151*, 676–688. [CrossRef]
9. Davies, D.S.; Ma, J.; Jegathees, T.; Goldsbury, C. Microglia show altered morphology and reduced arborization in human brain during aging and Alzheimer's disease. *Brain Pathol.* **2017**, *27*, 795–808. [CrossRef]
10. Stellwagen, D.; Malenka, R.C. Synaptic scaling mediated by glial TNF-α. *Nature* **2006**, *440*, 1054–1059. [CrossRef]
11. Yirmiya, R.; Goshen, I. Immune modulation of learning, memory, neural plasticity and neurogenesis. *Brain Behav. Immun.* **2011**, *25*, 181–213. [CrossRef] [PubMed]
12. Rizzo, F.R.; Musella, A.; De Vito, F.; Fresegna, D.; Bullitta, S.; Vanni, V.; Guadalupi, L.; Bassi, M.S.; Buttari, F.; Mandolesi, G.; et al. Tumor Necrosis Factor and Interleukin-1β Modulate Synaptic Plasticity during Neuroinflammation. *Neural Plast.* **2018**, *2018*, 8430123. [CrossRef] [PubMed]
13. Barrientos, R.M.; Frank, M.G.; Hein, A.M.; Higgins, E.A.; Watkins, L.R.; Rudy, J.W.; Maier, S.F. Time course of hippocampal IL-1 beta and memory consolidation impairments in aging rats following peripheral infection. *Brain Behav. Immun.* **2009**, *23*, 46–54. [CrossRef] [PubMed]
14. Griffin, R.; Nally, R.; Nolan, Y.; McCartney, Y.; Linden, J.; Lynch, M.A. The age-related attenuation in long-term potentiation is associated with microglial activation. *J. Neurochem.* **2006**, *99*, 1263–1272. [CrossRef] [PubMed]
15. Cunningham, A.J.; Murray, C.A.; ONeill, L.A.J.; Lynch, M.A.; OConnor, J.J. Interleukin-1 beta (IL-1 beta) and tumour necrosis factor (TNF) inhibit long-term potentiation in the rat dentate gyrus in vitro. *Neurosci. Lett.* **1996**, *203*, 17–20. [CrossRef] [PubMed]
16. Cargnello, M.; Roux, P.P. Activation and Function of the MAPKs and Their Substrates, the MAPK-Activated Protein Kinases. *Microbiol. Mol. Biol. Rev.* **2011**, *75*, 50–83. [CrossRef] [PubMed]

17. Pearson, G.; Robinson, F.; Beers Gibson, T.; Xu, B.E.; Karandikar, M.; Berman, K.; Cobb, M.H. Mitogen-activated protein (MAP) kinase pathways: Regulation and physiological functions. *Endocr. Rev.* **2001**, *22*, 153–183. [CrossRef]
18. Kelly, A.; Vereker, E.; Nolan, Y.; Brady, M.; Barry, C.; Loscher, C.E.; Mills, K.H.; Lynch, M.A. Activation of p38 plays a pivotal role in the inhibitory effect of lipopolysaccharide and interleukin-1 beta on long term potentiation in rat dentate gyrus. *J. Biol. Chem.* **2003**, *278*, 19453–19462. [CrossRef]
19. Abbasi, A.; Forsberg, K.; Bischof, F. The role of the ubiquitin-editing enzyme A20 in diseases of the central nervous system and other pathological processes. *Front. Mol. Neurosci.* **2015**, *8*, 21. [CrossRef]
20. Kempuraj, D.; Thangavel, R.; Natteru, P.A.; Selvakumar, G.P.; Saeed, D.; Zahoor, H.; Zaheer, S.; Iyer, S.S.; Zaheer, A. Neuroinflammation Induces Neurodegeneration. *J. Neurol Neurosurg. Spine* **2016**, *1*, 1003.
21. Sharman, M.J.; Verdile, G.; Kirubakaran, S.; Parenti, C.; Singh, A.; Watt, G.; Karl, T.; Chang, D.; Li, C.G.; Münch, G. Targeting Inflammatory Pathways in Alzheimer's Disease: A Focus on Natural Products and Phytomedicines. *CNS Drugs* **2019**, *33*, 457–480. [CrossRef] [PubMed]
22. Panarsky, R. Investigation of the Potential Antioxidant and Anti-Inflammatory Effect of Ladostigil and Its Metabolites. Ph.D. Thesis, The Hebrew University, Jerusalem, Israel, 2012.
23. Panarsky, R.; Luques, L.; Weinstock, M. Anti-inflammatory effects of ladostigil and its metabolites in aged rat brain and in microglial cells. *J. Neuroimmune Pharmacol.* **2012**, *7*, 488–498. [CrossRef] [PubMed]
24. Weinstock, M.; Bejar, C.; Schorer-Apelbaum, D.; Panarsky, R.; Luques, L.; Shoham, S. Dose-dependent Effects of Ladostigil on Microglial Activation and Cognition in Aged Rats. *J. Neuroimmune Pharmacol.* **2013**, *8*, 345–355. [CrossRef]
25. Linial, M.; Stern, A.; Weinstock, M. Effect of ladostigil treatment of aging rats on gene expression in four brain areas associated with regulation of memory. *Neuropharmacology* **2020**, *177*, 108229. [CrossRef] [PubMed]
26. Kettenmann, H.; Hanisch, U.K.; Noda, M.; Verkhratsky, A. Physiology of microglia. *Physiol. Rev.* **2011**, *91*, 461–553. [CrossRef]
27. Walz, W.; Ilschner, S.; Ohlemeyer, C.; Banati, R.; Kettenmann, H. Extracellular Atp Activates a Cation Conductance and a K+ Conductance in Cultured Microglial Cells from Mouse-Brain. *J. Neurosci.* **1993**, *13*, 4403–4411. [CrossRef] [PubMed]
28. Shieh, C.H.; Heinrich, A.; Serchov, T.; van Calker, D.; Biber, K. P2X7-dependent, but differentially regulated release of IL-6, CCL2, and TNF-α in cultured mouse microglia. *Glia* **2014**, *62*, 592–607. [CrossRef]
29. Chakfe, Y.; Seguin, R.; Antel, J.P.; Morissette, C.; Malo, D.; Henderson, D.; Séguéla, P. ADP and AMP induce interleukin-1β release from microglial cells through activation of ATP-primed P2X receptor channels. *J. Neurosci.* **2002**, *22*, 3061–3069. [CrossRef]
30. Reichert, F.; Rotshenker, S. Complement-receptor-3 and scavenger-receptor-AI/II mediated myelin phagocytosis in microglia and macrophages. *Neurobiol. Dis.* **2003**, *12*, 65–72. [CrossRef]
31. Shamash, S.; Reichert, F.; Rotshenker, S. The cytokine network of Wallerian degeneration: Tumor necrosis factor-α, interleukin-1α, and interleukin-1β. *J. Neurosci.* **2002**, *22*, 3052–3060. [CrossRef]
32. Latour, S.; Tanaka, H.; Demeure, C.; Mateo, V.; Rubio, M.; Brown, E.J.; Maliszewski, C.; Lindberg, F.P.; Oldenborg, A.; Ullrich, A.; et al. Bidirectional negative regulation of human T and dendritic cells by CD47 and its cognate receptor signal-regulator protein-alpha: Down-regulation of IL-12 responsiveness and inhibition of dendritic cell activation. *J. Immunol.* **2001**, *167*, 2547–2554. [CrossRef] [PubMed]
33. Denizot, F.; Lang, R. Rapid Colorimetric Assay for Cell-Growth and Survival—Modifications to the Tetrazolium Dye Procedure Giving Improved Sensitivity and Reliability. *J. Immunol. Methods* **1986**, *89*, 271–277. [CrossRef] [PubMed]
34. Moradov, D.; Finkin-Groner, E.; Bejar, C.; Sunita, P.; Schorer-Apelbaum, D.; Barasch, D.; Nemirovski, A.; Cohen, M.; Weinstock, M. Dose-limiting inhibition of acetylcholinesterase by ladostigil results from the rapid formation and fast hydrolysis of the drug-enzyme complex formed by its major metabolite, R-MCPAI. *Biochem. Pharmacol.* **2015**, *94*, 164–172. [CrossRef] [PubMed]
35. Bolger, A.M.; Lohse, M.; Usadel, B. Trimmomatic: A flexible trimmer for Illumina sequence data. *Bioinformatics* **2014**, *30*, 2114–2120. [CrossRef]
36. Dobin, A.; Davis, C.A.; Schlesinger, F.; Drenkow, J.; Zaleski, C.; Jha, S.; Batut, P.; Chaisson, M.; Gingeras, T.R. STAR: Ultrafast universal RNA-seq aligner. *Bioinformatics* **2013**, *29*, 15–21. [CrossRef] [PubMed]
37. Frankish, A.; Diekhans, M.; Ferreira, A.M.; Johnson, R.; Jungreis, I.; Loveland, J.; Mudge, J.M.; Sisu, C.; Wright, J.; Armstrong, J.; et al. GENCODE reference annotation for the human and mouse genomes. *Nucleic Acids Res.* **2019**, *47*, D766–D773. [CrossRef] [PubMed]
38. Robinson, M.D.; McCarthy, D.J.; Smyth, G.K. edgeR: A Bioconductor package for differential expression analysis of digital gene expression data. *Bioinformatics* **2010**, *26*, 139–140. [CrossRef]
39. Ritchie, M.E.; Phipson, B.; Wu, D.; Hu, Y.F.; Law, C.W.; Shi, W.; Smyth, G.K. powers differential expression analyses for RNA-sequencing and microarray studies. *Nucleic Acids Res.* **2015**, *43*, e47. [CrossRef]
40. Friedle, S.A.; Brautigam, V.M.; Nikodemova, M.; Wright, M.L.; Watters, J.J. The P2X7-Egr pathway regulates nucleotide-dependent inflammatory gene expression in microglia. *Glia* **2011**, *59*, 1–13. [CrossRef]
41. Davalos, D.; Grutzendler, J.; Yang, G.; Kim, J.V.; Zuo, Y.; Jung, S.; Littman, D.R.; Dustin, M.L.; Gan, W.B. ATP mediates rapid microglial response to local brain injury. *Nat. Neurosci.* **2005**, *8*, 752–758. [CrossRef]
42. Orr, A.G.; Orr, A.L.; Li, X.J.; Gross, R.E.; Traynelis, S.F. Adenosine A(2A) receptor mediates microglial process retraction. *Nat. Neurosci.* **2009**, *12*, 872–878. [CrossRef] [PubMed]

43. Angulo, E.; Casado, V.; Mallol, J.; Canela, E.I.; Vinals, F.; Ferrer, I.; Lluis, C.; Franco, R. A1 adenosine receptors accumulate in neurodegenerative structures in Alzheimer disease and mediate both amyloid precursor protein processing and tau phosphorylation and translocation. *Brain Pathol.* **2003**, *13*, 440–451. [CrossRef]
44. Illes, P. P2X7 Receptors Amplify CNS Damage in Neurodegenerative Diseases. *Int. J. Mol. Sci.* **2020**, *21*, 5996. [CrossRef] [PubMed]
45. Yan, Y.H.; Tan, X.; Wu, X.M.; Shao, B.; Wu, X.H.; Cao, J.H.; Xu, J.; Jin, W.; Li, L.; Xu, W.; et al. Involvement of early growth response-2 (Egr-2) in lipopolysaccharide-induced neuroinflammation. *J. Mol. Histol.* **2013**, *44*, 249–257. [CrossRef] [PubMed]
46. Ravetti, M.G.; Rosso, O.A.; Berretta, R.; Moscato, P. Uncovering Molecular Biomarkers That Correlate Cognitive Decline with the Changes of Hippocampus' Gene Expression Profiles in Alzheimer's Disease. *PLoS ONE* **2010**, *5*, e10153. [CrossRef]
47. Qin, X.; Wang, Y.; Paudel, H.K. Inhibition of Early Growth Response 1 in the Hippocampus Alleviates Neuropathology and Improves Cognition in an Alzheimer Model with Plaques and Tangles. *Am. J. Pathol.* **2017**, *187*, 1828–1847. [CrossRef]
48. Sagare, A.P.; Sweeney, M.D.; Makshanoff, J.; Zlokovic, B.V. Shedding of soluble platelet-derived growth factor receptor-β from human brain pericytes. *Neurosci. Lett.* **2015**, *607*, 97–101. [CrossRef]
49. Cicognola, C.; Mattsson-Carlgren, N.; van Westen, D.; Zetterberg, H.; Blennow, K.; Palmqvist, S.; Ahmadi, K.; Strandberg, O.; Stomrud, E.; Janelidze, S.; et al. Associations of CSF PDGFRbeta With Aging, Blood-Brain Barrier Damage, Neuroinflammation, and Alzheimer Disease Pathologic Changes. *Neurology* **2023**, *101*, e30–e39. [CrossRef]
50. Nuttall, R.K.; Silva, C.; Hader, W.; Bar-Or, A.; Patel, K.D.; Edwards, D.R.; Yong, V.W. Metalloproteinases are enriched in microglia compared with leukocytes and they regulate cytokine levels in activated microglia. *Glia* **2007**, *55*, 516–526. [CrossRef]
51. Boone, D.L.; Turer, E.E.; Lee, E.G.; Ahmad, R.C.; Wheeler, M.T.; Tsui, C.; Hurley, P.; Chien, M.; Chai, S.; Hitotsumatsu, O.; et al. The ubiquitin-modifying enzyme A20 is required for termination of Toll-like receptor responses. *Nat. Immunol.* **2004**, *5*, 1052–1060. [CrossRef]
52. Duong, B.H.; Onizawa, M.; Oses-Prieto, J.A.; Advincula, R.; Burlingame, A.; Malynn, B.A.; Ma, A. A20 restricts ubiquitination of pro-interleukin-1beta protein complexes and suppresses NLRP3 inflammasome activity. *Immunity* **2015**, *42*, 55–67. [CrossRef] [PubMed]
53. Priem, D.; van Loo, G.; Bertrand, M.J.M. A20 and Cell Death-driven Inflammation. *Trends Immunol.* **2020**, *41*, 421–435. [CrossRef] [PubMed]
54. Mohebiany, A.N.; Ramphal, N.S.; Karram, K.; Di Liberto, G.; Novkovic, T.; Klein, M.; Marini, F.; Kreutzfeldt, M.; Hartner, F.; Lacher, S.M.; et al. Microglial A20 Protects the Brain from CD8 T-Cell-Mediated Immunopathology. *Cell Rep.* **2020**, *30*, 1585–1597.e6. [CrossRef] [PubMed]
55. Deng, H.J.; Deji, Q.; Zhaba, W.; Liu, J.Q.; Gao, S.Q.; Han, Y.L.; Zhou, M.L.; Wang, C.X. A20 Establishes Negative Feedback With TRAF6/NF-kappaB and Attenuates Early Brain Injury After Experimental Subarachnoid Hemorrhage. *Front. Immunol.* **2021**, *12*, 623256. [CrossRef]
56. Schneider, L.S.; Geffen, Y.; Rabinowitz, J.; Thomas, R.G.; Schmidt, R.; Ropele, S.; Weinstock, M.; Grp, L.S. Low-dose ladostigil for mild cognitive impairment A phase 2 placebo-controlled clinical trial. *Neurology* **2019**, *93*, E1474–E1484. [CrossRef]

biomolecules

MDPI

Review

Neuroprotective Properties of Clove (*Syzygium aromaticum*): State of the Art and Future Pharmaceutical Applications for Alzheimer's Disease

Tatevik Sargsyan [1,2,†], Hayarpi M. Simonyan [2,†], Lala Stepanyan [1], Avetis Tsaturyan [1,2], Caterina Vicidomini [3,*], Raffaele Pastore [4], Germano Guerra [4] and Giovanni N. Roviello [3,*]

1 Scientific and Production Center "Armbiotechnology" NAS RA, 14 Gyurjyan Str., Yerevan 0056, Armenia; tatev-sargsyan@ysu.am (T.S.); lala_stepanyan@rambler.ru (L.S.)
2 Institute of Pharmacy, Yerevan State University, 1 Alex Manoogian Str., Yerevan 0025, Armenia
3 Institute of Biostructures and Bioimaging, Italian National Council for Research (IBB-CNR), Area di Ricerca Site and Headquarters, Via Pietro Castellino 111, 80131 Naples, Italy
4 Department of Medicine and Health Sciences "Vincenzo Tiberio", University of Molise, Via F. De Santis, 86100 Campobasso, Italy
* Correspondence: caterina.vicidomini@ibb.cnr.it (C.V.); giovanni.roviello@cnr.it (G.N.R.); Tel.: +39-0812203415 (C.V.)
† These authors contributed equally to this work.

Abstract: This study explores the neuropharmacological potential of various molecular and amino acid components derived from *Syzygium aromaticum* (clove), an aromatic spice with a long history of culinary and medicinal use. Key bioactive compounds such as eugenol, α-humulene, β-caryophyllene, gallic acid, quercetin, and luteolin demonstrate antioxidant, anti-inflammatory, and neuroprotective properties by scavenging free radicals, modulating calcium channels, and reducing neuroinflammation and oxidative stress. Moreover, gallic acid and asiatic acid may exhibit protective effects, including neuronal apoptosis inhibition, while other useful properties of clove phytocompounds include NF-κB pathway inhibition, membrane stabilization, and suppression of pro-inflammatory pathways, possibly in neurons or other relevant cell types, further contributing to neuroprotection and cognitive enhancement. Amino acid analysis revealed essential and non-essential amino acids such as aspartic acid, serine, glutamic acid, glycine, histidine, and arginine in various clove parts (buds, fruits, branches, and leaves). These amino acids play crucial roles in neurotransmitter synthesis, immune modulation, antioxidant defense, and metabolic regulation. Collectively, these bioactive molecules and amino acids contribute to clove's antioxidant, anti-inflammatory, neurotrophic, and neurotransmitter-modulating effects, highlighting its potential as a preventive and therapeutic candidate for neurodegenerative disorders. While preliminary preclinical studies support these neuroprotective properties, further research, including clinical trials, is needed to validate the efficacy and safety of clove-based interventions in neuroprotection.

Keywords: cloves; herbal medicine; nutrition; Alzheimer's disease; neurodegeneration

1. Introduction

Neurodegenerative diseases are a global health challenge, affecting millions and imposing significant burdens on healthcare systems, families, and economies. They are particularly prevalent in Western societies, where factors like aging populations may contribute to their high incidence [1]. These disorders, including Alzheimer's disease (AD), Parkinson's disease (PD), Huntington's disease (HD), and amyotrophic lateral sclerosis (ALS), progressively deteriorate nerve cells in the brain and spinal cord, leading to cognitive decline, motor impairments, and often fatal consequences. The economic implications of neurodegenerative disorders are substantial, with immense costs associated with patient care, medication, long-term assistance, and loss of productivity [2]. These expenses strain healthcare systems and exert financial pressure on families, significantly affecting their quality of life [3–6]. While advancements in research and healthcare have significantly improved the diagnosis and management of neurodegenerative diseases, the absence of effective disease-modifying treatments remains a critical challenge. This dichotomy reflects the progress made in understanding the pathology of neurodegenerative diseases and the ongoing difficulties in translating this knowledge into therapies that alter their trajectory. The complex nature of these diseases demands better therapeutic strategies, emphasizing the urgency for innovative interventions, as AD is a progressive neurodegenerative disorder that represents one of the most pressing healthcare challenges of our time. It is believed to be a disorder characterized by the aggregation of amyloid-beta (Aβ) plaques and neurofibrillary tangles, which are associated with synaptic dysfunction, neuronal loss, and cognitive impairment [7]. Genetic factors, such as mutations in the amyloid precursor protein (APP) and presenilin genes, play a role in early-onset AD, while aging, environmental factors, and lifestyle contribute to late-onset AD [8,9]. Current AD therapies face significant limitations. Cholinesterase inhibitors (donepezil, rivastigmine, galantamine) and the N-methyl-D-aspartate (NMDA) receptor antagonist (memantine [10]) are approved drugs for AD symptomatic treatment [1]. However, these medications do not modify the underlying disease pathology. Recently, drugs such as aducanumab and lecanemab have been approved, although their benefits appear to be modest, primarily slowing cognitive decline rather than reversing or halting the progression of Alzheimer's disease [11,12]. These therapies target amyloid plaques, a hallmark of Alzheimer's, but their impact on long-term disease progression remains uncertain. Additionally, the most of therapies are often invasive (e.g., intravenous administration) and can have severe side effects, such as brain swelling and bleeding [13]. Symptomatic treatments for memory loss and neuropsychiatric symptoms provide temporary relief, but do not address underlying neurodegeneration. Donanemab, an amyloid β-directed antibody, received approval in the USA for the treatment of adults with early symptomatic Alzheimer's disease, but the National Institute for Health and Care Excellence has recently not recommended it for use in the National Health Service (NHS) in the United Kingdom due to its relatively small benefits and high rollout costs, including regular infusions and side effect monitoring [14]. Advances in disease-modifying therapies remain critical to improving outcomes [15,16]. In this context, nature-inspired and plant-based therapies [17–20] are gaining attention for their potential in treating AD, offering a natural and holistic approach to a condition with limited pharmaceutical solutions [21,22]. Addressing socially relevant diseases drives the search for new therapeutic strategies, which are based on molecular systems that include not only natural, but also synthetic compounds, such as peptidic and oligonucleotidic molecules, as well as hybrid structures like nucleopeptides [23–33]. However, plants like Ginkgo biloba, Huperzine A, ginseng, and turmeric contain bioactive compounds that address several key aspects of AD's pathology, including oxidative stress, neuroinflammation, and Aβ plaque formation [34]. Ginkgo biloba, one of the most studied herbs for AD, is known for its flavonoids and terpenoids that act as powerful antioxidants and anti-inflammatory agents. Clinical trials have shown that extracts like EGb 761 can improve cognitive functions in mild-to-moderate dementia and alleviate neuropsychiatric symptoms, though results vary, highlighting the need for more rigorous studies [35,36]. Similarly, Huperzine A, derived from the Chinese club moss, shows promise as it enhances cholinergic signaling in the brain by inhibiting acetylcholinesterase, though long-term efficacy remains to be fully understood [37]. Ginseng, which is rich in ginsenosides, has demonstrated the ability to reduce Aβ accumulation and oxidative damage both critical in AD progression [38,39]. Meanwhile, curcumin, the active compound in turmeric, has been noted for its anti-inflammatory properties and potential to reduce plaque burden although its poor bioavailability poses a challenge [40–42]. Sage (*Salvia officinalis*) is another herb that has shown benefits in improving memory and cognitive performance in clinical trials alone and in addiction with *Hypericum perforatum*, also known as St. John's Wort [43,44]. The strength of herbal therapies lies in their ability to target multiple pathways involved in AD's complex mechanisms. Among the various underlying mechanisms of AD, mitochondrial dysfunction plays a crucial role. This includes factors such as increased production of reactive oxygen species (ROS) [45,46], disrupted calcium balance, and disturbances in mitochondrial dynamics [47]. In this context, the clove (*Syzygium aromaticum*) [48] and its primary bioactive compound, eugenol [49], provided very interesting results for their potential therapeutic effects on AD. These effects are primarily linked to clove's antioxidant and anti-inflammatory properties, which address two critical pathological processes in AD: oxidative stress and neuroinflammation [50]. Eugenol has been shown to neutralize free radicals and reduce inflammatory cytokines, thereby protecting neurons from damage associated with Aβ plaques and tau protein aggregation, hallmarks of AD [51]. Preclinical studies suggest that clove essential oil can attenuate cognitive decline and improve memory functions in animal models of AD [52]. It achieves this by modulating oxidative pathways and preventing neurodegeneration. Additionally, the anti-inflammatory effects of eugenol help suppress microglial activation, which plays a crucial role in the progression of neurodegenerative disorders [53]. Clove is widely used in both culinary and traditional medicinal practices across the world. In cooking, clove is a staple spice in many cuisines, particularly in Asia, the Middle East, and Africa, where it is used to flavor dishes, teas, and desserts [54]. Its aromatic and slightly sweet, yet pungent, taste enhances a variety of foods, including curries, baked goods, and beverages like chai tea. Medically, clove has been an integral part of traditional systems like Ayurveda, Unani, and Chinese medicine. It is used for its antiseptic, analgesic, and digestive properties. Clove oil is applied to relieve toothache and oral infections, while clove teas are consumed to soothe digestive discomfort and boost immunity [55]. In Armenia, clove oil is a key ingredient in an ointment called Yubivaks, which is believed to heal burns. The ointment is currently undergoing preclinical testing for its anti-burn activity [56]. Epidemiological studies on the use of clove and its association with AD incidence in the area where the spice was extensively used are limited but promising. While clinical data in humans are sparse, preliminary findings suggest that clove has potential neuroprotective effects that could lower the risk or progression of AD in populations where clove is a dietary staple or traditional medicine [57]. Furthermore, traditional usage of clove in regions like Southeast Asia, where its consumption is common, aligns with its observed health benefits in experimental studies. Although direct epidemiological evidence linking clove consumption to lower AD incidence is not yet well-established, ongoing pharmacological and biochemical research supports its potential for inclusion in therapeutic strategies against neurodegenerative disease. Owing to its molecular composition, clove contains numerous bioactive compounds with significant neuropharmacological potential (Table 1). These molecules demonstrate antioxidant, anti-inflammatory, and neuroprotective properties. The amounts of molecular components in clove can vary depending on factors, such as the part of the plant used (buds, leaves, or stems) and the method of extraction. For instance, thirty-six constituents were identified from the essential oil of clove buds, and twenty-nine from the essential oil of clove leaves using gas chromatography–mass spectrometry. Major classes of compounds include sesquiterpenes, phenyl propanoids, oxygenated sesquiterpenes, and esters. The composition of major constituents varied between the two oils, with the clove bud essential oil containing eugenol (65.29%), trans-caryophyllene (20.06%), and α-humulene (3.38%), while the clove leaf essential oil contained eugenol (64.47%), trans-caryophyllene (27.19%), and α-humulene (3.62%) [58]. Remarkably, clove components, like eugenol and β-caryophyllene, have been found to be able to cross the blood–brain barrier and have been subjected to studies against glioblastoma [59]. The already mentioned eugenol exhibits neuroprotective, antioxidant, and anti-inflammatory effects by scavenging free radicals, inhibiting neuroinflammation, and modulating calcium channels (Table 1). β-Caryophyllene functions as a CB2 receptor agonist, reducing oxidative stress and neuroinflammation, while gallic acid and quercetin enhance memory and cognitive function by reducing oxidative stress and inhibiting acetylcholinesterase. Luteolin and kaempferol offer neuroprotection through anti-inflammatory pathways, while tannic acid and paeoniflorin mitigate oxidative stress and stabilize cell membranes. Compounds such as isorhamnetin, ellagic acid, and rhamnocitrin demonstrate free radical scavenging and anti-inflammatory properties. Eugenin, oleanolic acid, and asiatic acid contribute to antioxidant defenses and modulate inflammatory pathways, protecting against neurodegenerative diseases. Arjunolic acid also protects against oxidative stress. Together, these phytocompounds highlight clove's potential as a neuroprotective agent.

Table 1. Key molecular components of clove, their potential neuropharmacological effects, mechanisms of action, and supporting references.

Molecular Component	Potential Neuropharmacological Effects	Mechanism of Action	Reference
Eugenol	Neuroprotective, antioxidant, anti-inflammatory	Scavenges free radicals, inhibits neuroinflammation, modulates calcium channels	[60–62]
α-Humulen	Antioxidant, anti-inflammatory	NF-κB inhibition, ROS neutralization, COX-2 suppression, membrane disruption.	[55,63–65]
β-Caryophyllene	Neuroprotective, anti-inflammatory, anti-anxiety	CB2 receptor agonist, modulates neuroinflammation, reduces oxidative stress	[66–68]
Gallic Acid	Antioxidant, anti-apoptotic, memory enhancer	Reduces oxidative stress, prevents neuronal apoptosis	[69–71]
Quercetin	Neuroprotective, cognitive enhancer, anti-neuroinflammatory	Inhibits acetylcholinesterase, reduces pro-inflammatory mediators	[66,72–74]
Luteolin	Neuroprotective	Inhibits inflammation, promotes neuroprotection, and reduces oxidative stress	[75–77]
Tannic acid	Antioxidant, anti-inflammatory, anti-neuroinflammatory	Free radical scavenging, metal chelation, lipid protection, NF-κB inhibition, cytokine reduction, modulates cytokines, inhibits microglial activation	[61,78,79]

Table 1. *Cont.*

Molecular Component	Potential Neuropharmacological Effects	Mechanism of Action	Reference
Casuarinin	Antioxidant	Scavenges free radicals, reducing oxidative stress	[80]
Paeoniflorin	Antioxidant, anti-inflammatory, neuroprotective	Inhibits pro-inflammatory cytokines, reduces ROS, stabilizes cell membranes	[81]
Kaempferol	Neuroprotective, anti-inflammatory	Suppresses pro-inflammatory pathways, protects against neuronal degeneration	[69,82]
Ellagic Acid	Antioxidant, neuroprotective	Scavenges free radicals, inhibits inflammation, regulates cell cycle	[69,83]
Rhamnocitrin	Antioxidant, neuroprotective	Free radical scavenging, reduction in neuroinflammation	[69,84,85]
Isorhamnetin	Antioxidant, anti-inflammatory	Free radical scavenging, inhibition of pro-inflammatory cytokines	[86,87]
Eugenin	Anti-inflammatory, antioxidant, neuroprotective	Neutralizes reactive oxygen species and reactive nitrogen species (RNS), inhibits the production of pro-inflammatory mediators, interferes with neuroinflammatory pathways	[87–90]
Oleanolic Acid	Antioxidant, anti-inflammatory	Scavenges free radicals and boosts cellular antioxidant defenses, inhibits the NF-κB pathway and reduces pro-inflammatory cytokines, modulates oxidative stress and inflammation	[87,91,92]
Asiatic Acid	Neuroprotective, anti-inflammatory	Protects neurons from oxidative stress and apoptosis, potentially benefiting neurodegenerative diseases like Alzheimer's and Parkinson's, suppresses pro-inflammatory mediators like IL-6 and TNF-α.	[87,91,92]
Arjunolic Acid	antioxidant, anti-inflammatory	Reduces oxidative stress, chelates metal ions and scavenges reactive oxygen species, reduces inflammation in various disease models.	[87,91,92]

Overall, clove contains a range of bioactive compounds with significant neuropharmacological potential and structural diversity (Figure 1).

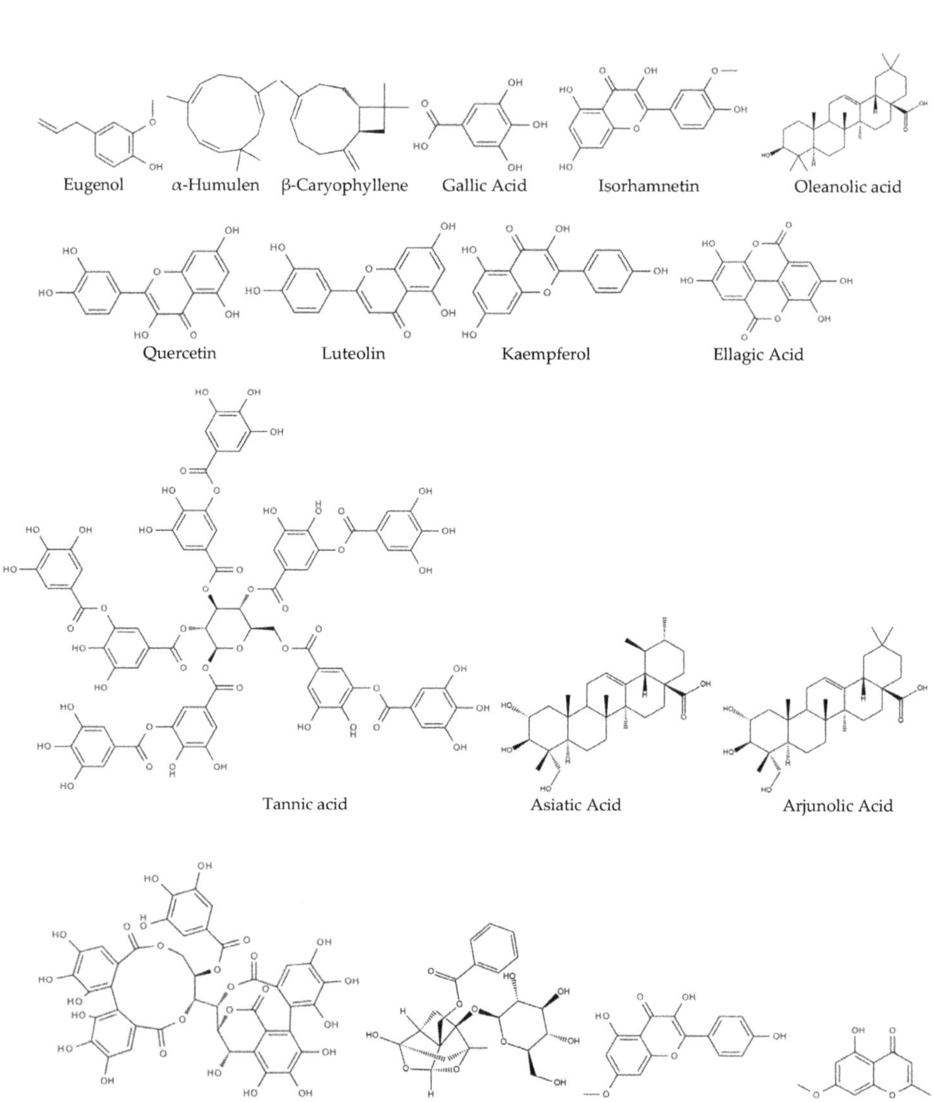

Figure 1. Chemical representations of the phytocompounds found in clove discussed in this section for their neuroprotective properties.

2. Cloves and Mechanisms of Neuroprotection

2.1. Clove Antioxidant Effects

Alzheimer's disease is a neurodegenerative disease that causes a gradual loss of normal motor and cognitive function. The complex AD pathophysiology involves various factors such as oxidative stress, neuroinflammation, Aβ aggregation, disturbed neurotransmission, and apoptosis. and are not able to cover different aspects of the disease [93]. Clove extracts were evaluated for their effects on hydrogen peroxide-induced oxidative stress in human neuroblastoma SH-SY5Y cell lines, which served as the experimental model. The results demonstrated that both the extracts and key bioactive compounds of *Syzygium aromaticum* effectively reduced ROS, restored mitochondrial membrane potential (MMP), and provided neuroprotection against H_2O_2-induced oxidative damage. This protective effect was attributed to the antioxidant properties of the extracts. Additionally, clove extracts significantly diminished lipid peroxidation and restored glutathione levels. The extracts also exhibited anti-acetylcholinesterase activity, anti-glycation effects, and the ability to inhibit Aβ aggregation and fibril formation. The multifaceted neuroprotective mechanisms of clove suggested its potential as a promising candidate for drug development in Alzheimer's disease [94]. Other studies have indicated that *Syzygium aromaticum* oil may minimize the neurotoxicity caused by acrylamide by reducing oxidative brain damage [95]. As previously mentioned in this work, clove contains bioactive compounds such as phenolic acids, flavonoids, and volatile oils, including eugenol, which demonstrate potent antioxidant properties. These antioxidants play a pivotal role in neuroprotection by scavenging free radicals and mitigating oxidative stress-induced damage in neuronal cells, a key factor in the pathogenesis of AD [96]. Several studies have reinforced the notion that clove extracts exhibit significant antioxidant activity, which may offer neuroprotection by mitigating oxidative damage in the brain, thereby reducing the risk of neurodegenerative diseases [97,98]. Eugenol has been reported to alleviate neuropathic pain [99] and demonstrate anti-amnestic activity in animal models of Alzheimer's disease, potentially through its antioxidant mechanism [100]. An interesting study demonstrated that chronic administration of clove essential oil improved memory and learning in rats, suggesting its potential role in cognitive enhancement [101]. Another study demonstrated the safety and antidepressant-like effects of *Syzygium aromaticum* essential oil after both acute and long-term treatment. Pronounced antidepressant effects were observed when the oil was administered intragastrically at a dose of 200 mg/kg. The toxicological profile was evaluated through prolonged administration at doses of 100, 200, and 400 mg/kg. Notably, only the highest dose (400 mg/kg) resulted in a significant reduction in body weight, while no significant changes were detected in organ weight ratios or cellular-level markers at any dose. These findings suggest that clove essential oil is a highly effective antidepressant with low toxicity when administered intragastrically [102]. As previously discussed, oxidative stress plays a central role in the pathogenesis of Alzheimer's disease, contributing to cellular damage through the accumulation of reactive oxygen species. Aging results from the accumulation of damage to cellular proteins and membranes, with ROS-induced oxidative stress being a significant factor in geriatric syndromes and various neurodegenerative diseases. In this context, Sirtuin 1 (SIRT1), an NAD^+-dependent deacetylase, emerged as a critical mediator in mitigating oxidative damage. SIRT1 is a pivotal regulator of cellular functions associated with aging and neurodegenerative disorders, influencing key signaling pathways related to autophagy, oxidative stress response, and mitochondrial activity—processes central to the development and progression of neurodegenerative diseases, like AD [103]. Expanding on this understanding, Shekhar et al. explored the neuroprotective properties of clove in addressing $A\beta_{25-35}$-induced neurotoxicity in neuronal cells. Their key findings confirmed that *Syzygium aromaticum* demonstrates substantial antioxidative capacity, as indicated by its ability to scavenge ROS and enhance the activity of key antioxidant enzymes, including superoxide dismutase, catalase, and glutathione peroxidase. Furthermore, the compound was found to upregulate both recombinant and endogenous SIRT1 activity while simultaneously downregulating γ-secretase, a protein complex involved in amyloid plaque formation. This activation of SIRT1 and reduction in γ-secretase suggested a holistic approach for addressing neurodegenerative diseases with *Syzygium aromaticum*. Clove inhibited the fibrillation and oligomerization of Aβ with high efficacy, and exhibited significant antioxidant activity to protect nerve cells. These findings highlighted its potential use in the treatment of neurodegenerative diseases, particularly AD. In fact, clove, as an ayurvedic product, promises healthy aging with minimal side effects and cost-effectiveness, offering a potential solution to current unmet medical needs [100]. Supporting these observations, other studies have further emphasized the multifaceted neuroprotective properties of *Syzygium aromaticum*. For example, research has shown that clove extract possesses anti-acetylcholinesterase activity, anti-glycation potential, and inhibits amyloid-beta aggregation, all of which contribute to its potential therapeutic benefits in Alzheimer's disease [104].

Additionally, the activation of SIRT1 by *Syzygium aromaticum* is consistent with findings that SIRT1 plays a crucial protective role against neurodegeneration, enhancing mitochondrial function and reducing oxidative stress [105]. Another study demonstrated that the compounds in *Syzygium aromaticum* effectively inhibited both acetylcholinesterase and butyrylcholinesterase, with stronger inhibition observed for the former enzyme. This suggests the potential of clove oils as an early therapeutic approach for brain dysfunction, particularly in neurodegenerative conditions such as Alzheimer's disease [106]. A gas chromatography–mass spectrometry analysis identified 58 volatile compounds in clove essential oil. To investigate its antioxidant and anti-aging effects, researchers employed the nematode *Caenorhabditis elegans* as a model organism. Chronic treatment with clove essential oil significantly extended the lifespan and improved the reproductive health of these nematodes. The oil demonstrated antioxidant activity by reducing levels of ROS and by upregulating key antioxidant enzymes, including superoxide dismutase 3 and glutathione S-transferase 4. Additionally, clove essential oil induced the translocation of the DAF-16/FOXO transcription factor from the cytoplasm to the nucleus. DAF-16, the *Caenorhabditis elegans* homolog of the FOXO transcription factor, plays a central role in the insulin/insulin-like growth factor 1 signaling pathway, which regulates longevity, stress resistance, and metabolism. Upon activation, DAF-16 moves into the nucleus and binds specific DNA sequences to promote the expression of genes involved in stress response and lifespan extension [107]. The treatment with clove essential oil leads to germ cell apoptosis in an *acep-1-* and *daf-16*-dependent manner, underscoring the intricate regulatory mechanisms that govern cell death in Caenorhabditis elegans. Although the precise role of *acep*-1 in this context requires further elucidation, the involvement of DAF-16 suggested that clove essential oil may modulate apoptotic pathways through its influence on this transcription factor. Overall, these findings indicate that clove essential oil possesses antioxidant and anti-aging properties, with DAF-16 playing a central role in mediating these effects [108]. The study demonstrated that combining endurance training with clove oil supplementation improved spatial memory in a rat model of Alzheimer's disease. This combined intervention increased the expression of the α7 nicotinic acetylcholine receptor in the hippocampus—an important receptor involved in cognitive function—while reducing levels of NLRP1 (NOD-like receptor protein 1), a key component of the inflammasome that mediates inflammatory responses. Additionally, the number of dark cells, which indicate cellular damage, decreased. These molecular and cellular changes are likely to contribute to enhanced spatial learning and memory [109].

2.2. *Clove Anti-Inflammatory Effects*

Inflammation plays a critical role in Alzheimer's disease, contributing significantly to the progression of the condition. Alzheimer's disease, like other proteinopathic neurodegenerative disorders, is characterized by the accumulation of amyloidogenic proteins. A neuroinflammatory component in Alzheimer's disease has been known for more than a decade, and although inflammation's contribution to the disease was initially underappreciated, recent genetic, bioinformatic, and preclinical data now confirm its importance in exacerbating the pathology of the disease. Neuroinflammation in Alzheimer's disease is primarily driven by the brain's intrinsic myeloid cells, known as microglia, and this inflammation intensifies as the disease progresses [110]. The effectiveness of some antidementia drugs in animal models of Alzheimer's disease has been linked to their anti-inflammatory properties. One such example is ellagic acid, a compound found in clove, which has demonstrated an ability to mitigate Alzheimer's-like behavior in 5xFAD mice. This antidementia effect is attributed to ellagic acid's ability to reduce inflammatory responses in the brain, decrease oxidative stress, lower amyloid beta deposition, reduce apoptosis, and promote neurogenesis, all of which contribute to the compound's potential as a therapeutic agent in Alzheimer's disease [111]. In the context of Alzheimer's disease, the anti-inflammatory effects of various molecular components found in clove further highlight the potential therapeutic benefits of this natural substance. As mentioned, ellagic acid has shown promise in mitigating Alzheimer's-like behavior in animal models by reducing inflammation, oxidative stress, and amyloid beta deposition. This anti-inflammatory property is not unique to ellagic acid, as clove is also known for its other molecular components that contribute to such effects. For example, clove essential oil, with its main component, eugenol, is renowned for its analgesic and anti-inflammatory properties. Eugenol has demonstrated significant anti-inflammatory effects [56], which may play a role in protecting the brain from the inflammatory processes that exacerbate Alzheimer's disease. *Helicobacter pylori* has been associated with an increased risk for various neurological diseases, including Alzheimer's disease, as well as other conditions like Parkinson's and multiple sclerosis. This link is primarily through mechanisms of chronic inflammation, systemic inflammation, and neuroinflammation. Given its role in promoting these inflammatory pathways, *Helicobacter pylori* is recognized as a contributing factor to the pathogenesis of neurological disorders. A growing body of evidence suggests that *Helicobacter pylori* infection plays a significant role in the development and progression of Alzheimer's disease through its impact on chronic inflammation and neuroinflammation. In particular, the presence of *H. pylori* has been associated with increased levels of specific anti-H. pylori antibodies in the cerebrospinal fluid and serum of Alzheimer's patients, which correlates with disease severity. Studies have also revealed that individuals carrying the ApoE4 polymorphism, the strongest genetic risk factor for Alzheimer's, are more susceptible to *H. pylori* infection, suggesting that this genetic variant may facilitate the entry of *H. pylori* into the brain. Furthermore, *H. pylori* infection induces systemic inflammation by releasing pro-inflammatory cytokines and toxins, which can cross the blood–brain barrier and disrupt its integrity. This breakdown of the blood–brain barrier, combined with the inflammatory responses triggered by the infection, likely contributes to the neurodegenerative processes observed in Alzheimer's disease. Thus, eradicating *Helicobacter pylori* can decrease systemic inflammation and improve endothelial function, potentially lowering the risk and severity of these conditions [112]. In this context, a report on the anti-inflammatory activity of clove essential oil in a *Helicobacter pylori* model fits seamlessly within the broader discussion of clove's potential in combating Alzheimer's disease [113] studied the anti-inflammatory effects of eugenol clove essential oil, specifically against *Helicobacter pylori*. Their results demonstrated that the essential oil inhibited human erythrocyte hemolysis at concentrations of 4, 8, 16, and 32 μg/L, with inhibition rates ranging from 53.04% to 63.64% (Table 2). Interestingly, sodium diclofenac, a well-established anti-inflammatory drug, showed similar inhibition rates; [114] evaluated the anti-inflammatory effect of nanoemulsion-based gels containing clove and cinnamon essential oils. Even if moderately, a nanogel based on clove showed activity in reducing paw edema, which is a model for inflammation; [115] found that clove essential oil reduced paw swelling by 40–60% in rats, highlighting its anti-inflammatory potential. The gel demonstrated comparable effectiveness to both diclofenac and neomycin, emphasizing the role of clove essential oil in inflammation reduction.

Table 2. Anti-inflammatory effects of clove essential oil with potential neuroprotective properties.

Model	Effect	Rate	Reference
In vitro (human erythrocyte)	Inhibited human erythrocyte hemolysis	Inhibition by 53.04–63.64%	[113]
In vivo (rat model)	Reduced paw swelling	Reduction by 40–60%	[115]

2.3. Neurotrophic and Neuropharmacological Effects

Some studies have highlighted the potential of clove extracts in promoting nerve cell growth and survival. Components within clove, particularly eugenol, have shown the ability to induce neurotrophic factors such as brain-derived neurotrophic factor (BDNF) and nerve growth factor (NGF). These neurotrophic factors support neuronal health, aiding in neurogenesis and neuronal regeneration [116,117]. Clove extracts have been suggested to modulate neurotransmitter levels in the brain. In particular, eugenol may influence neurotransmitter systems, including acetylcholine and dopamine pathways, potentially impacting cognitive function and mood regulation. This modulation could contribute to neuroprotective effects [118–120]. Several preclinical studies have explored the neuropharmacological effects of clove extracts in animal models. These investigations have reported improvements in cognitive function, memory enhancement, and neurobehavioral outcomes upon administration of clove extract or its constituents. Such effects imply a potential for clove-based interventions in neurodegenerative conditions [121–123]. Eugenol has been shown to protect against neuronal death in PC12 cells treated with Aβ. It also mitigated the transcriptional upregulation of the pro-apoptotic protein Bax and the down-regulation of the anti-apoptotic protein B-cell lymphoma (Bcl), which are typically induced by Aβ in these cells [124]. Furthermore, eugenol has demonstrated its potential therapeutic effects by improving memory impairment and reducing the number of amyloid plaques, indicating its potential to influence and alter the underlying processes of Alzheimer's disease [125]. However, the exact mechanisms by which eugenol exerts these effects remain unclear. To further explore these mechanisms, Jung et al. investigated the impact of eugenol on Alzheimer's disease pathologies and its therapeutic action using a 5×FAD mouse model. Eugenol was found to improve cognitive function, reduce necroptotic cell death, and decrease Aβ accumulation in the 5×FAD mouse model. The therapeutic effects of eugenol may vary depending on the specific brain region targeted. In particular, eugenol exhibited anti-inflammatory properties in the cortex and facilitated microglia-mediated phagocytosis of Aβ in the hippocampus. These findings suggest that eugenol could offer therapeutic benefits for Alzheimer's disease by modulating the inflammatory response and addressing amyloid-related pathologies [126]. Overall, the effects reported to date in the scientific literature of the different clove extracts and oils, which are due to their molecular components, in in vitro and in vivo models are summarized in the following Table 3.

Table 3. Clove extract and oil effects in in vitro and in vivo models.

In Vitro/In Vivo Model	Biological Activity	Plant Part	Extract/Oil	Reference
Antioxidant tests (DPPH, FRAP) (in vitro)	Antioxidant	Buds	Supercritical extract	[94]
Antioxidant analysis (ABTS, DPPH) (in vitro)	Antioxidant	Buds	Essential oil	[95]

Table 3. *Cont.*

In Vitro/In Vivo Model	Biological Activity	Plant Part	Extract/Oil	Reference
Neuron culture, Aβ-induced damage (in vitro)	Neuroprotection (Alzheimer's disease)	Buds	Ethanol extract	[97]
Primary neuronal cells, scopolamine-induced memory impairment model (in vitro)	Memory enhancement, neuroprotection	Buds	Oil	[98]
Neuronal cell line PC12, stress-induced damage (in vitro)	Neurogenesis, memory improvement	Buds	Eugenol (oil component)	[118]
Neuropathic pain model, eugenol injection in cerebrospinal fluid (in vivo; **rats**)	Pain relief in neuropathic pain	Buds	Eugenol	[96]
Alzheimer's disease model, Aβ-induced memory impairment (in vivo; **mice**)	Neuroprotection (SIRT1 pathway)	Buds	Extract	[101]
Acrylamide-induced neurotoxicity model (in vivo; **rats**)	Neuroprotection in toxic brain damage	Buds	Oil	[92]
Alzheimer's disease model, effect of physical exercise (in vivo; **rats**)	Memory enhancement, reduction in damaged cells	Buds	Oil	[106]
Alzheimer's disease model, mitochondrial function analysis (in vivo; **rats**)	Memory restoration, apoptosis reduction, improved mitochondria	Buds	Extract	[49]

3. Amino Acid and Peptide Components in Cloves with Neuroprotective Potential

Clove is highly valued for its diverse bioactive compounds, including those found in clove essential oil, flavonoids, and phenolic compounds. While plant oils and extracts are commonly used in neuropharmacological studies, the consumption of whole plant material in the case of clove is believed to be linked to neuroprotective effects. In this context, recent studies have highlighted the significant amino acid content across various parts of the plant, such as the buds, fruits, branches, and leaves, which contribute to its nutritional and medicinal properties as summarized in Table 4. Remarkably, amino acids play a crucial role in metabolic processes, immune function, and antioxidant activity, making clove an essential resource for health and wellness applications.

Table 4. Amino acid composition of clove in its different parts like buds, fruits, and branches [127,128], as well as their biological activities connected to their neuroprotective role [87].

Amino Acid	Buds (mg/kg)	Buds (mg/kg)	Fruits (mg/kg)	Branches (mg/kg)	Leaves (mg/kg)	Biological Properties	Reference
Aspartic Acid	111.6	42.8	105.4	-	-	Supports metabolism and neurotransmission	[129]
Serine	69.8	80.5	41.5	57.9	37.9	Supports protein synthesis and acts as a precursor for neurotransmitters	[130]
Glutamic Acid	93.8	91.3	74.1	64.2	66.4	Functions as an excitatory neurotransmitter and antioxidant	[131]
Glycine	61.2	-	42.3	40.5	41.4	Neurotransmitter, anti-inflammatory, cytoprotective, immunomodulatory, metabolic precursor	[132–134]
Histidine	121.6	-	118.8	121.2	120.6	Encompasses neurotransmitter synthesis, enzymatic catalysis, metal ion chelation, and plays a role in the modulation of immune responses and growth	[135,136]
Arginine	133.1	113.7	96.1	250.1	89.9	Encompasses nitric oxide production, immune enhancement, antimicrobial action, and metabolic regulation	[137,138]
Threonine *	38.4	260.4	40.1	-	34.8	Plays a critical role in protein synthesis, immune function, and various metabolic pathways	[139]
Alanine	94.5	-	93.8	52.3	55.2	Supports gluconeogenesis, insulin secretion, immune function, and longevity	[140]
Tyrosine	77.5	40.0	69.3	64.1	66.7	Precursor for hormones like dopamine and adrenaline. Affects cognition, thermoregulation, neurotransmission, and may influence lifespan at varying doses	[141]

Table 4. *Cont.*

Amino Acid	Buds (mg/kg)	Buds (mg/kg)	Fruits (mg/kg)	Branches (mg/kg)	Leaves (mg/kg)	Biological Properties	Reference
Valine *	65.9	106.1	50.2	45.7	44.9	Contributes to muscle growth, tissue repair, has antioxidant properties, and activates NRF2 to improve cellular health and growth	[142,143]
Methionine *	63.3	14.1	62.8	-	-	Acts as an antioxidant and supports liver detoxification	[144,145]
Lysine *	68.9	-	68.5	68.2	66.8	Essential for protein synthesis, longevity, metabolism, and tissue repair; plays a significant role in antioxidant and anti-inflammatory activities	[146,147]
Isoleucine *	59.8	16.8	53.1	-	-	Branched-chain amino acid, affects glucose metabolism, insulin resistance, and may play a role in aging	[143,148]
Leucine *	61.8	27.7	56.8	-	-	Influences lifespan, metabolism, muscle function, and longevity regulation pathways	[143,149]
Phenylalanine *	75.0	21.1	74.8	-	83.3	Precursor for neurotransmitters like dopamine and norepinephrine, antioxidant.	[150]
Proline	154.9	-	203.7	97.6	63.7	Enhances collagen synthesis and cellular repair, supports antioxidant activity, and contributes to metabolic regulation.	[151]
Tryptophan *	-	12.1	-	-	-	Recursor to melatonin, serotonin, and vitamin B3. It influences aging, neurotransmitter synthesis, mood regulation, and sleep cycles.	[152,153]

Table 4. *Cont.*

Amino Acid	Buds (mg/kg)	Buds (mg/kg)	Fruits (mg/kg)	Branches (mg/kg)	Leaves (mg/kg)	Biological Properties	Reference
Total Amino Acids (TAA)	1351.1	830.2	1251.2	861.8	771.6		
Essential Amino Acids (EAA)	433.1	461.9	406.9	113.9	229.8		

* The asterisk denotes essential amino acids.

As observed from the data in Table 4, clove is a significant source of both essential and non-essential amino acids, which contribute to its remarkable biological activities. The plant's various parts, including the buds, fruits, branches, and leaves, contain an array of amino acids that play key roles in human health. Amino acids in clove contribute to energy metabolism, immune system support, tissue repair, neurotransmission, and antioxidant properties, making it a valuable resource for nutritional and pharmacological applications. The buds and fruits of *Syzygium aromaticum* exhibit similar levels of essential amino acids, with total contents of 433.1 and 406.2 mg/kg, respectively. These values are higher than those found in the branches (113.9 mg/kg) and leaves (229.8 mg/kg). The essential amino acids found in clove are of particular interest, as they cannot be synthesized by the human body and must be obtained through diet. These essential amino acids, such as histidine, threonine, valine, and methionine, are critical for various physiological functions including immune modulation, muscle growth, collagen production, and neurotransmitter synthesis. Their presence in clove enhances its potential as a health-promoting ingredient in functional foods and medicinal products. On the other hand, non-essential amino acids like glutamic acid, proline, and alanine also contribute to the plant's bioactivity. For instance, glutamic acid, a key neurotransmitter, plays an important role in brain function [154]. Similarly, proline supports collagen synthesis, contributing to tissue repair and skin health. These amino acids, though not required through the diet, provide significant benefits to the body, especially in terms of maintaining homeostasis and overall well-being. The presence of these amino acids underscores clove's therapeutic potential in managing a variety of conditions, from wound healing to inflammation and oxidative stress reduction. The amino acid profile of clove also highlights its versatility as a natural source of bioactive compounds. In addition to these amino acids, clove contains numerous other secondary metabolites, such as flavonoids, phenolic compounds, and essential oils, which further enhance its biological activities, as mentioned in the previous sections of this work. These properties collectively contribute to clove's reputation as a multifunctional medicinal plant, ideal for both preventive health and therapeutic purposes, including neuroprotection. It is clear that the inclusion of clove in various health and wellness products, particularly in nutritional supplements, could significantly contribute to improving human health.

The peptide composition of clove is also noteworthy, particularly the identification of ghrelin, an endogenous ligand of the growth hormone secretagogue receptor (Figure 2). In clove, ghrelin was found to have concentrations of 4070.75 $\pm$ 664.67 pg/mg in the flower bud [155].

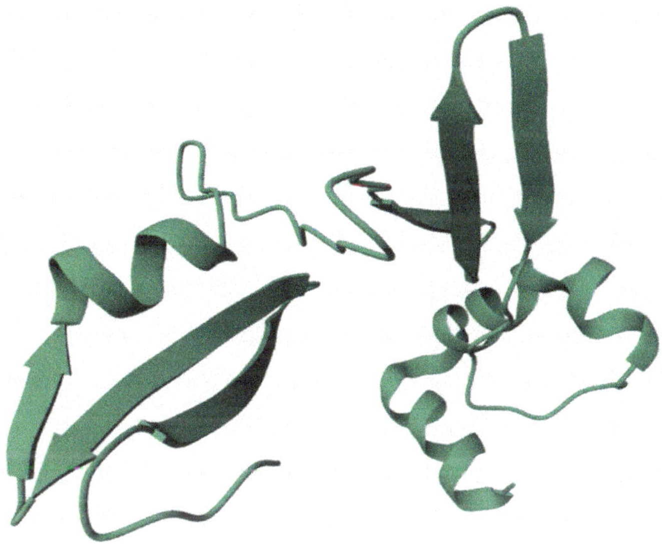

Figure 2. Three-dimensional structure of human ghrelin within its precursor protein, as accessed from https://www.modelarchive.org/doi/10.5452/ma-cfzyt (accessed on 15 February 2025) [156].

Known for its roles in regulating food intake, energy homeostasis, and insulin release, ghrelin has recently drawn attention for its potential therapeutic effects in neurological disorders, particularly AD. In AD, ghrelin or its receptor agonists have shown promise in attenuating pathology related to amyloid-beta accumulation, tau hyperphosphorylation, neuroinflammation, and cognitive decline [157].

4. Conclusions

Clove has been traditionally valued not only for its culinary uses, but also for its medicinal properties. Recent studies have drawn attention to its potential role in managing neurodegenerative diseases like Alzheimer's disease, with evidence suggesting that its bioactive compounds, particularly eugenol, may offer neuroprotective effects. Although epidemiological studies directly linking clove consumption to AD prevention are still limited, preclinical research demonstrates significant promise. Clove contains a range of bioactive compounds, including flavonoids, phenolic compounds, and amino acids, all of which contribute to its health benefits. Clove is a rich source of both essential and non-essential amino acids, which support numerous biological activities such as energy metabolism, neurotransmission, immune function, and antioxidant action. These amino acids play a crucial role in maintaining metabolic processes and immune function, which are essential for brain health. The amino acid profile and the presence of peptides, like ghrelin, in clove support the role of this spice in brain health and cognitive function. Clove also contains significant levels of eugenol, which is the main compound responsible for its neuroprotective properties. Eugenol has demonstrated antioxidant, anti-inflammatory, and neuroprotective effects in several studies. Eugenol's ability to modulate inflammation and oxidative stress makes it a promising candidate for AD therapy. Overall, clove's amino acid content enhances its versatility in supporting cognitive health, alongside its other bioactive compounds, such as flavonoids and phenolic compounds, all of which work synergistically to improve cognitive function, support antioxidant defenses, and reduce neuroinflammation. The presence of essential amino acids, particularly in the buds and fruits of clove, make it a valuable ingredient for both traditional and modern medicinal applications. The combination of antioxidant and anti-inflammatory properties, along with the amino acids' role in neurotransmission and tissue repair, position clove as a potentially valuable adjunct to AD management, while some of clove's molecular constituents can contribute to managing other conditions related to oxidative stress, such as cardiovascular disease, by supporting vascular health and reducing inflammation. In conclusion, eugenol and other clove phytocompounds, along with the peptide ghrelin and the specific amino acid composition of clove—particularly its high levels of neuroprotective compounds like glutamic acid—enhance its potential as a therapeutic agent for Alzheimer's disease. Further clinical research is necessary to fully understand the therapeutic potential of clove, particularly in combination with other neuroprotective agents, and to optimize its use for neurodegenerative diseases. With its diverse bioactive profile, clove holds significant promise as a natural remedy in managing cognitive decline and other neurodegenerative conditions.

Author Contributions: All authors contributed to the writing, investigation, and preparation of this manuscript. All authors have read and agreed to the published version of the manuscript.

Funding: This research did not have external funding.

Data Availability Statement: Not applicable.

Acknowledgments: We are grateful to CNR and University of Cassino for supporting this investigation. G.N. Roviello and T. Sargsyan would like to express their sincere gratitude to the Higher Education and Science Committee of the RA Ministry of Education, Science, Culture, and Sports of Armenia for their support through the Adjunct Research Professorship Program 2024. This opportunity has significantly contributed to the advancement of this research.The authors express their gratitude to the Science Committee of the Republic of Armenia for supporting their research within the framework of the project № 24RL-1D014.

Conflicts of Interest: The authors declare no conflicts of interest.

References

1. Prince, M.; Bryce, R.; Albanese, E.; Wimo, A.; Ribeiro, W.; Ferri, C.P. The Global Prevalence of Dementia: A Systematic Review and Meta-Analysis. *Alzheimers Dement.* **2013**, *9*, 63.e2–75.e2. [CrossRef] [PubMed]
2. Wimo, A.; Guerchet, M.; Ali, G.-C.; Wu, Y.-T.; Prina, A.M.; Winblad, B.; Jönsson, L.; Liu, Z.; Prince, M. The worldwide costs of dementia 2015 and comparisons with 2010. *Alzheimer's Dement.* **2017**, *13*, 1–7. [CrossRef] [PubMed]
3. Zahra, W.; Rai, S.N.; Birla, H.; Singh, S.S.; Dilnashin, H.; Rathore, A.S.; Singh, S.P. The Global Economic Impact of Neurodegenerative Diseases: Opportunities and Challenges. *Bioeconomy Sustain. Dev.* **2019**, *17*, 333–345. [CrossRef]
4. Roland, K.P.; Chappell, N.L. Caregiver Experiences Across Three Neurodegenerative Diseases: Alzheimer's, Parkinson's, and Parkinson's With Dementia. *J. Aging Health* **2019**, *31*, 256–279. [CrossRef]
5. Aza, A.; Gómez-Vela, M.; Badia, M.; Orgaz, M.B.; González-Ortega, E.; Vicario-Molina, I.; Montes-López, E. Listening to families with a person with neurodegenerative disease talk about their quality of life: Integrating quantitative and qualitative approaches. *Health Qual. Life Outcomes* **2022**, *20*, 76. [CrossRef]
6. Dokholyan, N.V.; Mohs, R.C.; Bateman, R.J. Challenges and progress in research, diagnostics, and therapeutics in Alzheimer's disease and related dementias. *Alzheimers Dement.* **2022**, *8*, e12330. [CrossRef]
7. Sehar, U.; Rawat, P.; Reddy, A.P.; Kopel, J.; Reddy, P.H. Amyloid Beta in Aging and Alzheimer's Disease. *Int. J. Mol. Sci.* **2022**, *23*, 12924. [CrossRef]
8. Lemche, E. Early Life Stress and Epigenetics in Late-Onset Alzheimer's Dementia: A Systematic Review. *Curr. Genomics* **2018**, *19*, 522–602. [CrossRef]
9. Andrade-Guerrero, J.; Santiago-Balmaseda, A.; Jeronimo-Aguilar, P.; Vargas-Rodríguez, I.; Cadena-Suárez, A.R.; Sánchez-Garibay, C.; Pozo-Molina, G.; Méndez-Catalá, C.F.; Cardenas-Aguayo, M.-d.-C.; Diaz-Cintra, S.; et al. Alzheimer's Disease: An Updated Overview of Its Genetics. *Int. J. Mol. Sci.* **2023**, *24*, 3754. [CrossRef]

10. Zemek, F.; Drtinova, L.; Nepovimova, E.; Sepsova, V.; Korabecny, J.; Klimes, J.; Kuca, K. Outcomes of Alzheimer's Disease Therapy with Acetylcholinesterase Inhibitors and Memantine. *Expert Opin. Drug Saf.* **2014**, *13*, 759–774. [CrossRef]
11. Kmietowicz, Z.; Mahase, E. Lecanemab: Benefits of Alzheimer's Drug Are "Just Too Small" to Justify Cost, Says NICE. *BMJ* **2024**, *386*, q1853. [CrossRef] [PubMed]
12. Beveridge, J.; Kaniecki, E.; Naidu, A.; Silverglate, B.D.; Grossberg, G. How Promising Are the Latest Monoclonal Antibodies Targeting Amyloid-β for the Treatment of Early Alzheimer's Disease? *Expert Opin. Emerg. Drugs* **2024**, *29*, 35–43. [CrossRef] [PubMed]
13. Cummings, J.; Zhou, Y.; Lee, G.; Zhong, K.; Fonseca, J.; Cheng, F. Alzheimer's disease drug development pipeline: 2024. *Alzheimers Dement.* **2024**, *10*, e12465. [CrossRef]
14. Mahase, E. Lecanemab and donanemab: NICE reconsiders controversial Alzheimer's drugs. *BMJ* **2025**, *388*, r463. [CrossRef]
15. Cummings, J. Corection to: New Approaches to Symptomatic Treatments for Alzheimer's Disease. *Mol. Neurodegener.* **2021**, *16*, 2. [CrossRef]
16. Lyketsos, C.G.; Szekely, C.A.; Mielke, M.M.; Rosenberg, P.B.; Zandi, P.P. Developing New Treatments for Alzheimer's Disease: The Who, What, When, and How of Biomarker-Guided Therapies. *Int. Psychogeriatr.* **2008**, *20*, 871–889. [CrossRef]
17. Roviello, V.; Gilhen-Baker, M.; Roviello, G.N.; Lichtfouse, E. River Therapy. *Environ. Chem. Lett.* **2022**, *20*, 2729–2734. [CrossRef]
18. Costanzo, M.; De Giglio, M.A.R.; Gilhen-Baker, M.; Roviello, G.N. The Chemical Basis of Seawater Therapies: A Review. *Environ. Chem. Lett.* 2024. [CrossRef]
19. Fik-Jaskółka, M.; Mittova, V.; Motsonelidze, C.; Vakhania, M.; Vicidomini, C.; Roviello, G.N. Antimicrobial Metabolites of Caucasian Medicinal Plants as Alternatives to Antibiotics. *Antibiotics* **2024**, *13*, 487. [CrossRef]
20. Ricci, A.; Roviello, G.N. Exploring the Protective Effect of Food Drugs against Viral Diseases: Interaction of Functional Food Ingredients and SARS-CoV-2, Influenza Virus, and HSV. *Life* **2023**, *13*, 402. [CrossRef]
21. Ding, H.; Reiss, A.B.; Pinkhasov, A.; Kasselman, L.J. Plants, Plants, and More Plants: Plant-Derived Nutrients and Their Protective Roles in Cognitive Function, Alzheimer's Disease, and Other Dementias. *Medicina* **2022**, *58*, 1025. [CrossRef] [PubMed]
22. Cooper, E.L.; Ma, M.J. Alzheimer Disease: Clues from Traditional and Complementary Medicine. *J. Tradit. Complement Med.* **2017**, *7*, 380–385. [CrossRef] [PubMed]
23. Fik-Jaskółka, M.A.; Mkrtchyan, A.F.; Saghyan, A.S.; Palumbo, R.; Belter, A.; Hayriyan, L.A.; Simonyan, H.; Roviello, V.; Roviello, G.N. Spectroscopic and SEM Evidence for G4-DNA Binding by a Synthetic Alkyne-Containing Amino Acid with Anticancer Activity. *Spectrochim. Acta A Mol. Biomol. Spectrosc.* **2020**, *229*, 117884. [CrossRef]
24. Roviello, V.; Musumeci, D.; Mokhir, A.; Roviello, G.N. Evidence of Protein Binding by a Nucleopeptide Based on a Thyminedecorated L-Diaminopropanoic Acid through CD and In Silico Studies. *Curr. Med. Chem.* **2021**, *28*, 5004–5015. [CrossRef]
25. Marzano, M.; Falanga, A.P.; Marasco, D.; Borbone, N.; D'Errico, S.; Piccialli, G.; Roviello, G.N.; Oliviero, G. Evaluation of an Analogue of the Marine ε-PLL Peptide as a Ligand of G-quadruplex DNA Structures. *Mar. Drugs* **2020**, *18*, 49. [CrossRef]
26. Roviello, G.N.; Ricci, A.; Buccia, E.M.; Pedone, C. Synthesis, Biological Evaluation and Supramolecular Assembly of Novel Analogues of Peptidyl Nucleosides. *Mol. BioSyst.* **2011**, *7*, 1115–1123. [CrossRef]
27. Roviello, G.N.; Musumeci, D.; Buccia, E.M.; Pedone, C. Evidences for Supramolecular Organization of Nucleopeptides: Synthesis, Spectroscopic and Biological Studies of a Novel Dithymine L-Serine Tetrapeptide. *Mol. BioSyst.* **2011**, *7*, 624–633. [CrossRef]
28. Roviello, G.; Musumeci, D.; Pedone, C.; Bucci, E.M. Synthesis, Characterization and Hybridization Studies of an Alternate Nucleo-Epsilon/Gamma-Peptide: Complexes Formation with Natural Nucleic Acids. *Amino Acids* **2010**, *38*, 103–111. [CrossRef]
29. Musumeci, D.; Oliviero, G.; Roviello, G.N.; Bucci, E.M.; Piccialli, G. G-Quadruplex-Forming Oligonucleotide Conjugated to Magnetic Nanoparticles: Synthesis, Characterization, and Enzymatic Stability Assays. *Bioconjug. Chem.* **2012**, *23*, 382–391. [CrossRef]
30. Roviello, G.N.; Di Gaetano, S.; Capasso, D.; Cesarani, A.; Bucci, E.M.; Pedone, C. Synthesis, Spectroscopic Studies and Biological Activity of a Novel Nucleopeptide with Moloney Murine Leukemia Virus Reverse Transcriptase Inhibitory Activity. *Amino Acids* **2010**, *38*, 1489–1496. [CrossRef]
31. Sargsyan, T.; Stepanyan, L.; Panosyan, H.; Hakobyan, H.; Israyelyan, M.; Tsaturyan, A.; Hovhannisyan, N.; Vicidomini, C.; Mkrtchyan, A.; Saghyan, A.; et al. Synthesis and Antifungal Activity of Fmoc-Protected 1,2,4-Triazolyl-α-Amino Acids and Their Dipeptides Against *Aspergillus* Species. *Biomolecules* **2025**, *15*, 61. [CrossRef] [PubMed]
32. Simonyan, H.; Palumbo, R.; Vicidomini, C.; Scognamiglio, P.L.; Petrosyan, S.; Sahakyan, L.; Melikyan, G.; Saghyan, A.; Roviello, G.N. Exploring the Binding of c-Myc G-Quadruplex and the Structural Impact of Synthetic Non-Proteinogenic Amino Acids on Serum Albumins: Implications for Potential Intrinsic c-Myc-Associated Anticancer Activity and Drug Delivery Systems. *Mol. Ther. Nucleic Acids* **2025**, 102478. [CrossRef]

33. Falanga, A.P.; Piccialli, I.; Greco, F.; D'Errico, S.; Nolli, M.G.; Borbone, N.; Oliviero, G.; Roviello, G.N. Nanostructural Modulation of G-Quadruplex DNA in Neurodegeneration: Orotate Interaction Revealed Through Experimental and Computational Approaches. *J. Neurochem.* **2025**, *169*, e16296. [CrossRef] [PubMed]
34. Yash, R.; Menghani, D.M.; Bhattad, D.M.; Chandak, K.K.; Taksande, J.R.; Umekar, M.J. A Review: Pharmacological and Herbal Remedies in the Management of Neurodegenerative Disorder (Alzheimer's). *Int. J. Pharmacogn. Life Sci.* **2021**, *2*, 18–27. [CrossRef]
35. McKeage, K.; Lyseng-Williamson, K.A. Ginkgo Biloba Extract EGb 761® in the Symptomatic Treatment of Mild-to-Moderate Dementia: A Profile of Its Use. *Drugs Ther. Perspect.* **2018**, *34*, 358–366. [CrossRef]
36. Kandiah, N.; Chan, Y.F.; Chen, C.; Dasig, D.; Dominguez, J.; Han, S.H.; Jia, J.; Kim, S.; Limpawattana, P.; Ng, L.L.; et al. Treatment of Dementia and Mild Cognitive Impairment with or without Cerebrovascular Disease: Expert Consensus on the Use of Ginkgo Biloba Extract, EGb 761®. *CNS Neurosci. Ther.* **2019**, *25*, 288–298. [CrossRef]
37. Yan, Y.-P.; Chen, J.-Y.; Lu, J.-H. Disease-Modifying Activity of Huperzine A on Alzheimer's Disease: Evidence from Preclinical Studies on Rodent Models. *Int. J. Mol. Sci.* **2022**, *23*, 15238. [CrossRef]
38. Shan, M.; Bai, Y.; Fang, X.; Lan, X.; Zhang, Y.; Cao, Y.; Zhu, D.; Luo, H. American Ginseng for the Treatment of Alzheimer's Disease: A Review. *Molecules* **2023**, *28*, 5716. [CrossRef]
39. Lee, B.C.; Choe, Y.M.; Suh, G.H.; Choi, I.G.; Kim, H.S.; Hwang, J.; Yi, D.; Jhoo, J.H.; Kim, J.W. Ginseng Intake and Alzheimer Disease-Specific Cognition in Older Adults According to Apolipoprotein ε4 Allele Status. *Front. Aging Neurosci.* **2023**, *15*, 1152626. [CrossRef]
40. Mittal, S.; Prajapati, K.P.; Ansari, M.; Anand, B.G.; Kar, K. Autooxidation of Curcumin in Physiological Buffer Causes an Enhanced Synergistic Anti-Amyloid Effect. *Int. J. Biol. Macromol.* **2023**, *235*, 123629. [CrossRef]
41. Sallaberry, C.A.; Voss, B.J.; Stone, W.B.; Estrada, F.; Bhatia, A.; Soto, J.D.; Griffin, C.W.; Vander Zanden, C.M. Curcumin Reduces Amyloid Beta Oligomer Interactions with Anionic Membranes. *ACS Chem. Neurosci.* **2023**, *14*, 4026–4038. [CrossRef] [PubMed]
42. Goozee, K.G.; Shah, T.M.; Sohrabi, H.R.; Rainey-Smith, S.R.; Brown, B.; Verdile, G.; Martins, R.N. Examining the Potential Clinical Value of Curcumin in the Prevention and Diagnosis of Alzheimer's Disease. *Br. J. Nutr.* **2016**, *115*, 449–465. [CrossRef] [PubMed]
43. Mohamed, I.E.; Osman, E.E.; Saeed, A.; Ming, L.C.; Goh, K.W.; Razi, P.; Abdullah, A.D.I.; Dahab, M. Plant Extracts as Emerging Modulators of Neuroinflammation and Immune Receptors in Alzheimer's Pathogenesis. *Heliyon* **2024**, *10*, e35943. [CrossRef] [PubMed]
44. Lee, J.; Lee, S.; Jo, W.; Ji, H.W.; Pyeon, M.; Moon, M.; Yun, J.; Lee, J.H.; Sohn, S.O. Effect of a *Salvia officinalis* and *Hypericum perforatum* Mixture on Improving Memory and Cognitive Decline. *Adv. Tradit. Med.* **2023**, *23*, 633–649. [CrossRef]
45. Vicidomini, C.; Palumbo, R.; Moccia, M.; Roviello, G.N. Oxidative Processes and Xenobiotic Metabolism in Plants: Mechanisms of Defense and Potential Therapeutic Implications. *J. Xenobiot.* **2024**, *14*, 1541–1569. [CrossRef]
46. Pirtskhalava, M.; Mittova, V.; Tsetskhladze, Z.R.; Palumbo, R.; Pastore, R.; Roviello, G.N. Georgian Medicinal Plants as Rich Natural Sources of Antioxidant Derivatives: A Review on the Current Knowledge and Future Perspectives. *Curr. Med. Chem.* **2024**, *31*, 4407–4424. [CrossRef]
47. Moawad, M.H.E.; Serag, I.; Alkhawaldeh, I.M.; Abbas, A.; Sharaf, A.; Alsalah, S.; Sadeq, M.A.; Shalaby, M.M.M.; Hefnawy, M.T.; Abouzid, M.; et al. Exploring the Mechanisms and Therapeutic Approaches of Mitochondrial Dysfunction in Alzheimer's Disease: An Educational Literature Review. *Mol. Neurobiol.* **2024**, *61*, 1–22. [CrossRef]
48. Vicidomini, C.; Roviello, V.; Roviello, G.N. Molecular Basis of the Therapeutical Potential of Clove (*Syzygium aromaticum* L.) and Clues to Its Anti-COVID-19 Utility. *Molecules* **2021**, *26*, 1880. [CrossRef]
49. Momo, E.J.; Nguimatsia, F.; Ateufouet Ngouango, L.; Lunga, P.K.; Pone Kamdem, B.; Jazet Dongmo, P.M. Eugenol-Rich Essential Oils from Flower Buds and Leaves of *Syzygium aromaticum* Show Antifungal Activity against Candida and Cryptococcus Species. *Future Pharmacol.* **2024**, *4*, 449–465. [CrossRef]
50. Hickey, J.P.; Collins, A.E.; Nelson, M.L.; Chen, H.; Kalisch, B.E. Modulation of Oxidative Stress and Neuroinflammation by Cannabidiol (CBD): Promising Targets for the Treatment of Alzheimer's Disease. *Curr. Issues Mol. Biol.* **2024**, *46*, 4379–4402. [CrossRef]
51. Saxena, B. Eugenol as Neuro-Phytomedicine: Recent Trends Pertaining to the Treatment of Neurological Disorders. In *NeuroPhytomedicine*, 1st ed.; CRC Press: Boca Raton, FL, USA, 2024; p. 18. [CrossRef]
52. Panahzadeh, F.; Mirnasuri, R.; Rahmati, M. Exercise and *Syzygium aromaticum* Reverse Memory Deficits, Apoptosis, and Mitochondrial Dysfunction of the Hippocampus in Alzheimer's Disease. *J. Ethnopharmacol.* **2022**, *286*, 114871. [CrossRef] [PubMed]
53. Lee, J.; Hong, S.; Ahn, M.; Kim, J.; Moon, C.; Matsuda, H.; Tanaka, A.; Nomura, Y.; Jung, K.; Shin, T. Eugenol Alleviates the Symptoms of Experimental Autoimmune Encephalomyelitis in Mice by Suppressing Inflammatory Responses. *Int. Immunopharmacol.* **2024**, *128*, 111479. [CrossRef] [PubMed]

54. Idowu, S.; Adekoya, A.E.; Igiehon, O.O.; Idowu, A.T. Clove (*Syzygium aromaticum*) Spices: A Review on Their Bioactivities, Current Use, and Potential Application in Dairy Products. *J. Food Meas. Charact.* **2021**, *15*, 3419–3435. [CrossRef]
55. Liñán-Atero, R.; Aghababaei, F.; García, S.R.; Hasiri, Z.; Ziogkas, D.; Moreno, A.; Hadidi, M. Clove Essential Oil: Chemical Profile, Biological Activities, Encapsulation Strategies, and Food Applications. *Antioxidants* **2024**, *13*, 488. [CrossRef]
56. Akopian, K.A.; Poghosyan, Y.M.; Poghosyan, S.B.; Matinyan, S.V.; Ter-Zakaryan, S.O.; Muradyan, S.A.; Movsisyan, M.R. Study of a Possible Allergic Effect of Ubivaks Ointment. *New Armen. Med. J.* **2015**, *9*, 32–35.
57. Grant, W.B.; Blake, S.M. Diet's Role in Modifying Risk of Alzheimer's Disease: History and Present Understanding. *J. Alzheimer's Dis.* **2023**, *96*, 1353–1382. [CrossRef]
58. Muderawan, I.W.; Laksmi, P.P.D.S.; Mudianta, I.W.; Martiningsih, N.W. Chemical Constituent and Antioxidant Activity of Clove (*Syzygium aromaticum*) Bud and Leaf Essential Oils from Bali. *Indones. J. Chem. Res.* **2025**, *12*, mud. [CrossRef]
59. Spigarelli, R.; Spisni, E.; Magalhães, M.; Cabral, C.; Gonçalves, A.C.; Saracino, I.M.; Botti, G.; Dalpiaz, A.; Beggiato, S.; Valerii, M.C. Clove Essential Oil as a Source of Antitumoral Compounds Capable of Crossing the Blood–Brain Barrier: A Focus on the Effects of β-Caryophyllene and Eugenol in a Glioblastoma Cell Line. *Int. J. Mol. Sci.* **2025**, *26*, 238. [CrossRef]
60. Nagababu, E.; Lakshmaiah, N. Inhibitory Effect of Eugenol on Non-Enzymatic Lipid Peroxidation in Rat Liver Mitochondria. *Biochem. Pharmacol.* **1992**, *43*, 2393–2400. [CrossRef]
61. Pandey, V.K.; Srivastava, S.; Ashish; Farooqui, A.; Shaikh, A.M.; Kovacs, B. Bioactive Properties of Clove (*Syzygium aromaticum*) Essential Oil Nanoemulsion: A Comprehensive Review. *Heliyon* **2024**, *10*, e22437. [CrossRef]
62. Silva, M.V.; Lima, A.C.A.; Silva, M.G.; Caetano, V.F.; Andrade, M.F.; Silva, R.G.C.; Moraes Filho, L.E.P.T.; Silva, I.D.L.; Vinhas, G.M. Clove Essential Oil and Eugenol: A Review of Their Significance and Uses. *Food Bioscience* **2024**, *62*, 105112. [CrossRef]
63. Stojanović, N.M.; Ranđelović, P.J.; Simonović, M.; Radić, M.; Todorović, S.; Corrigan, M.; Harkin, A.; Boylan, F. Essential Oil Constituents as Anti-Inflammatory and Neuroprotective Agents: An Insight through Microglia Modulation. *Int. J. Mol. Sci.* **2024**, *25*, 5168. [CrossRef] [PubMed]
64. Viveiros, M.M.H.; Silva, M.G.; da Costa, J.G.M.; de Oliveira, A.G.; Rubio, C.; Padovani, C.R.; Rainho, C.A.; Schellini, S.A. Anti-inflammatory Effects of α-Humulene and β-Caryophyllene on Pterygium Fibroblasts. *Int. J. Ophthalmol.* **2022**, *15*, 1903–1907. [CrossRef] [PubMed]
65. Dalavaye, N.; Nicholas, M.; Pillai, M.; Erridge, S.; Sodergren, M.H. The Clinical Translation of α-humulene—A Scoping Review. *Planta Med.* **2024**, *90*, 664–674. [CrossRef]
66. Alberti, T.B.; Barbosa, W.L.R.; Vieira, J.L.F.; Raposo, N.R.B.; Dutra, R.C. (−)-β-Caryophyllene, a CB2 Receptor-Selective Phytocannabinoid, Suppresses Motor Paralysis and Neuroinflammation in a Murine Model of Multiple Sclerosis. *Int. J. Mol. Sci.* **2017**, *18*, 691. [CrossRef]
67. Bahi, A.; Al Mansouri, S.; Al Memari, E.; Al Ameri, M.; Nurulain, S.M.; Ojha, S. β-Caryophyllene, a CB2 Receptor Agonist Produces Multiple Behavioral Changes Relevant to Anxiety and Depression in Mice. *Physiol. Behav.* **2014**, *135*, 119–124. [CrossRef]
68. Cho, J.Y.; Chang, H.J.; Lee, S.K.; Kim, H.J.; Hwang, J.K.; Chun, H.S. Amelioration of Dextran Sulfate Sodium-Induced Colitis in Mice by Oral Administration of β-Caryophyllene, a Sesquiterpene. *Life Sci.* **2007**, *80*, 932–939. [CrossRef]
69. Batiha, G.E.; Alkazmi, L.M.; Wasef, L.G.; Beshbishy, A.M.; Nadwa, E.H.; Rashwan, E.K. *Syzygium aromaticum* L. (Myrtaceae): Traditional Uses, Bioactive Chemical Constituents, Pharmacological and Toxicological Activities. *Biomolecules* **2020**, *10*, 202. [CrossRef]
70. Mansouri, M.T.; Farbood, Y.; Sameri, M.J.; Sarkaki, A.; Naghizadeh, B.; Rafeirad, M. Neuroprotective Effects of Oral Gallic Acid Against Oxidative Stress Induced by 6-Hydroxydopamine in Rats. *Food Chem.* **2013**, *138*, 1028–1033. [CrossRef]
71. Hajipour, S.; Sarkaki, A.; Farbood, Y.; Eidi, A.; Mortazavi, P.; Valizadeh, Z. Effect of Gallic Acid on Dementia Type of Alzheimer Disease in Rats: Electrophysiological and Histological Studies. *Basic Clin. Neurosci.* **2016**, *7*, 97–106. [CrossRef]
72. Costa, L.G.; Garrick, J.M.; Roquè, P.J.; Pellacani, C. Mechanisms of Neuroprotection by Quercetin: Counteracting Oxidative Stress and More. *Oxid. Med. Cell. Longev.* **2016**, *2016*, 2986796. [CrossRef] [PubMed]
73. Ansari, M.A.; Abdul, H.M.; Joshi, G.; Opii, W.O.; Butterfield, D.A. Protective Effect of Quercetin in Primary Neurons Against Aβ(1–42): Relevance to Alzheimer's Disease. *J. Nutr. Biochem.* **2009**, *20*, 269–275. [CrossRef] [PubMed]
74. Sabogal-Guáqueta, A.M.; Muñoz-Manco, J.I.; Ramírez-Pineda, J.R.; Lamprea-Rodriguez, M.; Osorio, E.; Cardona-Gómez, G.P. The Flavonoid Quercetin Ameliorates Alzheimer's Disease Pathology and Protects Cognitive and Emotional Function in Aged Triple Transgenic Alzheimer's Disease Model Mice. *Neuropharmacology* **2015**, *93*, 134–145. [CrossRef] [PubMed]
75. Ryu, B.; Kim, H.M.; Lee, J.S.; Lee, C.K.; Sezirahiga, J.; Woo, J.H.; Choi, J.H.; Jang, D.S. New Flavonol Glucuronides from the Flower Buds of *Syzygium aromaticum* (Clove). *J. Agric. Food Chem.* **2016**, *64*, 3048–3053. [CrossRef]
76. Kempuraj, D.; Thangavel, R.; Kempuraj, D.D.; Ahmed, M.E.; Selvakumar, G.P.; Raikwar, S.P.; Zaheer, S.A.; Iyer, S.S.; Govindarajan, R.; Chandrasekaran, P.N.; et al. Neuroprotective Effects of Flavone Luteolin in Neuroinflammation and Neurotrauma. *Biofactors* **2021**, *47*, 190–197. [CrossRef]

77. Ahmad, S.; Jo, M.H.; Ikram, M.; Khan, A.; Kim, M.O. Deciphering the Potential Neuroprotective Effects of Luteolin against Aβ1–42-Induced Alzheimer's Disease. *Int. J. Mol. Sci.* **2021**, *22*, 9583. [CrossRef]
78. Jing, W.; Xiaolan, C.; Yu, C.; Feng, Q.; Haifeng, Y. Pharmacological Effects and Mechanisms of Tannic Acid. *Biomed. Pharmacother.* **2022**, *154*, 113561. [CrossRef]
79. Wu, Y.; Zhong, L.; Yu, Z.; Qi, J. Anti-Neuroinflammatory Effects of Tannic Acid Against Lipopolysaccharide-Induced BV2 Microglial Cells via Inhibition of NF-κB Activation. *Drug Dev. Res.* **2019**, *80*, 262–268. [CrossRef]
80. Souza-Moreira, T.M.; Severi, J.A.; Lee, K.; Preechasuth, K.; Santos, E.; Gow, N.A.; Munro, C.A.; Vilegas, W.; Pietro, R.C. Anti-Candida Targets and Cytotoxicity of Casuarinin Isolated from *Plinia cauliflora* Leaves in a Bioactivity-Guided Study. *Molecules* **2013**, *18*, 8095–8108. [CrossRef]
81. Ali, A.; Wu, H.; Ponnampalam, E.N.; Cottrell, J.J.; Dunshea, F.R.; Suleria, H.A.R. Comprehensive Profiling of Most Widely Used Spices for Their Phenolic Compounds through LC-ESI-QTOF-MS2 and Their Antioxidant Potential. *Antioxidants* **2021**, *10*, 721. [CrossRef]
82. Alrumaihi, F.; Almatroodi, S.A.; Alharbi, H.O.A.; Alwanian, W.M.; Alharbi, F.A.; Almatroudi, A.; Rahmani, A.H. Pharmacological Potential of Kaempferol, a Flavonoid in the Management of Pathogenesis via Modulation of Inflammation and Other Biological Activities. *Molecules* **2024**, *29*, 2007. [CrossRef] [PubMed]
83. Cortés-Rojas, D.F.; de Souza, C.R.; Oliveira, W.P. Clove (*Syzygium aromaticum*): A Precious Spice. *Asian Pac. J. Trop. Med.* **2014**, *4*, 90–96. [CrossRef]
84. Cai, L.; Wu, C.D. Compounds from *Syzygium aromaticum* Possessing Growth Inhibitory Activity against Oral Pathogens. *J. Nat. Prod.* **1996**, *59*, 987–990. [CrossRef] [PubMed]
85. Yessirita, N.; Verawati, R.; Purnamasari, D.; Rollando, R.; Mandeli, R.S.; Albari, M.T.; Azhari, P.; Zainul, R.; Kharisma, V.D.; Jakhmola, V.; et al. In Silico Study of Rhamnocitrin Extract from Clove (*Syzygium Aromaticum*) in Inhibiting Adenosine A1 Adenylate Cyclase Interaction. *Pharmacogn. J.* **2023**, *15*, 512–517. [CrossRef]
86. Gong, G.; Guan, Y.Y.; Zhang, Z.L.; Rahman, K.; Wang, S.J.; Zhou, S.; Luan, X.; Zhang, H. Isorhamnetin: A Review of Pharmacological Effects. *Biomed. Pharmacother.* **2020**, *128*, 110301. [CrossRef]
87. Xue, Q.; Xiang, Z.; Wang, S.; Cong, Z.; Gao, P.; Liu, X. Recent Advances in Nutritional Composition, Phytochemistry, Bioactive, and Potential Applications of *Syzygium aromaticum* L. (Myrtaceae). *Front. Nutr.* **2022**, *9*, 1002147. [CrossRef]
88. Han, A.R. Identification and PEP Inhibitory Activity of Acetophenone Glucosides from the Clove Buds (*Syzygium aromaticum*). *J. Korean Soc. Appl. Biol. Chem.* **2010**, *53*, 847–851. [CrossRef]
89. Kumar Pandey, V.; Shams, R.; Singh, R.; Dar, A.H.; Pandiselvam, R.; Rusu, A.V.; Trif, M. A Comprehensive Review on Clove (*Caryophyllus aromaticus* L.) Essential Oil and Its Significance in the Formulation of Edible Coatings for Potential Food Applications. *Front. Nutr.* **2022**, *9*, 987674. [CrossRef]
90. Abdul Aziz, A.H.; Rizkiyah, D.N.; Qomariyah, L.; Irianto, I.; Che Yunus, M.A.; Putra, N.R. Unlocking the Full Potential of Clove (*Syzygium aromaticum*) Spice: An Overview of Extraction Techniques, Bioactivity, and Future Opportunities in the Food and Beverage Industry. *Processes* **2023**, *11*, 2453. [CrossRef]
91. Sara, A.; Ali, S.N.; Begum, S.; Siddiqui, B.S. Chemical Constituents of *Syzygium aromaticum*. *Chem. Nat. Comp.* **2018**, *54*, 1192–1193. [CrossRef]
92. Sen, A. Prophylactic and Therapeutic Roles of Oleanolic Acid and Its Derivatives in Several Diseases. *World J. Clin. Cases* **2020**, *8*, 1767–1792. [CrossRef] [PubMed]
93. Benninghoff, J.; Perneczky, R. Anti-Dementia Medications and Anti-Alzheimer's Disease Drugs: Side Effects, Contraindications, and Interactions. In *NeuroPsychopharmacotherapy*; Springer: Cham, Switzerland, 2022; pp. 1–10. [CrossRef]
94. Sharma, H.; Kim, D.Y.; Shim, K.H.; Sharma, N.; An, S.S.A. Multi-Targeting Neuroprotective Effects of *Syzygium aromaticum* Bud Extracts and Their Key Phytocompounds against Neurodegenerative Diseases. *Int. J. Mol. Sci.* **2023**, *24*, 8148. [CrossRef] [PubMed]
95. Aboubakr, M.; Ibrahim, S.S.; Said, A.M.; Elgendey, F.; Anis, A. Neuroprotective Effects of Clove Oil in Acrylamide-Induced Neurotoxicity in Rats. *Pak. Vet. J.* **2019**, *39*, 111–115. [CrossRef]
96. Gandara, J.; Barreto, G.E.; Martins, N.; Sharifi-Rad, J. Oxidative Stress in Alzheimer's Disease: Current Knowledge of Signaling Pathways and Therapeutics. *Mol. Biol. Rep.* **2023**, *50*, 1–17. [CrossRef]
97. Ivanović, J.; Dimitrijević-Branković, S.; Mišić, D. Evaluation and Improvement of Antioxidant and Antibacterial Activities of Supercritical Extracts from Clove Buds. *J. Funct. Foods* **2013**, *5*, 416–423. [CrossRef]
98. Kiki, M.J. In Vitro Antiviral Potential, Antioxidant, and Chemical Composition of Clove (*Syzygium aromaticum*) Essential Oil. *Molecules* **2023**, *28*, 2421. [CrossRef]
99. Lionnet, L.; Beaudry, F.; Vachon, P. Intrathecal Eugenol Administration Alleviates Neuropathic Pain in Male Sprague-Dawley Rats. *Phytother. Res.* **2010**, *24*, 1645–1653. [CrossRef]

100. Shekhar, S.; Yadav, Y.; Singh, A.P.; Pradhan, R.; Desai, G.R.; Dey, A.B.; Dey, S. Neuroprotection by Ethanolic Extract of *Syzygium aromaticum* in Alzheimer's Disease-Like Pathology via Maintaining Oxidative Balance through SIRT1 Pathway. *Exp. Gerontol.* **2018**, *110*, 277–283. [CrossRef]
101. Halder, S.; Mehta, A.K.; Kar, R.; Mustafa, M.; Mediratta, P.K.; Sharma, K.K. Clove Oil Reverses Learning and Memory Deficits in Scopolamine-Treated Mice. *Planta Med.* **2011**, *77*, 830–834. [CrossRef]
102. Liu, B.B.; Luo, L.; Liu, X.L.; Geng, D.; Li, C.F.; Chen, S.M.; Chen, X.M.; Yi, L.T.; Liu, Q. Essential Oil of *Syzygium aromaticum* Reverses the Deficits of Stress-Induced Behaviors and Hippocampal p-ERK/p-CREB/Brain-Derived Neurotrophic Factor Expression. *Planta Medica* **2015**, *81*, 185–192. [CrossRef]
103. Thapa, R.; Moglad, E.; Afzal, M.; Kumar, G.; Bhat, A.A.; Khan, M.A.; Khan, M.A.; Khan, M.I.; Khan, M.I.; Khan, M.I.; et al. The Role of Sirtuin 1 in Ageing and Neurodegenerative Disease: A Molecular Perspective. *Ageing Res. Rev.* **2024**, *90*, 102545. [CrossRef] [PubMed]
104. Amir Rawa, M.S.; Mazlan, M.K.N.; Ahmad, R.; Nogawa, T.; Wahab, H.A. Roles of Syzygium in Anti-Cholinesterase, Anti-Diabetic, Anti-Inflammatory, and Antioxidant: From Alzheimer's Perspective. *Plants* **2022**, *11*, 1476. [CrossRef] [PubMed]
105. Zhang, Y.; Zhang, Y.; Zhang, Y.; Zhang, Y.; Zhang, Y.; Zhang, Y.; Zhang, Y.; Zhang, Y.; Zhang, Y.; Zhang, Y.; et al. The Neuroprotective Effects of SIRT1 in Mice Carrying the APP/PS1 Transgene. *Aging Cell* **2014**, *13*, 808–817. [CrossRef]
106. Kiki, M.J. Chemical Composition, In Vivo, and In Silico Molecular Docking Studies of the Effect of *Syzygium aromaticum* (Clove) Essential Oil on Ochratoxin A-Induced Acute Neurotoxicity. *Plants* **2025**, *14*, 130. [CrossRef]
107. Sun, X.; Chen, W.D.; Wang, Y.D. DAF-16/FOXO Transcription Factor in Aging and Longevity. *Front. Pharmacol.* **2017**, *8*, 548. [CrossRef]
108. Li, X.; Sun, L.; Wang, Y.; Chen, Q.; Qiao, X. Eugenol Alleviates Lipopolysaccharide-Induced Neuroinflammation by Activating the Nrf2/HO-1 Pathway in BV2 Microglial Cells. *Comp. Biochem. Physiol. C Toxicol. Pharmacol.* **2020**, *234*, 108938. [CrossRef]
109. Gorgin Karaji, Z.; Fathi, M.; Mirnasori, R.; van der Zee, E.A. Swimming Exercise and Clove Oil Can Improve Memory by Molecular Responses Modification and Reduce Dark Cells in Rat Model of Alzheimer's Disease. *Exp. Gerontol.* **2023**, *174*, 112192. [CrossRef]
110. Heppner, F.L.; Ransohoff, R.M.; Becher, B. Immune Attack: The Role of Inflammation in Alzheimer Disease. *Nat. Rev. Neurosci.* **2015**, *16*, 358–372. [CrossRef]
111. Li, Y.; Zhang, J.; Zhang, L.; Hu, C.; Zhou, L.; Cheng, Y.; Liu, Q. Ellagic Acid (EA) Ameliorates Alzheimer's Disease by Reducing Aβ Levels, Oxidative Stress, and Attenuating Inflammation. *Eur. J. Pharmacol.* **2025**, *986*, 177099. [CrossRef]
112. Pădureanu, V.; Dop, D.; Caragea, D.C.; Rădulescu, D.; Pădureanu, R.; Forțofoiu, M.-C. Cardiovascular and Neurological Diseases and Association with Helicobacter Pylori Infection—An Overview. *Diagnostics* **2024**, *14*, 1781. [CrossRef]
113. Kim, K.H.; Rateb, M.; Hassan, H.; Elbestawy, M.K.M.; El-Sherbiny, G.M.; Moghannem, S.A. Antibacterial, Antibiofilm and Anti-Inflammatory Activities of Eugenol Clove Essential Oil against Resistant Helicobacter pylori. *Molecules* **2023**, *28*, 2448. [CrossRef] [PubMed]
114. Esmaeili, F.; Zahmatkeshan, M.; Yousefpoor, Y.; Alipanah, H.; Safari, E.; Osanloo, M. Anti-Inflammatory and Anti-Nociceptive Effects of Cinnamon and Clove Essential Oils Nanogels: An in Vivo Study. *BMC Complement. Med. Ther.* **2022**, *22*, 143. [CrossRef] [PubMed]
115. Banerjee, K.; Madhyastha, H.; Sandur, V.R.; Manikandanath, N.T.; Thiagarajan, N.; Thiagarajan, P. Anti-Inflammatory and Wound Healing Potential of a Clove Oil Emulsion. *Colloids Surf. B Biointerfaces* **2020**, *193*, 111102. [CrossRef] [PubMed]
116. Akbar, L.; Juliandi, B.; Boediono, A.; Batubara, I.; Subangkit, M. Effects of Eugenol on Memory Performance, Neurogenesis, and Dendritic Complexity of Neurons in Mice Analyzed by Behavioral Tests and Golgi Staining of Brain Tissue. *J. Stem Cells Regen. Med.* **2021**, *17*, 35–41. [CrossRef]
117. Revi, N.; Sankaranarayanan, S.A.; Rengan, A.K. A Study on the Role of Eugenol Encapsulated Liposomes in Facilitating Neuron-Microglia Mediated Wound Recovery. *Materialia* **2022**, *23*, 101454. [CrossRef]
118. Garabadu, D.; Sharma, M. Eugenol Attenuates Scopolamine-Induced Hippocampal Cholinergic, Glutamatergic, and Mitochondrial Toxicity in Experimental Rats. *Neurotoxicol. Res.* **2019**, *35*, 848–859. [CrossRef]
119. Prasad, S.N.; Bharath, M.M.; Muralidhara. Neurorestorative Effects of Eugenol, a Spice Bioactive: Evidence in Cell Model and Its Efficacy as an Intervention Molecule to Abrogate Brain Oxidative Dysfunctions in the Streptozotocin Diabetic Rat. *Neurochem. Int.* **2016**, *95*, 24–36. [CrossRef]
120. Soares, G.A.B.e.; Bhattacharya, T.; Chakrabarti, T.; Tagde, P.; Cavalu, S. Exploring Pharmacological Mechanisms of Essential Oils on the Central Nervous System. *Plants* **2022**, *11*, 21. [CrossRef]
121. Ahmad, A.; Husain, A.; Mujeeb, M.; Khan, S.A.; Najmi, A.K.; Siddique, N.A.; Damanhouri, Z.A.; Anwar, F. Eugenol Enhances Memory Performance and Neurogenesis in Mice. *J. Neuropharm.* **2021**, *9*, 8372414. Available online: https://pmc.ncbi.nlm.nih.gov/articles/PMC8372414.

122. Hussain, M.; Khattak, M.N.K.; Shaheen, M.; Zaman, W.; Khan, M.S.; Saeed, M.; Rauf, A.; Anwar, F.; Shamsi, M. Clove Oil Reverses Learning and Memory Deficits in Scopolamine-Treated Mice. *Brain Res.* **2015**, *1234*, 45–52.
123. Costa, J.G.; Ratti, A.L.; Filho, P.P.; Silva, L.M.; Martins, A.M.; Moreira, A.; Cavalcanti, R.P.; Araujo, S.; Rocha, A.C.; de Lima, G. Clove Extract Improves Cognitive Impairment in Septic Mice by Activating the SIRT1 Signaling Pathway. *Food Sci. Technol.* **2020**, *40*, 1234–1245.
124. Liang, Z.H.; Cheng, X.H.; Ruan, Z.G.; Wang, H.; Li, S.S.; Liu, J.; Li, G.Y.; Tian, S.M. Protective Effects of Components of the Chinese Herb Grassleaf Sweetflag Rhizome on PC12 Cells Incubated with Amyloid-Beta42. *Neural Regen. Res.* **2015**, *10*, 1292–1297. [CrossRef] [PubMed]
125. Taheri, P.; Yaghmaei, P.; Tehrani, H.S.; Ebrahim-Habibi, A. Effects of Eugenol on Alzheimer's Disease-like Manifestations in Insulin- and Aβ-Induced Rat Models. *Neurophysiology* **2019**, *51*, 114–119. [CrossRef]
126. Jung, H.A.; Kim, S.M.; Kim, J.Y.; Yang, H.O. Eugenol Ameliorates Alzheimer's Disease Pathologies in a 5×FAD Mouse Model. *Phytomedicine* 2023. [CrossRef]
127. Liu, J.G.; Liu, H.P. Total Nutrient Analysis of Various Parts of *Eugenia caryophyllata*. *Biotic Res.* **2021**, *43*, 357–362. [CrossRef]
128. Ma, S.S. Food safety standards and development of health food of clove. Ph.D. Thesis, Hainan University, Haikou, China, 2018.
129. Holeček, M. Aspartic Acid in Health and Disease. *Nutrients* **2023**, *15*, 4023. [CrossRef]
130. Ye, Y.; Xu, H.; Xiong, Y.; Tong, Z.; Li, X. L-Serine, an Endogenous Amino Acid, Is a Potential Neuroprotective Agent for Neurological Disease and Injury. *Front. Mol. Neurosci.* **2021**, *14*, 726665. [CrossRef]
131. Moldovan, O.L.; Sandulea, A.; Lungu, I.A.; Gâz, Ș.A.; Rusu, A. Identification of Some Glutamic Acid Derivatives with Biological Potential by Computational Methods. *Molecules* **2023**, *28*, 4123. [CrossRef]
132. Petrat, F.; Boengler, K.; Schulz, R.; de Groot, H. Glycine, a Simple Physiological Compound Protecting by Yet Puzzling Mechanism(s) Against Ischaemia-Reperfusion Injury: Current Knowledge. *Br. J. Pharmacol.* **2012**, *165*, 2059–2072. [CrossRef]
133. Razak, M.A.; Begum, P.S.; Viswanath, B.; Rajagopal, S. Multifarious Beneficial Effect of Nonessential Amino Acid, Glycine: A Review. *Oxid. Med. Cell Longev.* **2017**, *2017*, 1716701. [CrossRef]
134. Zhong, Z.; Wheeler, M.D.; Li, X.; Froh, M.; Schemmer, P.; Yin, M.; Bunzendaul, H.; Bradford, B.; Lemasters, J.J. L-Glycine: A Novel Anti-Inflammatory, Immunomodulatory, and Cytoprotective Agent. *Curr. Opin. Clin. Nutr. Metab. Care* **2003**, *6*, 229–240. [CrossRef] [PubMed]
135. Holeček, M. Histidine in Health and Disease: Metabolism, Physiological Importance, and Use as a Supplement. *Nutrients* **2020**, *12*, 848. [CrossRef] [PubMed]
136. Canfield, C.-A.; Bradshaw, P.C. Amino acids in the regulation of aging and aging-related diseases. *Transl. Med. Aging* **2019**, *3*, 70–89. [CrossRef]
137. Gupta, M.N.; Uversky, V.N. Biological Importance of Arginine: A Comprehensive Review of the Roles in Structure, Disorder, and Functionality of Peptides and Proteins. *Int. J. Biol. Macromol.* **2024**, *257 Pt 1*, 128646. [CrossRef]
138. Kurhaluk, N. The Effectiveness of L-arginine in Clinical Conditions Associated with Hypoxia. *Int. J. Mol. Sci.* **2023**, *24*, 8205. [CrossRef]
139. Feng, L.; Peng, Y.; Wu, P.; Hu, K.; Jiang, W.D.; Liu, Y.; Jiang, J.; Li, S.H.; Zhou, X.Q. Threonine Affects Intestinal Function, Protein Synthesis and Gene Expression of TOR in Jian Carp (*Cyprinus carpio* var. Jian). *PLoS ONE* **2013**, *8*, e69974. [CrossRef]
140. Holeček, M. Origin and Roles of Alanine and Glutamine in Gluconeogenesis in the Liver, Kidneys, and Small Intestine under Physiological and Pathological Conditions. *Int. J. Mol. Sci.* **2024**, *25*, 7037. [CrossRef]
141. Jongkees, B.J.; Hommel, B.; Kühn, S.; Colzato, L.S. Effect of Tyrosine Supplementation on Clinical and Healthy Populations under Stress or Cognitive Demands—A Review. *J. Psychiatr. Res.* **2015**, *70*, 50–57. [CrossRef]
142. Egbujor, M.C.; Olaniyan, O.T.; Emeruwa, C.N.; Saha, S.; Saso, L.; Tucci, P. An Insight into the Role of Amino Acids as Antioxidants via NRF2 Activation. *Amino Acids* **2024**, *56*, 23. [CrossRef]
143. Gorissen, S.H.M.; Phillips, S.M. Branched-Chain Amino Acids (Leucine, Isoleucine, and Valine) and Skeletal Muscle. In *Nutrition and Skeletal Muscle*; Elsevier: Amsterdam, The Netherlands, 2019; pp. 283–298. [CrossRef]
144. Li, Z.; Wang, F.; Liang, B.; Su, Y.; Sun, S.; Xia, S.; Shao, J.; Zhang, Z.; Hong, M.; Zhang, F.; et al. Methionine Metabolism in Chronic Liver Diseases: An Update on Molecular Mechanism and Therapeutic Implication. *Signal Transduct. Target Ther.* **2020**, *5*, 280. [CrossRef]
145. Derouiche, F.; Djemil, R.; Sebihi, F.Z.; Douaouya, L.; Maamar, H.; Benjemana, K. High Methionine Diet Mediated Oxidative Stress and Proteasome Impairment Causes Toxicity in Liver. *Sci. Rep.* **2024**, *14*, 5555. [CrossRef]
146. Xiao, C.W.; Hendry, A.; Kenney, L.; Bertinato, J. L-Lysine Supplementation Affects Dietary Protein Quality and Growth and Serum Amino Acid Concentrations in Rats. *Sci. Rep.* **2023**, *13*, 19943. [CrossRef]

147. Huang, D.; Maulu, S.; Ren, M.; Liang, H.; Ge, X.; Ji, K.; Yu, H. Dietary Lysine Levels Improved Antioxidant Capacity and Immunity via the TOR and p38 MAPK Signaling Pathways in Grass Carp, *Ctenopharyngodon idellus* Fry. *Front. Immunol.* **2021**, *12*, 635015. [CrossRef]
148. Doi, M.; Yamaoka, I.; Nakayama, M.; Mochizuki, S.; Sugahara, K.; Yoshizawa, F. Isoleucine, a Blood Glucose-Lowering Amino Acid, Increases Glucose Uptake in Rat Skeletal Muscle in the Absence of Increases in AMP-Activated Protein Kinase Activity. *J. Nutr.* **2005**, *135*, 2103–2108. [CrossRef]
149. Ely, I.A.; Phillips, B.E.; Smith, K.; Wilkinson, D.J.; Piasecki, M.; Breen, L.; Larsen, M.S.; Atherton, P.J. A Focus on Leucine in the Nutritional Regulation of Human Skeletal Muscle Metabolism in Ageing, Exercise, and Unloading States. *Clin. Nutr.* **2023**, *42*, 1849–1865. [CrossRef]
150. Fernstrom, J.D.; Fernstrom, M.H. Tyrosine, Phenylalanine, and Catecholamine Synthesis and Function in the Brain. *J. Nutr.* **2007**, *137*, 1539S–1547S. [CrossRef]
151. Vettore, L.A.; Westbrook, R.L.; Tennant, D.A. Proline Metabolism and Redox; Maintaining a Balance in Health and Disease. *Amino Acids* **2021**, *53*, 1779–1788. [CrossRef]
152. Richard, D.M.; Dawes, M.A.; Mathias, C.W.; Acheson, A.; Hill-Kapturczak, N.; Dougherty, D.M. L-Tryptophan: Basic Metabolic Functions, Behavioral Research, and Therapeutic Indications. *Int. J. Tryptophan Res.* **2009**, *2*, 45–60. [CrossRef]
153. Kikuchi, A.M.; Tanabe, A.; Iwahori, Y. A Systematic Review of the Effect of L-Tryptophan Supplementation on Mood and Emotional Functioning. *J. Diet. Suppl.* **2021**, *18*, 316–333. [CrossRef]
154. Bon, L.I.; Maksimovich, N.Y.; Burak, I.N. Amino Acids that Play an Important Role in the Functioning of the Nervous System: A Review. *Clin. Trials Clin. Res.* **2023**, *2*, 3. [CrossRef]
155. Aydin, S.; Dagli, A.F.; Ozkan, Y.; Kendir, Y.; Sahin, I.; Aksoy, A.; Ozercan, I.H. Immunohistochemical and Quantitative Analysis of Ghrelin in *Syzygium aromaticum*. *Cell Biol. Int.* **2011**, *35*, 437–441. [CrossRef]
156. Sehnal, D.; Rose, A.S.; Koca, J.; Burley, S.K.; Velankar, S. Mol*: Towards a Common Library and Tools for Web Molecular Graphics. In *Workshop on Molecular Graphics and Visual Analysis of Molecular Data*; Byška, J., Krone, M., Sommer, B., Eds.; The Eurographics Association: Limassol, Cyprus, 2018. [CrossRef]
157. Jeon, S.G.; Hong, S.B.; Nam, Y.; Tae, J.; Yoo, A.; Song, E.J.; Kim, K.I.; Lee, D.; Park, J.; Lee, S.M.; et al. Ghrelin in Alzheimer's Disease: Pathologic Roles and Therapeutic Implications. *Ageing Res. Rev.* **2019**, *55*, 100945. [CrossRef]

biomolecules

MDPI

Review

Deciphering the Role of Adrenergic Receptors in Alzheimer's Disease: Paving the Way for Innovative Therapies

Androulla N. Miliotou [1], Andria Kotsoni [1] and Lefteris C. Zacharia [1,2,*]

[1] Department of Health Sciences, School of Life and Health Sciences, University of Nicosia, 46 Makedonitissas Avenue, 2417 Nicosia, Cyprus; miliotou.a@unic.ac.cy (A.N.M.)

[2] Bioactive Molecules Research Center, School of Life and Health Sciences, University of Nicosia, 46 Makedonitissas Avenue, 2417 Nicosia, Cyprus

* Correspondence: zacharia.l@unic.ac.cy

Abstract: Neurodegenerative diseases are currently among the most devastating diseases with no effective disease-modifying drugs in the market, with Alzheimer's disease (AD) being the most prevalent. AD is a complex multifactorial neurodegenerative disorder characterized by progressive and severe cognitive impairment and memory loss. It is the most common cause of progressive memory loss (dementia) in the elderly, and to date, there is no effective treatment to cure or slow disease progression substantially. The role of adrenergic receptors in the pathogenesis of Alzheimer's disease and other tauopathies is poorly understood or investigated. Recently, some studies indicated a potential benefit of drugs acting on the adrenergic receptors for AD and dementias, although due to the heterogeneity of the drug classes used, the results on the whole remain inconclusive. The scope of this review article is to comprehensively review the literature on the possible role of adrenergic receptors in neurodegenerative diseases, stemming from the use of agonists and antagonists including antihypertensive and asthma drugs acting on the adrenergic receptors, but also from animal models and in vitro models where these receptors have been studied. Ultimately, we hope to obtain a better understanding of the role of these receptors, identify the gaps in knowledge, and explore the possibility of repurposing such drugs for AD, given their long history of use and safety.

Keywords: adrenergic receptors; Alzheimer's disease; neurodegeneration; neuroinflammation; tauopathies; beta amyloid; adrenergic agonists; adrenergic antagonists

1. Introduction

The most widespread and common type of dementia is Alzheimer's disease (AD), which poses a significant challenge to global health and inflicts devastating consequences on patients, caregivers, and health systems. It is a progressive illness that leads to cognitive impairment, loss of memory, and inability to perform even simple daily living tasks in the end stages [1,2]. AD affects millions, and that number is expected to increase substantially owing to the aging population [3]. Despite the heavy research on AD, no disease-modifying treatments are available that could alter the course of AD. Current treatment options are largely symptomatic in nature and result in slight and short-lived benefits in cognitive function without modifying the mechanisms responsible for neurodegeneration [4].

AD is a complex disorder that is associated with multiple pathological features such as amyloid-beta (Aβ) accumulation, hyperphosphorylation of tau protein, neuroinflammation, loss of synapse connectivity and cardio-vascular abnormalities [5–7]. All of these processes are interrelated, which eventually result in a loss of neurons. Recently, research has started focusing on the noradrenergic system, more specifically on adrenergic receptors (ARs), and their role in the development of Alzheimer's disease as a significant contributing factor [8–14]. Adrenergic receptors are GPCRs whose primary function is to respond to norepinephrine (NE) or epinephrine in the brain where such cognitive, vasculature and inflammatory activities are regulated. Such dysregulation, with subsequent loss of the non adrenergic system and innervation, is caused by the early degeneration of the locus coeruleus (LC), which is one of the few regions that produces NE and is affected in AD [11,15–17].

Adrenergic receptors, divided into α and β, each have a different function and signaling pathways. Such receptors are present in most parts of the Central Nervous System (CNS) and are important in the control of synaptic plasticity, neuroinflammation, neurovascular coupling and various metabolic activities. Degeneration of the LC also occurs in AD, leading to a dramatic hypofunction of NE signaling, whose neural effects are likely to potentiate the toxicity of Aβ, promote the accumulation of tau, and damage neurons. What is more, adrenergic receptors can both protect from AD and increase its risk, which is dependent on the subtype of the receptor, its activated form, and the place where it is situated [18,19]. This presents opportunities for drug design but also points to problems associated with using them effectively in practice.

Given the extensive expression of adrenergic receptors and their multifunctional roles, they are considered promising targets for drug development for AD. There are many drugs targeting adrenergic receptors that are currently being used in clinical practice for hypertension, bronchial asthma, and some psychiatric disorders. This is advantageous since it allows for the repurposing of such drugs for the treatment of AD without extensive clinical trials. However, whether they can be beneficial is still under debate, as in some cases they seem to be beneficial, but not always. Thus, potential modulation of these receptors should be carried out in a very specific manner based on receptor subtypes, brain regions, and disease stage.

However, whether activation or blockade of the adrenergic receptors is beneficial is not clear. This review evaluates the evidence regarding the role of adrenergic receptors in AD in a consolidated and comprehensive manner, drawing on insights from preclinical models, genetic studies, and clinical trials. An overview of the role of the dual role of adrenergic receptors is found in Figure 1. By synthesizing these findings, the review aims to elucidate the critical importance of adrenergic receptor modulation in the pathophysiology of AD and its potential as a therapeutic target. In particular, it aims to delineate whether evidence points to whether activation or inhibition of adrenergic receptors is likely protective, and the possibility that such drugs can be designed or existing ones repurposed. In doing so, it will explore how adrenergic signaling influences key pathological processes, including Aβ accumulation, tau hyperphosphorylation, neuroinflammation, and synaptic dysfunction, while also addressing the dual protective and pathological roles of these receptors. The overarching goal is to pave the way for innovative, mechanism-driven therapies that can harness the therapeutic potential of adrenergic receptor modulation, while mitigating its detrimental effects, to provide transformative solutions for patients suffering from AD and related neurodegenerative disorders.

1.1. α-Adrenergic Receptors

Among the adrenergic receptor subtypes, α-ARs stand out for their nuanced roles in neurodegeneration, particularly in AD and related tauopathies. This is a group of G-protein-coupled receptors which are classified into α1 and α2 adrenergic receptors. They are naturally activated by norepinephrine (NE) and are important for synaptic plasticity, neurovascular activities, and inflammation [20]. Due to their presence in dense innervation and modulating effect on Aβ processing and tau pathology, they are positioned at the center of neurodegeneration associated with Alzheimer's disease pathology.

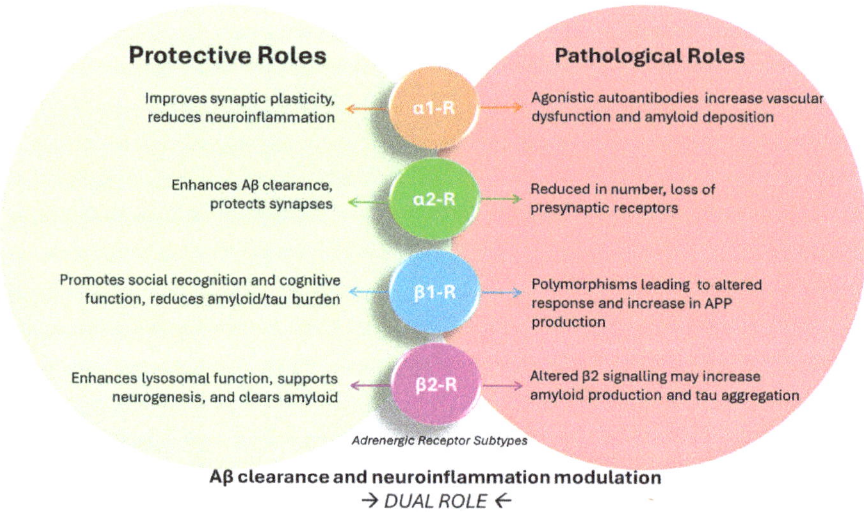

Figure 1. A visual summary of the dual roles of adrenergic receptors in Alzheimer's disease.

The α-ARs therapeutic efficacy is complicated by their dualism. α1-ARs have the most pronounced effect and have shown to be neuroprotective in preclinical studies, where activation has been associated with enhancement of synaptic function, reduced neuroinflammation, and enhanced cognitive outcomes [9]. Likewise, selective α2-AR agonists decrease the amyloidogenic processes while increasing the neurogenic processes, both very useful in terms of treatment. Still, those receptors pose some issues. In some patients with AD, there are circulating agonistic α1-AR autoantibodies that cause chronic overstimulation of those receptors, which leads to neurotoxic and vasculotoxic effects [21]. With respect to α2-ARs, however, the side effects may include the promotion of amyloid production and pathological amyloid precursor protein (APP) activation upon receptor stimulation [22].

In recent years, studies on genetic, molecular, and pharmacological aspects of α-ARs advances have greatly enhanced understanding of their roles in the pathophysiology of AD. This section examines α-ARs in AD and discusses their positive and negative effects, addressing how they can mediate the control of dementia.

1.2. Positive Roles of α-Adrenergic Receptors in Alzheimer's Disease and Other Neurodegenerative Disorders

1.2.1. α1-Adrenergic Receptors

The α1 adrenergic receptors (α1-ARs) have emerged as a focal point of interest in the context of AD and other neurodegenerative disorders due to their potential to influence key mechanisms underlying neuronal function and pathology. The therapeutic potential of these receptors is underscored by studies demonstrating both the protective effects of their activation and the benefits of selective inhibition. These seemingly contradictory findings emphasize the complexity of α1-AR signaling and its role in neurodegeneration.

One promising development in this area is the discovery of positive allosteric modulators (PAMs) for α1A-ARs. Notably, Compound 3 (Cmpd-3), a PAM for α1A-ARs, has been shown to rescue LTP defects and normalize amyloid β (Aβ)-40 and –42 levels in AD mouse models. These effects were achieved without impacting blood pressure, which is a common concern in adrenergic receptor-targeted therapies. Furthermore, oral administration of Cmpd-3 at doses ranging from 3 to 9 mg/kg once daily for three months led to significant cognitive improvements that outperformed donepezil, which is the standard acetyl cholinesterase inhibitor treatment for AD. These findings highlight the potential for PAM like Cmpd-3 as a disease-modifying agent with robust therapeutic efficacy [23]. Furthermore, mice with constitutively active α_{1A}AR showed enhancements in learning and memory compared to α_{1A}AR knockout. WT mice treated with the α_{1A}AR-selective agonist cirazoline also showed enhanced cognitive functions, suggesting that long-term α_{1A}AR stimulation improves synaptic plasticity, cognitive function, mood, and longevity [24]. Further supporting the neuroprotective role of α1-ARs, avenanthramide-C, a compound derived from oats, has demonstrated the ability to reverse memory impairments in AD models Tg2576 and 5XFAD. This compound improves recognition and spatial memory and reduces neuroinflammation, effects that are mediated by its interaction with α1A-ARs. The beneficial effects of avenanthramide-C were abolished by prazosin, a specific α1A-AR inhibitor, confirming the critical role of these receptors in protecting [25]. Interestingly, studies on α1B-AR knockout mice have highlighted the critical role of this receptor subtype in cognitive processes. Mice deficient in α1B-ARs exhibited significant impairments in memory consolidation and exploratory behavior, indicating that α1B-ARs may support cognitive function when activated [26]. These findings suggest that modulation of specific α1-AR subtypes could yield differential therapeutic outcomes depending on the context of receptor activity and the disease state.

Despite these data, pharmacological inhibition of α1-ARs has also revealed neuroprotective effects in certain contexts. Doxazosin, an α1-adrenergic blocker, has been shown to protect hippocampal slices from amyloid-β toxicity by preventing glycogen synthase kinase-3β (GSK-3β) activation and tau hyperphosphorylation [27]. Similarly, prazosin reduced Aβ generation, induced anti-inflammatory responses, and prevented memory deficits in transgenic APP23 mice despite not affecting amyloid plaque load. These findings underscore the anti-inflammatory and neuroprotective benefits of α1-AR antagonists in AD [28]. Another α1-AR antagonist, terazosin, has demonstrated a novel mechanism for mitigating neurodegenerative pathology. By increasing ATP levels and enhancing autophagy, terazosin reduced pathological protein aggregates in AD and other neurodegenerative disease models, highlighting its potential as a treatment strategy [29]. Terazosin was also shown to reduce amyloid plaque burden, tau hyperphosphorylation, and glial activation. Behavioral deficits in AD models were also significantly improved with terazosin treatment. These findings suggest that selective inhibition of α1-ARs may counteract pathological processes underlying AD, offering a potential therapeutic pathway [30].

1.2.2. α2-Adrenergic Receptors

The α2 adrenergic receptors (α2-ARs) have also been implicated in neuroprotection and cognitive enhancement in AD and other neurodegenerative disorders with mixed results, as both activating and blockading have been shown to be neuroprotective, with most studies favoring blockade rather than activation for protection. Genetic studies have provided further evidence for the protective role of α2-ARs. A deletion variant of the α2b-adrenergic receptor has been associated with enhanced memory formation and a reduced risk of developing AD in 311 Greek subjects. This variant was more prevalent in control subjects compared to individuals with AD or mild cognitive impairment, suggesting its potential as a protective genetic factor [31].

Pharmacological modulation of α2C-ARs (α2 subtype C) has shown promise. ORM-10921, a selective α2C-AR antagonist, improved cognitive function (measured by water maze) and alleviated symptoms in CNS disorders, including AD. Its high selectivity for the α2C-AR subtype supports its potential as a targeted therapeutic for neurodegenerative diseases [32]. In addition to these pharmacological findings, the α2-adrenoceptor antagonist dexefaroxan has been found to enhance neuron survival, offering a novel approach for preserving cognitive function in neurodegenerative conditions. Dexefaroxan also reduced cholinergic degeneration and improved memory performance in rodent models, further supporting its therapeutic potential [33–35]. Another α2-adrenoblocker, mesedin, has demonstrated multiple neuroprotective effects in vivo, including anti-amyloidogenic action, increasing choline acetyltransferase levels, enhancing Aβ degradation, and reducing neuroinflammation. These findings highlight the multifaceted role of α2-AR modulation in mitigating AD pathology [36].

However, the α2A adrenergic receptor has also been implicated in the enhancement of Aβ generation, linking noradrenergic dysregulation to AD progression. Activation of α2A-ARs disrupts the interaction between APP and sorting-related receptors with A repeat (SorLA), a critical regulator of APP trafficking. This disruption facilitates the redistribution of APP to endosomes, where it is cleaved by β-secretase, leading to increased Aβ production. Blockade of the α2AR with Idazoxan reduced AD-related pathology and ameliorated cognitive deficits in an AD transgenic mode [37].

Agonists such as brimonidine and clonidine have demonstrated substantial neuroprotective effects, particularly by reducing retinal ganglion cell apoptosis. This effect is achieved via reduction in Aβ and APP processing, emphasizing the therapeutic relevance of α2-ARs in conditions characterized by amyloid pathology [38].

1.3. Negative Roles of α-Adrenergic Receptors in Alzheimer's Disease and Other Neurodegenerative Disorders

1.3.1. α1-Adrenergic Receptors

Despite the above evidence, the α1 adrenergic receptors (α1-ARs) have also been associated with several adverse outcomes in AD and other neurodegenerative disorders, underscoring their complex role in disease pathology. Research indicates that their persistent activation and the presence of autoantibodies targeting these receptors contribute significantly to vascular and neuronal dysfunctions.

Agonistic autoantibodies (agAABs) against α1-ARs have been identified in approximately 50% of AD patients. These autoantibodies mimic natural agonists, binding persistently to the receptors and causing prolonged activation. This aberrant signaling leads to non-physiological intracellular calcium elevations, triggering macrovascular and microvascular impairments. In animal models, the presence of these autoantibodies has been linked to significant reductions in blood flow and vessel density, impairing cerebral perfusion and contributing to neurodegeneration. Immunoadsorption has emerged as a potential strategy to mitigate these effects. In a study involving AD patients, the removal of α1-AR autoantibodies through immunoadsorption stabilized cognitive function over a follow-up period of 12–18 months, highlighting the therapeutic potential of this approach [39,40]. Further investigations revealed that 59% of patients with mild to moderate AD and vascular dementia harbor autoantibodies targeting both α1- and β2-adrenergic receptors. These autoantibodies preferentially bind to the extracellular loop of α1-ARs, mobilizing intracellular calcium and exacerbating vascular lesions. Such processes are believed to facilitate amyloid plaque formation and vascular lesions [21].

In addition, the detrimental effects of α1-AR activation extend to cerebral vasculature. Neuroimaging studies on rats demonstrated significant reductions in relative cerebral blood volume (rCBV) within the cerebrum, cortex, and hippocampus due to α1-AR autoantibody activity. These findings correlate with impaired neurovascular function, a hallmark of AD pathology [41]. The IMAD pilot trial (efficacy of immunoadsorption for treatment of Alzheimer's disease and agonistic autoantibodies against α1-AR) investigates the potential of immunoadsorption to remove α1-AR autoantibodies and mitigate their harmful effects as suggested by previously mentioned studies on autoantibodies [42].

Furthermore, Aβ peptides have been found to directly activate α1-ARs, contributing to vascular dysfunction. This interaction induces a positive chronotropic effect in cardiac assays and triggers intracellular calcium release in vascular smooth muscle cells. Such effects are attenuated by α1-AR blockers, including prazosin, emphasizing the receptor's involvement in Aβ-induced vascular pathology [43]. This highlights the detrimental role of α1-AR activation in AD-associated cardiovascular and cerebrovascular abnormalities.

The pathological effects of α1-ARs are not limited to vascular changes but extend to behavioral outcomes. Aggressive behavior in AD patients has been associated with upregulation of α1-ARs. Postmortem studies revealed that increased α1 receptors were correlated with aggressiveness in AD patients, implicating heightened α1-AR activity in behavioral dysregulation [44].

In conclusion, α1-ARs play a multifaceted role in AD and other neurodegenerative disorders, with their pathological activation contributing to vascular dysfunction, amyloid pathology, neuroinflammation, and behavioral changes. The presence of agonistic autoantibodies further exacerbates these effects, while receptor inhibition or immunoadsorption offers potential therapeutic benefits. Continued research into α1-AR modulation and the development of targeted interventions are crucial for addressing the complex pathophysiology of AD and improving patient outcomes.

1.3.2. α2-Adrenergic Receptors

The negative roles of α2AR in neurodegenerative diseases are much less implicated and not widely supported by the literature. In AD brains, a significant reduction in α2-AR density has been observed in the prefrontal cortex, with decreases of approximately 50% compared to age-matched controls. This reduction is attributed to a loss of presynaptic receptors on noradrenergic synapses rather than changes in ligand affinity, highlighting the degeneration of neurons. These findings suggest that α2-AR deficits may contribute to the cognitive and behavioral impairments associated with AD [45,46].

1.4. β-Adrenergic Receptors

β-ARs represent a distinct and significant component of the adrenergic receptor family, with notable roles in AD pathophysiology. β-ARs are primarily associated with their ability to regulate intracellular signaling pathways that influence neuroinflammation, synaptic plasticity, and energy metabolism. Comprised of three subtypes—β1, β2, and β3—these G-protein-coupled receptors are activated by NE and epinephrine and are broadly distributed in the brain. They interact with cAMP-mediated pathways, making them key modulators of cellular functions that are directly implicated in AD pathology [47].

This section examines the multifaceted roles of β-ARs in AD, delineating their contributions to both disease progression and potential therapeutic interventions. By exploring the intricate mechanisms by which β-ARs influence AD pathology, this discussion aims to clarify their distinct functions relative to α-adrenergic receptors and highlight their potential as targets for precision medicine in neurodegenerative disorders.

1.5. Positive Roles of β-Adrenergic Receptors in Alzheimer's Disease and Other Neurodegenerative Disorders

1.5.1. β1 Adrenergic Receptors

The β1-ARs have emerged as a promising therapeutic target in AD and other neurodegenerative disorders due to their role in modulating cognitive functions, including memory and social learning. Importantly all studies indicate that the protective effects are mediated by activating the β1 receptor. One critical area of β1-AR activity is in the medial amygdala, where these receptors facilitate the learning and processing of social cues. Dysfunction in social recognition has been effectively addressed in preclinical studies through selective activation of β1-ARs. Xamoterol, a selective partial β1-AR agonist, was shown to rescue social recognition deficits in an AD mouse model (APP) by activating the protein kinase A (PKA)/phosphorylated cAMP-response element-binding protein (phospho-CREB) signaling cascade. This pathway is essential for synaptic plasticity and memory formation, highlighting the potential of β1-AR modulation to restore cognitive abilities specifically related to social interaction in AD [48]. Chronic administration of xamoterol has demonstrated significant reductions in key pathological features of AD, including Aβ and tau pathology. In the 5XFAD mouse model, a widely used AD model characterized by robust Aβ and tau accumulation, xamoterol treatment reduced neuroinflammation markers. It also lowered mRNA expression levels of inflammatory mediators, indicating its potential to modulate both systemic and localized neuroinflammatory processes. The impact of β1-AR activation on AD pathology is further underscored by xamoterol's ability to attenuate Aβ plaque deposition and tau hyperphosphorylation in vivo [49]. Additionally, chronic nebivolol treatment (a selective β1 adrenergic receptor antagonist) on Tg2576 mice with established amyloid neuropathology and cognitive impairments significantly reduced brain amyloid content but failed to improve cognitive function [50]. These outcomes suggest that β1-AR modulation not only protects against neuronal loss and cognitive decline but also addresses the underlying pathological mechanisms driving AD progression. The therapeutic potential of β1-ARs has also been studied with another selective partial agonist, STD-101-D1. This compound demonstrates selective partial agonistic activity on G-protein signaling with an EC50 value in the low nanomolar range, indicating high efficacy and potency. Functionally selective agonists like STD-101-D1 allow for targeted modulation of β1-AR activity, minimizing side effects associated with full receptor activation. STD-101-D1 has shown dual benefits in neuroprotection and inflammation reduction, both of which are critical in addressing AD pathology. In vitro and in vivo studies have demonstrated that this compound effectively inhibits the tumor necrosis factor-α (TNF-α) response triggered by lipopolysaccharide (LPS), a known inducer of neuroinflammation. Neuroinflammation, a significant contributor to synaptic dysfunction and neuronal loss in AD, is thus effectively mitigated by STD-101-D1, making it a strong candidate for further therapeutic development [51]. The ability of β1-AR-targeted therapies to modulate neuroinflammatory pathways positions these receptors as key targets not only for AD but also for other neurodegenerative and neuroinflammatory disorders. The neuroprotective effects of β1-AR modulation are closely tied to their influence on intracellular signaling cascades. Activation of β1-ARs leads to increased cAMP production, which activates PKA and downstream signaling molecules such as CREB. This cascade promotes the transcription of neuroprotective genes, including those involved in synaptic repair and neuronal survival. Furthermore, β1-ARs appear to regulate neuroinflammation through inhibition of pro-inflammatory cytokines like TNF-α, providing a dual mechanism of action that addresses both neuronal and immune components of neurodegeneration.

1.5.2. β2-Adrenergic Receptors

The β2-ARs have emerged as crucial modulators in the context of AD and other neurodegenerative conditions. By influencing cellular, immune, and synaptic processes, β2-AR activation offers a multifaceted approach to mitigating AD pathology and preserving cognitive function. Their ability to influence cellular clearance processes, immune regulation, and synaptic function positions them as promising therapeutic targets. Central to this is their role in restoring lysosomal function and autophagy, both of which are crucial in mitigating AD pathology. In particular, β2-AR agonists such as isoproterenol have been shown to reacidify lysosomes in presenilin-1 (PSEN1) KO fibroblasts from PEN1 familial AD patients, which restores lysosomal proteolysis, calcium homeostasis and normal autophagy flux [52]. Building on these cellular benefits, β2-ARs also play a significant role in modulating the brain's immune response. Microglia, the brain's immune cells, are highly responsive to β2-AR signaling, which protects them from Aβ-induced inflammation. This anti-inflammatory effect has been observed in both pharmacological treatments using β-adrenergic agonists and lifestyle interventions like environmental enrichment. Mice exposed to these interventions showed reduced microglial activation and a decline in pro-inflammatory cytokine levels, highlighting the ability of β2-AR signaling to mitigate harmful neuroinflammatory responses that exacerbate AD pathology [53].

Activation of β2-ARs significantly contributes to preserving synaptic health by preventing LTP inhibition by Aβ protein. Through the cAMP/PKA signaling pathway, activation of β2-ARs prevents Aβ oligomer-induced inhibition of LTP, a key mechanism underlying learning and memory. This synaptic protection underscores the potential of β2-AR-targeted therapies in maintaining cognitive function and lessening the effects of Aβ accumulation during aging [54]. β2-ARs are particularly significant in the context of AD. These receptors regulate lysosomal function and autophagy, which are essential for the clearance of Aβ, a hallmark of AD. Studies demonstrate that β2-AR activation enhances autophagy by restoring lysosomal acidity and promoting the degradation of Aβ aggregates. Furthermore, β2-AR stimulation protects against Aβ-induced synaptotoxicity by supporting synaptic integrity and plasticity through downstream signaling pathways, such as the cAMP/PKA pathway. β1-ARs are also critical in AD, with their activation enhancing learning and memory processes via modulation of LTP and synaptic plasticity [55].

These neuroprotective effects extend beyond synaptic plasticity to include reductions in amyloid pathology and neuritic damage. Chronic stimulation of β2-ARs in preclinical models, such as the 5xFAD mouse model of AD, has resulted in lower amyloid plaque burdens and improved neuritic integrity, suggesting that microglial β2-AR signaling not only limits amyloid deposition but also preserves neurons [10]. The evidence supporting β2-ARs' protective role is further strengthened by epidemiological studies. Retrospective analyses have suggested that individuals exposed to β2-AR agonists have a reduced risk of developing AD, with hazard ratios indicating a protective effect, while an increased risk was observed with patients receiving non selective β2 AR agonists. These findings align with preclinical data, supporting the idea that β2-AR activation can delay or prevent the onset of neurodegeneration [56]. Moreover, β2-ARs contribute to promoting neurogenesis and synaptic health, key factors in cognitive resilience. In APP/PS1 transgenic mice, β2-AR agonists such as clenbuterol enhanced neurogenesis, increased dendritic branching, and upregulated synaptic protein expression. These structural and functional improvements were accompanied by reductions in amyloid plaques, reduced APP phosphorylation, and amelioration of memory deficits, reinforcing the therapeutic promise of β2-AR-targeted treatments [57,58]. In addition to its effects on amyloid pathology, adrenergic signaling offers promise in targeting tau pathology. β-AR agonists such as salbutamol have been identified as potential inhibitors of tau aggregation. Salbutamol reduces tau filament formation and prevents the structural transition of tau into β-sheet-rich aggregates. This novel approach suggests that adrenergic signaling could address both major pathological pathways in AD, amyloid and tau, through distinct mechanisms [59].

All these positive effects of the β2 adrenergic receptor are corroborated by the fact that deletion of the β2-AR gene ameliorates pathological effects in senile PS1/APP mice, indicating that β2AR may represent a potential therapeutic target for preventing the progression of AD [60].

Adding to the therapeutic potential, non-pharmacological interventions such as aerobic exercise also leverage β2-AR activation to deliver benefits in AD. Aerobic exercise has been shown to reverse autophagy-lysosomal deficits and attenuate amyloid pathology through the upregulation of the AMPK-MTOR signaling pathway and VMA21 levels. These molecular changes enhance the clearance of Aβ and improve cognitive outcomes, illustrating the synergy between lifestyle interventions and β2-AR activation in managing AD [61].

1.6. Adrenergic Signaling, Beyond the Activity of Specific Receptor Subtypes

Adrenergic signaling, beyond the activity of specific receptor subtypes, has been increasingly associated with beneficial outcomes in neurodegenerative diseases, particularly AD. Activation of adrenergic pathways plays a critical role in preserving synaptic function and enhancing neuroprotective mechanisms, making it a promising area for therapeutic intervention. One key aspect of adrenergic signaling is its impact on synaptic plasticity, a process that is often impaired in AD. β-AR activation has been shown to enhance LTP, a mechanism underlying memory and learning. In the TgF344-AD rat model, the loss of LC-NA axons is compensated by increased β-AR, and this heightened β-AR function resulted in increased LTP magnitude and preserved learning and memory abilities. These findings suggest that β-AR activation could serve as a compensatory mechanism to counteract the synaptic dysfunction characteristic of AD [62]. Adding to this, vagus nerve stimulation, which indirectly activates β-ARs, has been found to modulate hippocampal synaptic transmission, offering potential cognitive benefits. This stimulation enhances noradrenaline release, leading to increased β-AR activation and improved synaptic transmission in the CA3 region of the hippocampus as the excitatory effect was diminished by the β-AR antagonist timolol. The ability to modulate hippocampal activity highlights the therapeutic potential of interventions that engage adrenergic pathways in addressing memory and learning deficits in AD [63]. Furthermore, β-adrenergic receptor activation plays a crucial role in mitigating Aβ-mediated synaptic impairment. Activation of both β1 and β2 adrenergic receptors by isoproterenol has been shown to prevent the inhibition of mossy fiber (axons) LTP caused by Aβ oligomers in mice. This neuroprotective effect underscores the potential of targeting specific β-AR subtypes to preserve synaptic function and combat Aβ-mediated damage in AD models [63].

Adrenergic signaling also extends its protective influence on microglial activity and amyloid clearance. Noradrenaline, acting through β1 and β2-ARs, has been demonstrated to prevent Aβ toxicity by stimulating NGF and BDNF in human neuronal cultures and primary rat hippocampal neurons [64,65]. Specific microglial β2-AR deletion worsened AD pathology in the 5xFAD mouse model, while chronic β2-AR stimulation resulted in attenuation of amyloid pathology and associated neuritic damage, suggesting microglial β2-AR might be used as a potential therapeutic target to modify AD pathology [10,66]. Additionally, NE can enhance amyloid clearance in murine microglia cell line N9 by involving upregulation of amyloid receptor expression and increased degradation of Aβ, driven by β2-AR activation. These findings highlight the dual role of adrenergic signaling in both protecting neurons from Aβ toxicity and reducing amyloid burden, addressing two critical aspects of AD pathology [64]. The anti-inflammatory properties of adrenergic signaling further strengthen its therapeutic potential. Studies on human THP-1 macrophages revealed that norepinephrine reduces Aβ-induced cytotoxicity and modulates cytokine secretion. This effect is mediated through β-AR activation, as the effects of NE were replicated by isoproterenol and blocked by propranolol. The effect is triggered by the cAMP/PKA signaling pathway and promotes CREB phosphorylation. By modulating immune responses, adrenergic signaling reduces inflammation and supports a more neuroprotective environment, further contributing to its benefits in AD [67].

Adrenergic signaling also plays a critical role in maintaining energy homeostasis in the brain, an often-overlooked aspect of neurodegenerative disease pathology. Noradrenaline, acting via β1-ARs, has been shown to regulate glycogen synthesis in astrocytes. This involves increased expression of protein targeting to glycogen (PTG) mRNA and enhanced glycogen synthesis, suggesting a role for adrenergic signaling in supporting metabolic demands and protecting neuronal health in energy-deficient states such as AD [68].

1.7. Negative Roles of β-Adrenergic Receptors in Alzheimer's Disease and Other Neurodegenerative Disorders

1.7.1. β1-Adrenergic Receptors

The β1-ARs, while often associated with therapeutic potential, have also been implicated in negative outcomes related to AD and other neurodegenerative disorders; however, with much less evidence stemming mainly from genetic studies. Polymorphisms in the β1-AR gene (*ADRB1*) and the G protein beta3 subunit (*GNB3*) gene have been identified as risk factors for AD. These genetic variations alter cell responsiveness to adrenergic stimulation, as evidenced by cAMP levels and mitogen-activated protein kinase (MAPK) activation and increased APP production in transfected human cell lines. These signaling alterations in specific polymorphisms may amplify pathological processes in AD, highlighting the genetic interplay between adrenergic receptor signaling and AD risk [69].

1.7.2. β2-Adrenergic Receptors

The β2-ARs, while often linked to protective roles, have also been associated with several negative effects in AD. These receptors are deeply involved in synaptic function and neuronal signaling, but their dysregulation can contribute to the progression of AD through multiple mechanisms, including Aβ and tau pathology, synaptic dysfunction, and genetic predispositions. One key negative effect of β2-ARs in AD is their vulnerability to Aβ-induced internalization and degradation. Aβ peptides bind directly to β2-ARs, triggering their internalization and subsequent desensitization. This leads to impaired adrenergic and glutamatergic signaling, both of which are critical for synaptic function and plasticity. By attenuating responses to subsequent adrenergic stimulation, Aβ disrupts the homeostasis of neurotransmitter systems vital for learning and memory [70].

In addition to their interaction with Aβ, β2-ARs play a role in exacerbating tau pathology. In tauopathy models, deletion of the β2-AR gene has been shown to reduce tau pathology and improve motor deficits. Mechanistically, this is associated with reduced activity of tau kinases such as glycogen synthase kinase-3β (GSK3β) and cyclin-dependent kinase 5 (CDK5), which are known to phosphorylate tau and drive its pathological aggregation.

These findings suggest that β2-AR signaling may contribute to tau hyperphosphorylation and accumulation, further linking β2-AR activity to neurodegeneration. This of course indicated that blockade of the receptors may be beneficial [71]. Altered β2-AR signaling also disrupts calcium homeostasis and APP processing, leading to additional synaptic dysfunction. Post-translational modifications of ryanodine receptors are regulated by β2-AR activity, resulting in abnormal calcium release that exacerbates βAPP processing. Blocking β2-AR signaling in cells has been shown to mitigate these effects, by reducing APP processing and Aβ production, indicating that β2-AR dysregulation may drive pathological APP cleavage and subsequent Aβ production, and β2 blockade could prove beneficial [72].

Further compounding its role in synaptic dysfunction, β2-AR activation contributes to Aβ-induced hyperactivity of AMPA receptors. This involves PKA-dependent phosphorylation of AMPA receptor subunits, which disrupts normal synaptic activity and exacerbates neuronal excitotoxicity. This effect is mediated by the binding of Aβ to β2-ARs [73]. The genetic component of β2-AR signaling dysfunction has also been implicated in AD. Polymorphisms in the β2-AR gene (Specifically Gly16Arg Gln27Glu) are associated with an increased risk of late-onset AD, particularly in conjunction with the APOE ε4 allele. These genetic interactions suggest that specific β2-AR variants may predispose individuals to AD by altering receptor function or adrenergic responsiveness [74].

The presence of agonistic autoantibodies targeting β2-ARs in AD patients further emphasizes their role in disease progression. These autoantibodies activate β2-ARs, mimicking adrenergic stimulation and potentially causing adrenergic overstimulation. Similar autoantibodies have been identified in other neurodegenerative disorders, such as glaucoma, suggesting a broader pathological mechanism linked to adrenergic overdrive. In addition, long-chain Aß-[Pyr]3-43, representing a major neurogenic plaque component, exerted an activation of the ß2-AR that could be blocked by antagonist ICI118.551 [75].

Another significant mechanism through which β2-ARs contribute to AD pathology is their interaction with the angiotensin II type 1 receptor in the production of Aβ, which seems to be more important than the β1-AR [76]. Activation of β2-ARs has been shown to enhance gamma-secretase activity, a key enzyme in Aβ production, thereby promoting amyloid plaque formation. This was also corroborated in vivo as chronic treatment with β2-AR agonist increased amyloid plaques in a mouse model of AD [77]. Stress-induced activation of β2-ARs further exacerbates this process, as evidenced by increased Aβ production in response to β2-AR agonists. Conversely, β2-AR antagonists have been shown to reduce Aβ production, highlighting a potential therapeutic approach for AD prevention [78,79]. However, β2-AR antagonism is not without risks. Inhibition of β2-ARs with ICI 118,551 has been linked to exacerbated cognitive deficits, tau hyperphosphorylation and amyloid pathology in 3XTg-AD mouse models. Specifically, β2-AR antagonism has been associated with increased Aβ levels and tau phosphorylation, suggesting that the timing and context of β2-AR modulation are critical in determining its therapeutic viability [80].

1.8. Negative Impacts of Adrenergic Signaling on Neurodegeneration

Neuroinflammation is another area where adrenergic signaling may have detrimental effects. While adrenergic pathways often exhibit anti-inflammatory properties, the chronic use of β-blockers in AD models has demonstrated pro-inflammatory effects in certain contexts. In APP mouse models, prolonged β-blocker administration was associated with increased markers of phagocytosis and heightened CNS inflammation. This was accompanied by impaired cognitive behavior, as evidenced by deficits in learning and memory tasks. These findings suggest that inappropriate modulation of adrenergic signaling can potentiate inflammatory processes, complicating efforts to target these pathways in neurodegenerative diseases [81].

The interaction between adrenergic and cholinergic systems also contributes to cognitive decline, particularly in aging and AD. Studies have shown that the combined blockade of muscarinic cholinergic and β-adrenergic receptors exacerbates learning and memory deficits in rodents. For instance, co-administration of scopolamine, a muscarinic cholinergic antagonist, and propranolol, a β-adrenergic antagonist, significantly impaired performance in learning tasks in rat models, while either scopolamine or propranolol alone had no effect. These findings underscore the interdependence of adrenergic and cholinergic systems in maintaining cognitive function and suggest that disruptions in either or both systems can synergistically exacerbate cognitive impairments [82].

2. Discussion

The role of adrenergic receptors in AD has emerged as a crucial yet complex area of investigation. The findings of this review highlight the duality of the adrenergic receptor role, showcasing both their therapeutic potential and their contributions to disease pathology. This nuanced understanding lays the groundwork for innovative therapeutic strategies, but also underscores the challenges inherent in targeting such a multifaceted system.

Adrenergic receptors, α1, α2, and β1 and β2 subtypes, are integral to various neural processes including synaptic plasticity, neurovascular regulation, and inflammation. Dysregulation of adrenergic signaling, often due to early degeneration of the locus coeruleus, is increasingly recognized as a driver of key AD pathologies such as Aβ accumulation, tau hyperphosphorylation, and chronic neuroinflammation. These findings align with the understanding that adrenergic receptors act at the intersection of cognitive, vascular, and immune pathways, making them critical modulators of both protective and pathological processes.

The evidence from the above in vitro and in vivo studies indicates that their role is naturally important; however, pharmacological modulation has not been always consistent as activation or blockade can have the same effect, making selective modulation difficult. Despite this, the evidence summarized in Tables 1 and 2 indicates that the positive effects of adrenergic receptors are derived by pharmacological modulation, highlighting the possibility of being used in the clinic, and Table 3 shows the negative roles. Interestingly, from the in vivo and in vitro models, it seems that β receptor activation is protective (Table 2), while for α1- AR, the results are inconsistent, as studies show that both activation and blockade can have beneficial effects (Table 1). The protective role of β2 AR is more evident a notion that is also supported by others [11].

While trying to delineate whether activation or blockade is beneficial, evidence is hampered by the lack of clinical data. In fact, strong evidence from clinical studies for the benefit of adrenoreceptors in AD is lacking. A number of studies have explored the potential protective effects of antihypertensive drugs acting on β receptors in AD, but the results have been mostly inconclusive. For instance, a study analyzing 849,378 antihypertensive users found that β-adrenoceptor blockers conferred small protective effects against dementia compared to other antihypertensives [83]. Another study involving 69,081 individuals suggested that highly blood–brain barrier (BBB)-permeable β blockers were associated with a reduced risk of AD compared to low BBB permeability β blockers [84]. However, the protective effect may not be solely due to blood pressure reduction, and the role of β1 receptors remains speculative. The role of asthma drugs acting on β2 adrenergic receptors (β2AR) has been studied a lot less. A recent cohort study suggested that selective β2AR agonists are associated with a decreased risk of developing AD, while non-selective AR antagonists are associated with an increased risk [56]. With regard to the α-AR, in a double-blind placebo-controlled parallel-group study, Prazosin improved behavioral symptoms in patients with agitation and aggression in AD [85]. In a randomized double-blind placebo-controlled exploratory phase 2a trial with 100 subjects with AD and neuropsychiatric symptoms, ORM-12741 (a selective antagonist of alpha-2C adrenoceptors, dosed at 30–60 mg or 100–200 mg, b.i.d for 12 weeks) administered in addition to standard therapy with cholinesterase inhibitors showed a statistically significant effect in Quality of Episodic Memory [86].

Table 1. Protective roles of α adrenergic receptors by pharmacological modulation.

Receptor	Action	Modulator	Effects	References
α1-AR	Positive Allosteric Modulator	Cmpd-3	• Improves cognitive functions without affecting blood pressure in AD mouse models. • Enhances LTP, normalizes Aβ levels in AD models	[23]
	Agonists/Activators	Cirazoline, Avenanthramide-C	• Enhances cognitive functions, synaptic plasticity • Improves memory, reduces, neuroinflammation in AD models (effects blocked by prazosin -α1-AR inhibitor).	[24,25]
	Antagonists	Doxazosin, Prazosin, Terazosin	• Reduce Aβ generation and induces anti-inflammatory responses, prevent memory deficits • Improve cognitive functions reduce protein aggregates, and amyloid plaque burden. • Protect against amyloid-β toxicity, prevent GSK-3β activation and tau hyperphosphorylation.	[27–30]
α2-AR	Agonists	Brimonidine, Clonidine	• Offer neuroprotective effects by reducing apoptosis of retinal ganglion cells • Reduce Aβ and APP processing in vitro	[38]
	Antagonists	ORM-10921, Dexefaroxan, Mesedin, Idazoxan	• Improve cognitive functions, enhance neuron survival, reduces cholinergic degeneration, • Exhibits anti-amyloidogenic effects and ameliorate cognitive deficits.	[32,33,35–37]

Table 2. Protective roles of β adrenergic receptors by pharmacological modulation.

Receptor	Action	Modulator	Effects	References
β1-AR	Agonist	Xamoterol, STD-101-D1	• Neuroprotection, reduction inflammation, and synaptic dysfunction. • Enhance cognitive functions, memory, and social learning • Reduce Aβ and tau pathology	[48,49,51]
β2-AR	Agonist	Isoproterenol (via β2), Clenbuterol, Salbutamol	• Restore lysosomal function and autophagy, promote degradation of Aβ. • Enhance neurogenesis, synaptic health, and cognitive function • Reduce amyloid plaques, mitigates Aβ-induced inflammation and synaptotoxicity, decrease neuroinflammatory responses.	[52–59]
Non-pharmacological	Leverage β2 AR Activation	Aerobic exercise	• Reverses autophagy-lysosomal deficits; attenuates amyloid pathology, improves cognitive outcomes	[61]

Table 3. Negative roles of α and β adrenergic receptors by pharmacological and non-modulation.

Receptor	Modulator/Action	Effects/Implications	References
α1-AR	Agonistic autoantibodies (agAABs)–Prolonged Activation	• Non-physiological calcium rise, reduced blood flow, neurodegeneration, amyloid plaques, vascular lesions	[21,39,41]
	Aβ peptides–Activation	• Vascular dysfunction, intracellular calcium release in smooth muscle cells	[43]
	α1-AR–Upregulation in AD	• Aggressive behavior in AD patients	[44]
α2-AR	α2-AR–Reduced density in AD brains	• Loss of presynaptic receptors, cognitive & behavioral impairments	[45,46]
β1-AR	β1-AR gene (*ADRB1*) and G protein beta3 subunit (*GNB3*) gene–Polymorphisms	• Altered adrenergic response, amplifies AD pathology	[69]
β2-AR	β2-AR Activation–Aβ peptides or other	• Impaired neurotransmitter signaling, disrupting learning and memory • Enhances gamma-secretase, promoting amyloid plaques • Contributes to tau hyperphosphorylation and neurodegeneration	[70,71,77]
	β2-AR–Altered signalling	• Disrupts calcium homeostasis, APP processing, synaptic dysfunction	[72]
	Gly16Arg, Gln27Glu–Polymorphisms	• Increased late-onset AD risk (APOE ε4 synergy)	[74]
	Inhibition–ICI 118,551	• Cognitive deficits, tau hyperphosphorylation, amyloid pathology	[80]

Therefore, it still remains uncertain if and which adrenergic receptor modulation may be protective. However, we do believe that with the available dataset from patients and big data analysis using artificial intelligence and machine learning, it will be possible to make a comprehensive analysis using clinical data to determine if modulating the AR is beneficial and under what circumstances. It will indeed be important to make such studies as there are many AR drugs that can easily be repurposed for AD, other dementias or neurodegenerative diseases [87–90], while repurposing for other antihypertensive drugs is also possible [91,92] A number of properties makes this class of drugs an attractive novel therapy for AD: their proven safety and their multi target drug ligand property, which is thought to be a very promising method for targeting complex neurodegenerative diseases [93]. Given the failures observed with current clinical trials [94], even a small indication that these drugs may have efficacy will be a giant leap forward. Importantly, since some of these drugs cross the blood–brain barrier (BBB) fairly easily they can prove to be more efficacious than the current drugs in clinical trials [84,95,96].

While the review underscores the potential of adrenergic receptors as therapeutic targets, several challenges remain. One significant hurdle is the dual nature of receptor activity, where the same pathway can yield both beneficial and harmful outcomes depending on the context. This necessitates the development of highly selective modulators capable of fine-tuning receptor activity in a disease- and region-specific manner. Genetic studies further complicate this landscape, revealing polymorphisms in adrenergic receptor genes that interact with AD risk factors such as APOE ε4 [97]. These findings suggest that patient-specific factors [98–100], including genetic background and disease stage, must be considered when designing adrenergic-based interventions [69,74]. Additionally, the review highlights the need for more comprehensive models that accurately reflect the complexities of adrenergic dysfunction in AD. Current preclinical models often fail to capture the interplay between adrenergic, cholinergic, and serotonergic systems, which collectively shape the neurochemical environment in AD [101,102].

The findings of this review reaffirm the central role of adrenergic signaling in the pathophysiology of AD and other tauopathies. Adrenergic receptors offer a promising yet challenging target for therapeutic development, capable of modulating processes as diverse as neuroinflammation, synaptic plasticity, and amyloid clearance. Future research should focus on developing receptor subtype-specific drugs, exploring combination therapies that integrate adrenergic modulation with other neuroprotective strategies, and tailoring interventions to individual patient profiles. By addressing these challenges, adrenergic receptor modulation could pave the way for transformative therapies in the treatment of neurodegenerative diseases.

3. Conclusions

The review emphasizes the intricate role of adrenergic receptors in AD, highlighting both their therapeutic potential and their contribution to disease pathology. Adrenergic receptors are physiologically integral to neural processes such as synaptic plasticity, neurovascular regulation, and inflammation. Despite promising preclinical evidence suggesting that these receptors could be viable therapeutic targets, some preclinical studies point to their negative effects. What is more, strong clinical evidence is lacking, and the limited clinical studies thus far have yielded inconclusive results regarding the benefits of drugs acting on the β1 or β2 receptors. The dual nature of receptor activity, where both activation and blockade can potentially produce similar effects, presents a significant challenge in developing selective modulators. Additionally, genetic factors, patient-specific variables, the complexity of the disease itself, and the lack of good animal models, further complicate research in this area and therapeutic approaches. Future research should prioritize the development of subtype-specific drugs, combination therapies, and more controlled clinical trials with appropriate covariate selection, while also leveraging big data and advanced analytical techniques to refine strategies for adrenergic modulation in AD. This could potentially lead to innovative treatments not only for AD but also for other neurodegenerative diseases.

Author Contributions: Conceptualization, L.C.Z.; methodology, L.C.Z.; resources, A.K.; writing—original draft preparation, A.N.M. and L.C.Z.; writing—review and editing, L.C.Z.; supervision, L.C.Z. All authors have read and agreed to the published version of the manuscript.

Funding: This research received no external funding.

Institutional Review Board Statement: Not applicable.

Informed Consent Statement: Not applicable.

Data Availability Statement: No new data were created in this study. Data sharing is not applicable to this article.

Acknowledgments: During the preparation of this manuscript, the author(s) used Accelerate, a University of Nicosia Chat GPT based tool for the purposes of sorting the papers into positive and negative effects and by adrenergic type. This was performed after papers were searched in PUBMED. In some cases AI was also used to summarize the findings. The authors have reviewed and edited the output and take full responsibility for the content of this publication.

Conflicts of Interest: The authors declare no conflict of interest.

References

1. Weller, J.; Budson, A. Current understanding of Alzheimer's disease diagnosis and treatment. *F1000Research* **2018**, *7*, 1161. [CrossRef] [PubMed]
2. Ossenkoppele, R.; van der Flier, W.M.; Verfaillie, S.C.; Vrenken, H.; Versteeg, A.; van Schijndel, R.A.; Sikkes, S.A.; Twisk, J.; Adriaanse, S.M.; Zwan, M.D.; et al. Long-term effects of amyloid, hypometabolism, and atrophy on neuropsychological functions. *Neurology* **2014**, *82*, 1768–1775. [CrossRef]

3. De la Rosa, A.; Olaso-Gonzalez, G.; Arc-Chagnaud, C.; Millan, F.; Salvador-Pascual, A.; García-Lucerga, C.; Blasco-Lafarga, C.; Garcia-Dominguez, E.; Carretero, A.; Correas, A.G.; et al. Physical exercise in the prevention and treatment of Alzheimer's disease. *J. Sport. Health Sci.* **2020**, *9*, 394–404. [CrossRef]
4. Passeri, E.; Elkhoury, K.; Morsink, M.; Broersen, K.; Linder, M.; Tamayol, A.; Malaplate, C.; Yen, F.T.; Arab-Tehrany, E. Alzheimer's Disease: Treatment Strategies and Their Limitations. *Int. J. Mol. Sci.* **2022**, *23*, 13954. [CrossRef] [PubMed]
5. Muralidar, S.; Ambi, S.V.; Sekaran, S.; Thirumalai, D.; Palaniappan, B. Role of tau protein in Alzheimer's disease: The prime pathological player. *Int. J. Biol. Macromol.* **2020**, *163*, 1599–1617. [CrossRef] [PubMed]
6. Pascoal, T.A.; Mathotaarachchi, S.; Shin, M.; Benedet, A.L.; Mohades, S.; Wang, S.; Beaudry, T.; Kang, M.S.; Soucy, J.; Labbe, A.; et al. Synergistic interaction between amyloid and tau predicts the progression to dementia. *Alzheimer's Dement.* **2017**, *13*, 644–653. [CrossRef] [PubMed]
7. Duyckaerts, C. Tau pathology in children and young adults: Can you still be unconditionally baptist? *Acta Neuropathol.* **2011**, *121*, 145–147. [CrossRef]
8. Bekdash, R.A. The Cholinergic System, the Adrenergic System and the Neuropathology of Alzheimer's Disease. *Int. J. Mol. Sci.* **2021**, *22*, 1273. [CrossRef] [PubMed]
9. Perez, D.M. alpha(1)-Adrenergic Receptors: Insights into Potential Therapeutic Opportunities for COVID-19, Heart Failure, and Alzheimer's Disease. *Int. J. Mol. Sci.* **2023**, *24*, 4188. [CrossRef] [PubMed]
10. Le, L.; Feidler, A.M.; Li, H.; Kara-Pabani, K.; Lamantia, C.; O'Banion, M.K.; Majewska, K.A. Noradrenergic signaling controls Alzheimer's disease pathology via activation of microglial beta2 adrenergic receptors. *bioRxiv* **2023**. [CrossRef]
11. Li, S. The beta-adrenergic hypothesis of synaptic and microglial impairment in Alzheimer's disease. *J. Neurochem.* **2023**, *165*, 289–302. [CrossRef] [PubMed]
12. Femminella, G.D.; Rengo, G.; Pagano, G.; De Lucia, C.; Komici, K.; Parisi, V.; Alessandroa, C.; Danielaa, L.; Carloa, V.; Pasquale Perrone, F.; et al. beta-adrenergic receptors and G protein-coupled receptor kinase-2 in Alzheimer's disease: A new paradigm for prognosis and therapy? *J. Alzheimer's Dis.* **2013**, *34*, 341–347. [CrossRef] [PubMed]
13. Yu, J.T.; Wang, N.D.; Ma, T.; Jiang, H.; Guan, J.; Tan, L. Roles of beta-adrenergic receptors in Alzheimer's disease: Implications for novel therapeutics. *Brain Res. Bull.* **2011**, *84*, 111–117. [CrossRef] [PubMed]
14. Luong, K.; Nguyen, L.T. The role of Beta-adrenergic receptor blockers in Alzheimer's disease: Potential genetic and cellular signaling mechanisms. *Am. J. Alzheimer's Dis. Other Demen.* **2013**, *28*, 427–439. [CrossRef]
15. Holland, N.; Robbins, T.W.; Rowe, J.B. The role of noradrenaline in cognition and cognitive disorders. *Brain* **2021**, *144*, 2243–2256. [CrossRef] [PubMed]
16. Gannon, M.; Wang, Q. Complex noradrenergic dysfunction in Alzheimer's disease: Low norepinephrine input is not always to blame. *Brain Res.* **2019**, *1702*, 12–16. [CrossRef] [PubMed]
17. Gannon, M.; Che, P.; Chen, Y.; Jiao, K.; Roberson, E.D.; Wang, Q. Noradrenergic dysfunction in Alzheimer's disease. *Front. Neurosci.* **2015**, *9*, 220. [CrossRef] [PubMed]
18. Yan, Z.; Rein, B. Mechanisms of synaptic transmission dysregulation in the prefrontal cortex: Pathophysiological implications. *Mol. Psychiatry* **2022**, *27*, 445–465. [CrossRef]
19. Shimohama, S.; Taniguchi, T.; Fujiwara, M.; Kameyama, M. Changes in beta-adrenergic receptor subtypes in Alzheimer-type dementia. *J. Neurochem.* **1987**, *48*, 1215–1221. [CrossRef] [PubMed]
20. O'donnell, J.; Zeppenfeld, D.; McConnell, E.; Pena, S.; Nedergaard, M. Norepinephrine: A neuromodulator that boosts the function of multiple cell types to optimize CNS performance. *Neurochem. Res.* **2012**, *37*, 2496–2512. [CrossRef] [PubMed]
21. Karczewski, P.; Hempel, P.; Kunze, R.; Bimmler, M. Agonistic autoantibodies to the α_1-adrenergic receptor and the β_2-adrenergic receptor in Alzheimer's and vascular dementia. *Scand. J. Immunol.* **2012**, *75*, 524–530. [CrossRef] [PubMed]
22. Zhang, F.; Gannon, M.; Chen, Y.; Zhou, L.; Jiao, K.; Wang, Q. The amyloid precursor protein modulates α_{2A}-adrenergic receptor endocytosis and signaling through disrupting arrestin 3 recruitment. *FASEB J.* **2017**, *31*, 4434–4446. [CrossRef]
23. Papay, R.S.; Stauffer, S.R.; Perez, D.M. A PAM of the α_{1A}-Adrenergic receptor rescues biomarker, long-term potentiation, and cognitive deficits in Alzheimer's disease mouse models without effects on blood pressure. *Curr. Res. Pharmacol. Drug Discov.* **2023**, *5*, 100160. [CrossRef] [PubMed]
24. Doze, V.A.; Papay, R.S.; Goldenstein, B.L.; Gupta, M.K.; Collette, K.M.; Nelson, B.W.; Lyons, M.J.; Davis, B.A.; Luger, E.J.; Wood, S.G.; et al. Long-term alpha1A-adrenergic receptor stimulation improves synaptic plasticity, cognitive function, mood, and longevity. *Mol. Pharmacol.* **2011**, *80*, 747–758. [CrossRef] [PubMed]
25. Ramasamy, V.S.; Samidurai, M.; Park, H.J.; Wang, M.; Park, R.Y.; Yu, S.Y.; Kang, H.K.; Hong, S.; Choi, W.-S.; Lee, Y.Y.; et al. Avenanthramide-C Restores Impaired Plasticity and Cognition in Alzheimer's Disease Model Mice. *Mol. Neurobiol.* **2020**, *57*, 315–330. [CrossRef]

26. Knauber, J.; Muller, W.E. Decreased exploratory activity and impaired passive avoidance behaviour in mice deficient for the alpha(1b)-adrenoceptor. *Eur. Neuropsychopharmacol.* **2000**, *10*, 423–427. [CrossRef] [PubMed]
27. Coelho, B.P.; Gaelzer, M.M.; Petry, F.d.S.; Hoppe, J.B.; Trindade, V.M.T.; Salbego, C.G.; Guma, F.T. Dual Effect of Doxazosin: Anticancer Activity on SH-SY5Y Neuroblastoma Cells and Neuroprotection on an In Vitro Model of Alzheimer's Disease. *Neuroscience* **2019**, *404*, 314–325. [CrossRef] [PubMed]
28. Katsouri, L.; Vizcaychipi, M.P.; McArthur, S.; Harrison, I.; Suárez-Calvet, M.; Lleo, A.; Lloyd, D.G.; Ma, D.; Sastre, M. Prazosin, an alpha(1)-adrenoceptor antagonist, prevents memory deterioration in the APP23 transgenic mouse model of Alzheimer's disease. *Neurobiol. Aging* **2013**, *34*, 1105–1115. [CrossRef]
29. Chen, H.; Li, Y.; Gao, J.; Cheng, Q.; Liu, L.; Cai, R. Activation of Pgk1 Results in Reduced Protein Aggregation in Diverse Neurodegenerative Conditions. *Mol. Neurobiol.* **2023**, *60*, 5090–5101. [CrossRef] [PubMed]
30. Yu, Z.; Yi, X.; Wang, Y.; Zeng, G.; Tan, C.; Cheng, Y.; Sun, P.; Liu, Z.; Wang, Y.; Liu, Y. Inhibiting alpha1-adrenergic receptor signaling pathway ameliorates AD-type pathologies and behavioral deficits in APPswe/PS1 mouse model. *J. Neurochem.* **2022**, *161*, 293–307. [CrossRef]
31. Koutroumani, M.; Daniilidou, M.; Giannakouros, T.; Proitsi, P.; Liapi, D.; Germanou, A.; Nikolakaki, E.; Tsolaki, M. The deletion variant of alpha2b-adrenergic receptor is associated with decreased risk in Alzheimer's disease and mild cognitive impairment. *J. Neurol. Sci.* **2013**, *328*, 19–23. [CrossRef]
32. Sallinen, J.; Holappa, J.; Koivisto, A.; Kuokkanen, K.; Chapman, H.; Lehtimäki, J.; Piepponen, P.; Mijatovic, J.; Tanila, H.; Virtanen, R.; et al. Pharmacological characterisation of a structurally novel alpha2C-adrenoceptor antagonist ORM-10921 and its effects in neuropsychiatric models. *Basic. Clin. Pharmacol. Toxicol.* **2013**, *113*, 239–249. [CrossRef] [PubMed]
33. Rizk, P.; Salazar, J.; Raisman-Vozari, R.; Marien, M.; Ruberg, M.; Colpaert, F.; Debeir, T. The alpha2-adrenoceptor antagonist dexefaroxan enhances hippocampal neurogenesis by increasing the survival and differentiation of new granule cells. *Neuropsychopharmacology* **2006**, *31*, 1146–1157. [CrossRef] [PubMed]
34. Debeir, T.; Marien, M.; Chopin, P.; Martel, J.-C.; Colpaert, F.; Raisman-Vozari, R. Protective effects of the alpha 2-adrenoceptor antagonist, dexefaroxan, against degeneration of the basalocortical cholinergic system induced by cortical devascularization in the adult rat. *Neuroscience* **2002**, *115*, 41–53. [CrossRef]
35. Chopin, P.; Colpaert, F.C.; Marien, M. Effects of acute and subchronic administration of dexefaroxan, an alpha(2)-adrenoceptor antagonist, on memory performance in young adult and aged rodents. *J. Pharmacol. Exp. Ther.* **2002**, *301*, 187–196. [CrossRef]
36. Melkonyan, M.M.; Hunanyan, L.; Lourhmati, A.; Layer, N.; Beer-Hammer, S.; Yenkoyan, K.; Schwab, M.; Danielyan, L. Neuroprotective, Neurogenic, and Amyloid Beta Reducing Effect of a Novel Alpha 2-Adrenoblocker, Mesedin, on Astroglia and Neuronal Progenitors upon Hypoxia and Glutamate Exposure. *Int. J. Mol. Sci.* **2017**, *19*, 9. [CrossRef] [PubMed]
37. Chen, Y.; Peng, Y.; Che, P.; Gannon, M.; Liu, Y.; Li, L.; Bu, G.; van Groen, T.; Jiao, K.; Wang, Q. alpha(2A) adrenergic receptor promotes amyloidogenesis through disrupting APP-SorLA interaction. *Proc. Natl. Acad. Sci. USA* **2014**, *111*, 17296–17301. [CrossRef] [PubMed]
38. Nizari, S.; Guo, L.; Davis, B.M.; Normando, E.M.; Galvao, J.; A Turner, L.; Bizrah, M.; Dehabadi, M.; Tian, K.; Cordeiro, M.F. Non-amyloidogenic effects of alpha2 adrenergic agonists: Implications for brimonidine-mediated neuroprotection. *Cell Death Dis.* **2016**, *7*, e2514. [CrossRef] [PubMed]
39. Karczewski, P.; Hempel, P.; Bimmler, M. Role of alpha1-adrenergic receptor antibodies in Alzheimer's disease. *Front. Biosci. (Landmark Ed.)* **2018**, *23*, 2082–2089. [PubMed]
40. Hempel, P.; Heinig, B.; Jerosch, C.; Decius, I.; Karczewski, P.; Kassner, U.; Kunze, R.; Steinhagen-Thiessen, E.; Bimmler, M. Immunoadsorption of Agonistic Autoantibodies Against alpha1-Adrenergic Receptors in Patients with Mild to Moderate Dementia. *Ther. Apher. Dial.* **2016**, *20*, 523–529. [CrossRef]
41. Pohlmann, A.; Karczewski, P.; Ku, M.-C.; Dieringer, B.; Waiczies, H.; Wisbrun, N.; Kox, S.; Palatnik, I.; Reimann, H.M.; Eichhorn, C.; et al. Cerebral blood volume estimation by ferumoxytol-enhanced steady-state MRI at 9.4 T reveals microvascular impact of alpha1 -adrenergic receptor antibodies. *NMR Biomed.* **2014**, *27*, 1085–1093. [CrossRef] [PubMed]
42. Stracke, S.; Lange, S.; Bornmann, S.; Kock, H.; Schulze, L.; Klinger-König, J.; Böhm, S.; Vogelgesang, A.; von Podewils, F.; Föel, A.; et al. Immunoadsorption for Treatment of Patients with Suspected Alzheimer Dementia and Agonistic Autoantibodies against Alpha1a-Adrenoceptor-Rationale and Design of the IMAD Pilot Study. *J. Clin. Med.* **2020**, *9*, 1919. [CrossRef] [PubMed]
43. Haase, N.; Herse, F.; Spallek, B.; Haase, H.; Morano, I.; Qadri, F.; Szijártó, I.A.; Roham, I.; Yilmaz, A.; Warrington, J.P.; et al. Amyloid-beta peptides activate alpha1-adrenergic cardiovascular receptors. *Hypertension* **2013**, *62*, 966–972. [CrossRef] [PubMed]
44. Sharp, S.I.; Ballard, C.G.; Chen, C.P.-H.; Francis, P.T. Aggressive behavior and neuroleptic medication are associated with increased number of alpha1-adrenoceptors in patients with Alzheimer disease. *Am. J. Geriatr. Psychiatry* **2007**, *15*, 435–437. [CrossRef] [PubMed]

45. Kalaria, R.N.; Andorn, A.C. Adrenergic receptors in aging and Alzheimer's disease: Decreased alpha 2-receptors demonstrated by [3H]p-aminoclonidine binding in prefrontal cortex. *Neurobiol. Aging* **1991**, *12*, 131–136. [CrossRef]
46. Pascual, J.; Grijalba, B.; García-Sevilla, J.A.; Zarranz, J.J.; Pazos, A. Loss of high-affinity alpha 2-adrenoceptors in Alzheimer's disease: An autoradiographic study in frontal cortex and hippocampus. *Neurosci. Lett.* **1992**, *142*, 36–40. [CrossRef]
47. Ippolito, M.; Benovic, J.L. Biased agonism at beta-adrenergic receptors. *Cell Signal* **2021**, *80*, 109905. [CrossRef]
48. Coutellier, L.; Ardestani, P.M.; Shamloo, M. beta1-adrenergic receptor activation enhances memory in Alzheimer's disease model. *Ann. Clin. Transl. Neurol.* **2014**, *1*, 348–360. [CrossRef]
49. Ardestani, P.M.; Evans, A.K.; Yi, B.; Nguyen, T.; Coutellier, L.; Shamloo, M. Modulation of neuroinflammation and pathology in the 5XFAD mouse model of Alzheimer's disease using a biased and selective beta-1 adrenergic receptor partial agonist. *Neuropharmacology* **2017**, *116*, 371–386. [CrossRef] [PubMed]
50. Wang, J.; Wright, H.M.; Vempati, P.; Li, H.; Wangsa, J.; Dzhuan, A.; Habbu, K.; Knable, L.A.; Ho, L.; Pasinetti, G.M. Investigation of nebivolol as a novel therapeutic agent for the treatment of Alzheimer's disease. *J. Alzheimer's Dis.* **2013**, *33*, 1147–1156. [CrossRef] [PubMed]
51. Yi, B.; Jahangir, A.; Evans, A.K.; Briggs, D.; Ravina, K.; Ernest, J.; Farimani, A.B.; Sun, W.; Rajadas, J.; Green, M.; et al. Discovery of novel brain permeable and G protein-biased beta-1 adrenergic receptor partial agonists for the treatment of neurocognitive disorders. *PLoS ONE* **2017**, *12*, e0180319. [CrossRef]
52. Lee, J.H.; Wolfe, D.M.; Darji, S.; McBrayer, M.K.; Colacurcio, D.J.; Kumar, A.; Stavrides, P.; Mohan, P.S.; Nixon, R.A. beta2-adrenergic Agonists Rescue Lysosome Acidification and Function in PSEN1 Deficiency by Reversing Defective ER to lysosome Delivery of ClC-7. *J. Mol. Biol.* **2020**, *432*, 2633–2650. [CrossRef] [PubMed]
53. Xu, H.; Rajsombath, M.M.; Weikop, P.; Selkoe, D.J. Enriched environment enhances beta-adrenergic signaling to prevent microglia inflammation by amyloid-beta. *EMBO Mol. Med.* **2018**, *10*, e8931. [CrossRef]
54. Li, S.; Jin, M.; Zhang, D.; Yang, T.; Koeglsperger, T.; Fu, H.; Selkoe, D.J. Environmental novelty activates beta2-adrenergic signaling to prevent the impairment of hippocampal LTP by Abeta oligomers. *Neuron* **2013**, *77*, 929–941. [CrossRef]
55. O'Dell, T.J.; Connor, S.A.; Gelinas, J.N.; Nguyen, P.V. Viagra for your synapses: Enhancement of hippocampal long-term potentiation by activation of beta-adrenergic receptors. *Cell Signal* **2010**, *22*, 728–736. [CrossRef]
56. Hutten, D.R.; Bos, J.H.; de Vos, S.; Hak, E. Targeting the Beta-2-Adrenergic Receptor and the Risk of Developing Alzheimer's Disease: A Retrospective Inception Cohort Study. *J. Alzheimer's Dis.* **2022**, *87*, 1089–1101. [CrossRef]
57. Zhao, P.; Chai, G.-S.; Wang, Y.-Y.; Yasheng, A. Beta 2-adrenergic receptor activation enhances neurogenesis in Alzheimer's disease mice. *Neural Regen. Res.* **2016**, *11*, 1617–1624. [CrossRef] [PubMed]
58. Chai, G.S.; Wang, Y.Y.; Zhu, D.; Yasheng, A.; Zhao, P. Activation of beta(2)-adrenergic receptor promotes dendrite ramification and spine generation in APP/PS1 mice. *Neurosci. Lett.* **2017**, *636*, 158–164. [CrossRef] [PubMed]
59. Townsend, D.J.; Mala, B.; Hughes, E.; Hussain, R.; Siligardi, G.; Fullwood, N.J.; Middleton, D.A. Circular Dichroism Spectroscopy Identifies the beta-Adrenoceptor Agonist Salbutamol As a Direct Inhibitor of Tau Filament Formation in Vitro. *ACS Chem. Neurosci.* **2020**, *11*, 2104–2116. [CrossRef]
60. Wang, D.; Fu, Q.; Zhou, Y.; Xu, B.; Shi, Q.; Igwe, B.; Matt, L.; Hell, J.W.; Wisely, E.V.; Oddo, S.; et al. beta2 adrenergic receptor, protein kinase A (PKA) and c-Jun N-terminal kinase (JNK) signaling pathways mediate tau pathology in Alzheimer disease models. *J. Biol. Chem.* **2013**, *288*, 10298–10307. [CrossRef]
61. Wu, J.J.; Yu, H.; Bi, S.G.; Wang, Z.X.; Gong, J.; Mao, Y.M.; Wang, F.Z.; Zhang, Y.Q.; Niu, Y.J.; Chai, G.S. Aerobic exercise attenuates autophagy-lysosomal flux deficits by ADRB2/beta2-adrenergic receptor-mediated V-ATPase assembly factor VMA21 signaling in APP-PSEN1/PS1 mice. *Autophagy* **2024**, *20*, 1015–1031. [CrossRef] [PubMed]
62. Goodman, A.M.; Langner, B.M.; Jackson, N.; Alex, C.; McMahon, L.L. Heightened Hippocampal beta-Adrenergic Receptor Function Drives Synaptic Potentiation and Supports Learning and Memory in the TgF344-AD Rat Model during Prodromal Alzheimer's Disease. *J. Neurosci.* **2021**, *41*, 5747–5761. [CrossRef]
63. Shen, H.; Fuchino, Y.; Miyamoto, D.; Nomura, H.; Matsuki, N. Vagus nerve stimulation enhances perforant path-CA3 synaptic transmission via the activation of beta-adrenergic receptors and the locus coeruleus. *Int. J. Neuropsychopharmacol.* **2012**, *15*, 523–530. [CrossRef] [PubMed]
64. Counts, S.E.; Mufson, E.J. Noradrenaline activation of neurotrophic pathways protects against neuronal amyloid toxicity. *J. Neurochem.* **2010**, *113*, 649–660. [CrossRef] [PubMed]
65. Kong, Y.; Ruan, L.; Qian, L.; Liu, X.; Le, Y. Norepinephrine promotes microglia to uptake and degrade amyloid beta peptide through upregulation of mouse formyl peptide receptor 2 and induction of insulin-degrading enzyme. *J. Neurosci.* **2010**, *30*, 11848–11857. [CrossRef] [PubMed]

66. Sugama, S.; Takenouchi, T.; Hashimoto, M.; Ohata, H.; Takenaka, Y.; Kakinuma, Y. Stress-induced microglial activation occurs through beta-adrenergic receptor: Noradrenaline as a key neurotransmitter in microglial activation. *J. Neuroinflamm.* **2019**, *16*, 266. [CrossRef]
67. Yang, J.H.; Lee, E.O.; Kim, S.E.; Suh, Y.H.; Chong, Y.H. Norepinephrine differentially modulates the innate inflammatory response provoked by amyloid-beta peptide via action at beta-adrenoceptors and activation of cAMP/PKA pathway in human THP-1 macrophages. *Exp. Neurol.* **2012**, *236*, 199–206. [CrossRef]
68. Allaman, I.; Pellerin, L.; Magistretti, P.J. Protein targeting to glycogen mRNA expression is stimulated by noradrenaline in mouse cortical astrocytes. *Glia* **2000**, *30*, 382–391. [CrossRef]
69. Bullido, M.J.; Ramos, M.C.; Ruiz-Gómez, A.; Tutor, A.S.; Sastre, I.; Frank, A.; Coria, F.; Gil, P.; Mayor, F.; Valdivieso, F. Polymorphism in genes involved in adrenergic signaling associated with Alzheimer's. *Neurobiol. Aging* **2004**, *25*, 853–859. [CrossRef]
70. Wang, D.; Yuen, E.Y.; Zhou, Y.; Yan, Z.; Xiang, Y.K. Amyloid beta peptide-(1-42) induces internalization and degradation of beta2 adrenergic receptors in prefrontal cortical neurons. *J. Biol. Chem.* **2011**, *286*, 31852–31863. [CrossRef]
71. Wisely, E.V.; Xiang, Y.K.; Oddo, S. Genetic suppression of beta2-adrenergic receptors ameliorates tau pathology in a mouse model of tauopathies. *Hum. Mol. Genet.* **2014**, *23*, 4024–4034. [CrossRef] [PubMed]
72. Bussiere, R.; Lacampagne, A.; Reiken, S.; Liu, X.; Scheuerman, V.; Zalk, R.; Martin, C.; Checler, F.; Marks, A.R.; Chami, M.; et al. Amyloid beta production is regulated by beta2-adrenergic signaling-mediated post-translational modifications of the ryanodine receptor. *J. Biol. Chem.* **2017**, *292*, 10153–10168. [CrossRef] [PubMed]
73. Wang, D.; Govindaiah, G.; Liu, R.; De Arcangelis, V.; Cox, C.L.; Xiang, Y.K. Binding of amyloid beta peptide to beta2 adrenergic receptor induces PKA-dependent AMPA receptor hyperactivity. *FASEB J.* **2010**, *24*, 3511–3521. [CrossRef]
74. Yu, J.T.; Tan, L.; Ou, J.R.; Zhu, J.X.; Liu, K.; Song, J.H.; Sun, Y.P. Polymorphisms at the beta2-adrenergic receptor gene influence Alzheimer's disease susceptibility. *Brain Res.* **2008**, *1210*, 216–222. [CrossRef] [PubMed]
75. Hohberger, B.; Prüss, H.; Mardin, C.; Lämmer, R.; Müller, J.; Wallukat, G. Glaucoma and Alzheimer: Neurodegenerative disorders show an adrenergic dysbalance. *PLoS ONE* **2022**, *17*, e0272811. [CrossRef] [PubMed]
76. Kikuchi, K.; Fujita, Y.; Shen, X.; Liu, J.; Terakawa, T.; Nishikata, D.; Niibori, S.; Ito, T.; Ashidate, K.; Kikuchi, T.; et al. Interaction between Angiotensin Receptor and beta-Adrenergic Receptor Regulates the Production of Amyloid beta-Protein. *Biol. Pharm. Bull.* **2020**, *43*, 731–735. [CrossRef]
77. Ni, Y.; Zhao, X.; Bao, G.; Zou, L.; Teng, L.; Wang, Z.; Song, M.; Xiong, J.; Bai, Y.; Pei, G. Activation of beta2-adrenergic receptor stimulates gamma-secretase activity and accelerates amyloid plaque formation. *Nat. Med.* **2006**, *12*, 1390–1396. [CrossRef]
78. Yu, N.N.; Wang, X.X.; Yu, J.T.; Wang, N.D.; Lu, R.C.; Miao, D.; Tian, Y.; Tan, L. Blocking beta2-adrenergic receptor attenuates acute stress-induced amyloid beta peptides production. *Brain Res.* **2010**, *1317*, 305–310. [CrossRef]
79. Wang, J.; Ono, K.; Dickstein, D.L.; Arrieta-Cruz, I.; Zhao, W.; Qian, X.; Lamparello, A.; Subnani, R.; Ferruzzi, M.; Pavlides, C.; et al. Carvedilol as a potential novel agent for the treatment of Alzheimer's disease. *Neurobiol. Aging* **2011**, *32*, 2321 e1-12. [CrossRef] [PubMed]
80. Branca, C.; Wisely, E.V.; Hartman, L.K.; Caccamo, A.; Oddo, S. Administration of a selective beta2 adrenergic receptor antagonist exacerbates neuropathology and cognitive deficits in a mouse model of Alzheimer's disease. *Neurobiol. Aging* **2014**, *35*, 2726–2735. [CrossRef]
81. Evans, A.K.; Ardestani, P.M.; Yi, B.; Park, H.H.; Lam, R.K.; Shamloo, M. Beta-adrenergic receptor antagonism is proinflammatory and exacerbates neuroinflammation in a mouse model of Alzheimer's Disease. *Neurobiol. Dis.* **2020**, *146*, 105089. [CrossRef]
82. Decker, M.W.; Gill, T.M.; McGaugh, J.L. Concurrent muscarinic and beta-adrenergic blockade in rats impairs place-learning in a water maze and retention of inhibitory avoidance. *Brain Res.* **1990**, *513*, 81–85. [CrossRef]
83. Walker, V.M.; Davies, N.M.; Martin, R.M.; Kehoe, P.G. Comparison of Antihypertensive Drug Classes for Dementia Prevention. *Epidemiology* **2020**, *31*, 852–859. [CrossRef] [PubMed]
84. Beaman, E.E.; Bonde, A.N.; Larsen SM, U.; Ozenne, B.; Lohela, T.J.; Nedergaard, M.; Hilmar Gíslason, G.; Knudsen, G.M.; Holst, S.C. Blood-brain barrier permeable beta-blockers linked to lower risk of Alzheimer's disease in hypertension. *Brain* **2022**, *146*, 1141–1151. [CrossRef] [PubMed]
85. Wang, L.Y.; Shofer, J.B.; Rohde, K.; Hart, K.L.; Hoff, D.J.; McFall, Y.H.; Raskind, M.A.; Peskind, E.R. Prazosin for the treatment of behavioral symptoms in patients with Alzheimer disease with agitation and aggression. *Am. J. Geriatr. Psychiatry* **2009**, *17*, 744–751. [CrossRef] [PubMed]
86. Rinne, J.O.; Wesnes, K.; Cummings, J.L.; Hakulinen, P.; Hallikainen, M.; Hänninen, J.; Murphy, M.; Riordan, H.; Scheinin, M.; Soininen, H.; et al. Tolerability of ORM-12741 and effects on episodic memory in patients with Alzheimer's disease. *Alzheimer's Dement.* **2017**, *3*, 1–9. [CrossRef]

87. Duraes, F.; Pinto, M.; Sousa, E. Old Drugs as New Treatments for Neurodegenerative Diseases. *Pharmaceuticals* **2018**, *11*, 44. [CrossRef] [PubMed]
88. Hernández, T.D.R.; Gore, S.V.; Kreiling, J.A.; Creton, R. Drug repurposing for neurodegenerative diseases using Zebrafish behavioral profiles. *Biomed. Pharmacother.* **2024**, *171*, 116096.
89. Walker, V.M.; Davies, N.M.; Jones, T.; Kehoe, P.G.; Martin, R.M. Can commonly prescribed drugs be repurposed for the prevention or treatment of Alzheimer's and other neurodegenerative diseases? Protocol for an observational cohort study in the UK Clinical Practice Research Datalink. *BMJ Open* **2016**, *6*, e012044. [CrossRef] [PubMed]
90. Tournissac, M.; Vu, T.-M.; Vrabic, N.; Hozer, C.; Tremblay, C.; Mélançon, K.; Planel, E.; Pifferi, F.; Calon, F. Repurposing beta-3 adrenergic receptor agonists for Alzheimer's disease: Beneficial effects in a mouse model. *Alzheimer's Res. Ther.* **2021**, *13*, 103. [CrossRef]
91. Lee, H.W.; Kim, S.; Jo, Y.; Kim, Y.; Ye, B.S.; Yu, Y.M. Neuroprotective effect of angiotensin II receptor blockers on the risk of incident Alzheimer's disease: A nationwide population-based cohort study. *Front. Aging Neurosci.* **2023**, *15*, 1137197. [CrossRef] [PubMed]
92. Giil, L.M.; Kristoffersen, E.K.; Vedeler, C.A.; Aarsland, D.; Nordrehaug, J.E.; Winblad, B.; Cedazo-Minguez, A.; Lund, A.; Reksten, T.R. Autoantibodies Toward the Angiotensin 2 Type 1 Receptor: A Novel Autoantibody in Alzheimer's Disease. *J. Alzheimer's Dis.* **2015**, *47*, 523–529. [CrossRef] [PubMed]
93. Hughes, R.E.; Nikolic, K.; Ramsay, R.R. One for All? Hitting Multiple Alzheimer's Disease Targets with One Drug. *Front. Neurosci.* **2016**, *10*, 177. [CrossRef] [PubMed]
94. Kim, C.K.; Lee, Y.R.; Ong, L.; Gold, M.; Kalali, A.; Sarkar, J. Alzheimer's Disease: Key Insights from Two Decades of Clinical Trial Failures. *J. Alzheimer's Dis.* **2022**, *87*, 83–100. [CrossRef] [PubMed]
95. Banks, W.A.; Rhea, E.M.; Reed, M.J.; Erickson, M.A. The penetration of therapeutics across the blood-brain barrier: Classic case studies and clinical implications. *Cell Rep. Med.* **2024**, *5*, 101760. [CrossRef] [PubMed]
96. Sun, Y.; Zabihi, M.; Li, Q.; Li, X.; Kim, B.J.; Ubogu, E.E.; Raja, S.N.; Wesselmann, U.; Zhao, C. Drug Permeability: From the Blood-Brain Barrier to the Peripheral Nerve Barriers. *Adv. Ther.* **2023**, *6*, 2200150. [CrossRef]
97. Di Battista, A.M.; Heinsinger, N.M.; Rebeck, G.W. Alzheimer's Disease Genetic Risk Factor. APOE-epsilon4 Also Affects Normal Brain Function. *Curr. Alzheimer Res.* **2016**, *13*, 1200–1207. [CrossRef]
98. Bengtsson Boström, K.; Hedner, J.; Grote, L.; Melander, O.; Von Wowern, F.; Råstam, L.; Groop, L.; Lindblad, U. Polymorphisms in alpha- and beta-Adrenergic Receptor Genes, Hypertension, and Obstructive Sleep Apnea: The Skaraborg Sleep Study. *Int. J. Hypertens.* **2010**, *2010*. [CrossRef] [PubMed]
99. Katsarou, M.-S.; Karathanasopoulou, A.; Andrianopoulou, A.; Desiniotis, V.; Tzinis, E.; Dimitrakis, E.; Lagiou, M.; Charmandari, E.; Aschner, M.; Tsatsakis, A.M.; et al. Beta 1, Beta 2 and Beta 3 Adrenergic Receptor Gene Polymorphisms in a Southeastern European Population. *Front. Genet.* **2018**, *9*, 560. [CrossRef] [PubMed]
100. Garland, E.M.; Biaggioni, I. Genetic polymorphisms of adrenergic receptors. *Clin. Auton. Res.* **2001**, *11*, 67–78. [CrossRef]
101. Chen, Z.Y.; Zhang, Y. Animal models of Alzheimer's disease: Applications, evaluation, and perspectives. *Zool. Res.* **2022**, *43*, 1026–1040. [CrossRef] [PubMed]
102. Padmanabhan, P.; Gotz, J. Clinical relevance of animal models in aging-related dementia research. *Nat. Aging* **2023**, *3*, 481–493. [CrossRef] [PubMed]

biomolecules

MDPI

Review

Immunoglobulin G and Complement as Major Players in the Neurodegeneration of Multiple Sclerosis

Peter G. E. Kennedy [1], Matthew Fultz [2], Jeremiah Phares [2] and Xiaoli Yu [2,*]

[1] Institute of Neuroscience and Psychology, University of Glasgow, Glasgow G61 1QH, UK; peter.kennedy@glasgow.ac.uk

[2] Department of Neurosurgery, University of Colorado Anschutz Medical Campus, Aurora, CO 80045, USA; matthew.fultz@cuanschutz.edu (M.F.); jeremiah.phares@cuanschutz.edu (J.P.)

* Correspondence: xiaoli.yu@cuanschutz.edu

Abstract: Multiple Sclerosis (MS) is an inflammatory, demyelinating, and neurodegenerative disease of the central nervous system (CNS) and is termed as one of the most common causes of neurological disability in young adults. Axonal loss and neuronal cell damage are the primary causes of disease progression and disability. Yet, little is known about the mechanism of neurodegeneration in the disease, a limitation that impairs the development of more effective treatments for progressive MS. MS is characterized by the presence of oligoclonal bands and raised levels of immunoglobulins in the CNS. The role of complement in the demyelinating process has been detected in both experimental animal models of MS and within the CNS of affected MS patients. Furthermore, both IgG antibodies and complement activation can be detected in the demyelinating plaques and cortical gray matter lesions. We propose here that both immunoglobulins and complement play an active role in the neurodegenerative process of MS. We hypothesize that the increased CNS IgG antibodies form IgG aggregates and bind complement C1q with high affinity, activating the classical complement pathway. This results in neuronal cell damage, which leads to neurodegeneration and demyelination in MS.

Keywords: multiple sclerosis; neurodegeneration; IgG; complement; neurons; disease progression

1. Introduction

Multiple sclerosis (MS) is the most common inflammatory demyelinating disease of the central nervous system (CNS), affecting young adults in Northern temperate zones with profound demyelination and neurodegeneration leading to long-term disability. More than 2.8 million people worldwide live with MS with no definitive cure in sight [1]. Axonal loss is a key pathological feature of MS [2], and it is well established that irreversible loss of axons and neurons is the primary cause of most MS patients' progressive neurological disability [3–5]. While it is clear that immunological abnormalities such as cerebrospinal fluid (CSF) oligoclonal bands and raised immunoglobulins are characteristic of MS, a key question is how the neurodegeneration typical of the disease is caused. We argue here that both immunoglobulins and complement play a significant role in the pathophysiology of MS [6] causing neurodegeneration, as the principal effector of an antibody-mediated immunological response is complement activation, and IgG in MS CSF is complement-fixing. We hypothesize that the increased CNS IgG antibodies form IgG aggregates and bind complement C1q with high affinity, causing neuronal cell damage, which leads to neurodegeneration (Figure 1). When referring to MS, we include all phenotypes of the disease, including primary progressive MS (PPMS), secondary progressive MS (SPMS), and relapsing and remitting MS (RRMS).

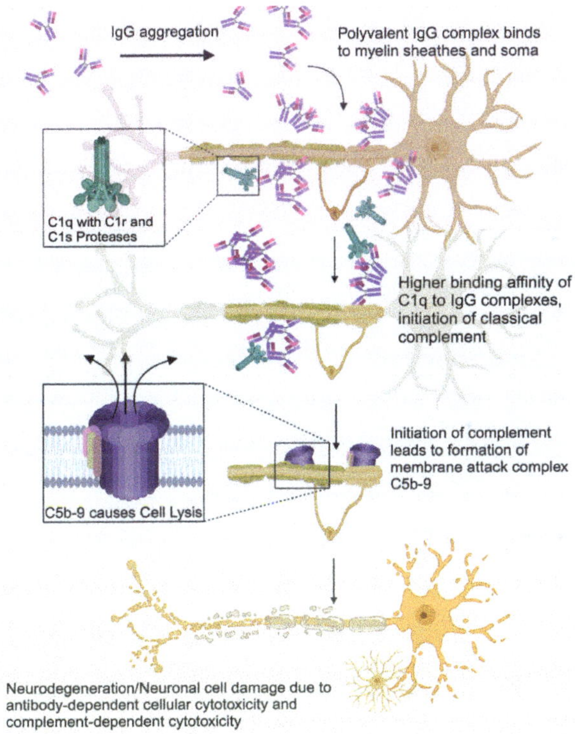

Figure 1. C1q exhibits increased binding to aggregated IgGs, which attach to the myelin and soma, leading to the C5b-9 membrane attack complex to form and lyse host-originating cells. Created with BioRender.com.

2. Neuron Damage and Neuronal Cell Loss Are the Primary Causes of MS Disease Progression and Disability

2.1. Grey Matter Pathology Correlates with Physical Disabilities and Cognitive Impairment in MS

MS has several destructive hallmarks, including acute damage to the neuroaxonal unit, progressive neuroaxonal degeneration, and brain atrophy, resulting in long-term disability. Pathological changes in the CNS gray matter, including the cortex, are central to the progression of MS [7]. Axonal injury correlates with demyelination [8], whereas gray matter pathology correlates with physical disabilities and cognitive impairment in MS [9–12]. We recently reported that plasma IgG aggregates in SPMS are elevated compared to both RRMS and PPMS [13], which caused significantly higher levels of neuronal apoptosis than in RRMS [14], supporting that neural cell apoptosis plays a vital role in MS pathogenesis [5]. The current hypothesis is predicated on the view that neurodegeneration is a leading cause of the disability seen in MS patients.

2.2. Demyelination and Neuron Loss May Occur Simultaneously in MS CNS

Demyelination of cerebral white matter is the pathological hallmark of MS. The classical notion is that demyelination is the consequence of inflammation, contributed by proinflammatory cytokines, CD8+ T cells, anti-myeline antibodies, and activated macrophages [14]. The disease process in MS causes the destruction of oligodendrocytes, which make myelin in the CNS, and this leads to demyelination [15]. Oligodendrocyte apoptosis was found in the new lesions of MS [16]. The pathological mechanism resulting in oligodendrocyte damage or death is not completely understood. Still, it seems highly likely there is a form of immune-mediated attack on these cells. Multiple factors have been shown to cause damage to oligodendrocytes, such as anti-myelin autoantibodies, cytotoxic T cells, and reactive oxygen species [17]. Demyelination is thought to drive neuronal degeneration and permanent neurological disability in individuals with MS [18]. However, oligodendrocyte and neuron damage resulting in neuronal cell loss may co-occur, with neuron loss being the leading event in the disease. Arguably, axonal damage may play a pivotal pathological role in MS, leading to secondary rather than primary demyelination. Recent studies suggest that cerebral white-matter demyelination and cortical neuronal degeneration can be independent events in the disease [18]. MS is conceptualized as a neurodegenerative rather than a pure neuroinflammatory disease with early involvement of axonal loss [2]. A recent study reported an inverse relationship between axon degeneration and demyelination [19].

2.3. Role of Complement in Animal Models of MS Neurodegeneration

Animal models of MS have yielded helpful information and insights into possible mechanisms of demyelination and neurodegeneration. An experimental allergic encephalomyelitis (EAE) model of MS showed that serum IgG antibodies from EAE rabbits cause demyelination [20]. Another earlier study showed that autoantibody depletion ameliorates disease in EAE [21]. An important finding in this area of complement-induced neurodegeneration was recently reported by Linzey et al. [22]. They studied two distinct murine models of MS, namely the proteolipid protein (PLP)-induced autoimmune EAE model (mimicking RRMS) and Theiler's murine encephalomyelitis virus-induced demyelinating disease (mimicking progressive MS). Interestingly, it was found that the chronic-progressive disease form was more dependent on the classical complement pathway and protected the mice from acute relapses, whereas, by contrast, the relapsing model was more dependent on the alternative complement pathway. These results show that complement may play distinct roles at the various stages of MS.

3. Lack of Understanding of the Disease Mechanism Is a Critical Barrier to Effective Treatments in Progressive MS

Irreversible loss of axons and neurons is the primary cause of the progressive neurological decline that most MS patients endure [5]. Although current therapeutic strategies (>20 disease-modifying therapies) are beneficial during RRMS by modulating or suppressing immune function, treatment options for progressive MS are limited [9], and no effective therapies are available to prevent neuronal damage and slow disease disability. Therefore, a better understanding of the underlying mechanisms of neuroaxonal damage leading to neurodegeneration is a critical step toward finding improved treatment options.

Often, drugs prescribed for the treatment of MS act as disease-modifying therapies (DMTs), which function to reduce relapse frequencies during RRMS and lessen symptoms [23]. However, options for treating PPMS and SPMS remain limited due to an incomplete understanding of their pathologies. As the pathogenesis of MS continues to be elucidated, treatments increase in efficacy, with more recent treatments targeting the sphingosine 1-phosphate (S1P) signaling pathway or acting as monoclonal antibodies against B-lymphocyte antigen CD20 or interleukin-2 receptor alpha protein CD25 [24]. S1P DMTs, such as Siponimod, enter the central nervous system via the blood-brain barrier (BBB), binding to S1P receptors and functioning to slow disability and promote the regeneration of myelin [24]. Conversely, monoclonal antibodies such as Ocrelizumab induce complement-dependent cytotoxicity (CDC) and antibody-dependent cellular toxicity (ADCC) onto CD20-positive B cells. However limited, current treatments show short-term improvement in easing the progression of the disease, with a recent clinical trial finding disease progression to be 9.1% lower in Beta-1a interferon versus 13.6% lower in Ocrelizumab after 12 weeks [25]. Despite their short-term efficacies, current DMTs only act to postpone disease progression, reaching eventual plateaus where further treatments hold little value when compared to their risk of adverse events [26].

4. IgG Antibodies and Complement Activation Are Consistently Found in Demyelinating Plaques and Cortical Grey Matter Lesions

4.1. Increased Intrathecal Synthesis of IgG and Oligoclonal Bands Are the Most Characteristic Features of MS, and Evidence Supports the Pathological Role of IgG in MS

A consistent observation in MS is the presence of CSF oligoclonal bands, which occur in over 90% of cases [24]. Once present, the intrathecal IgG, mainly consisting of IgG1 and IgG3 [27], remains stable over time. Accumulating evidence supports the pathological role of CSF immunoglobulins. CSF OCBs were associated with increased levels of disease activity and disability, with the conversion from a clinically isolated syndrome (CIS) to early RRMS, along with more significant brain atrophy and increased disease activity [28–35]. Further, CSF in MS patients induced inflammatory demyelination and axonal damage in mice [36]. It has also been shown that some myelin-specific antibodies derived from MS CSF can cause both demyelination and complement-dependent cytotoxicity in mouse cerebellar oligodendrocytes [36]. Alcázar et al. reported that the soluble factors in primary progressive MS (PPMS) CSF can induce axonal damage and neuronal apoptosis [37]. Additionally, caspase inhibitors are shown to be protective against neuronal apoptosis induced by MS CSF [38]. Most recent studies demonstrated that CSF IgG in PPMS is pathogenic and causes motor disability and spinal cord pathology, including demyelination, impaired remyelination, reactive astrogliosis, and axonal damage [39]. These studies support the pathogenic effects of CSF IgG in MS.

4.2. Evidence Supports the Pathological Role of Blood IgG Antibodies in MS

Lisak et al. reported that MS plasma B cells secreted a factor or factors that induced apoptosis in cultured rat neurons [40]. Our recent report confirms the presence of IgG aggregate-mediated complement activation and cytotoxicity in astrocytes and neurons [41]. The extensive evidence for the role of complement in MS has recently been reviewed in detail by Saez-Calveras and Stuve [42]. These results support the early studies that most MS sera (>80%) caused demyelination (with natural complement) in newborn rat cerebellum [43]. It is also of relevance that we have recently shown that MS plasma contains IgG aggregates that induce complement-dependent neuronal cytotoxicity [41]. These MS plasma IgG aggregates may also function as biomarkers for the disease [41].

4.3. IgG Effector Functions

While extensive research has failed to identify the antigen specificity of the IgG antibodies in MS, there is nevertheless evidence to support IgG effector functions and their role in disease pathogenesis. The IgG Fc domain mediates a wide range of effector functions, including anti-body-dependent cellular cytotoxicity (ADCC) and complement-dependent cytotoxicity (CDC). For CDC, binding of complement component 1q (C1q) triggers the activation of the complement cascade and leads to the formation of the membrane attack complex (MAC), creating pores in the cell membrane and causing the lysis of target cells [44].

4.4. A Brief Introduction of Complement Activation

The complement system is a unique bridge between innate and adaptive immunity. It contributes to the removal of invading pathogens and dead or dying cells [45]. Complement activation plays a critical role in synaptic pruning during brain development [46]. It hails from three unique methods of activation: the classical pathway, the alternative pathway, and the lectin pathway. The classical pathway starts with complement C1, which consists of serine proteases C1r and C1s and the hexameric molecule C1q. C1q binds to the Fc regions of IgGs, and IgM is bound to antigens, which triggers the serine proteases to cleave C2 and C4 into fragments, resulting in the formation of C3 convertase [47,48]. The lectin pathway involves soluble pattern-recognition molecules, most notably mannose-binding lectin (MBL), which binds to carbohydrate signatures on bacteria or fungi, activating MASP-2, and forming C3 convertase [49]. Finally, the alternative pathway involves the hydrolysis of C3 into an initial form of C3 convertase, C3 (H_2O) Bb, by Factor D. Further cleavage in the alternate pathway involving Factor B leads to the formation of C3 convertase and C5 convertase [47]. Though originating from different means of activation, the complement system always results in the production of anaphylatoxins such as C3a, opsonins such as C3b, and the C5b-9 membrane attack complex (MAC) [47]. The insertion of MAC causes necrosis of the cells, resulting in ruptures of the membrane and cell death. Ninomiya and Sims reported that inhibitor CD59 binds complement proteins C8 and C9 at the membrane to prevent insertion and polymerization of MAC pores [50].

4.5. IgG and Complement in MS Brain Lesions

Excessive amounts of IgG antibodies were reported in early studies of MS autopsy active lesions in both free/soluble and tissue-bound/particulate forms [51–53]. The IgG antibodies purified from corresponding soluble and particulate samples showed OCBs [53]. Over 20 times more IgG was extracted from MS plaques than control brain [54]. In the landmark paper by Lucchinetti et al., antibodies and complements were found extensively in pattern II, whereas patterns III and IV were suggestive of oligodendrocyte damage/loss [55]. The lack of detection of IgG antibodies in the latter patterns may indicate the limitation of the detection method, as only immunohistochemistry was used.

The co-localization of IgG antibodies, complement, and Fc gamma receptors (FcγR) in the active lesions was reported, suggesting the pathological role of these antibodies in the early stages of MS [56]. Furthermore, complement activation in cortical gray matter lesions of brain tissues in patients with progressive MS [57] implies that antibodies and complements may contribute to the pathological mechanism underlying the disease's irreversible progression. The dual roles of the C5b-9 complement complex in demyelination have been discussed extensively in early studies [58]. The activation of C5b-9 promotes demyelination, and the sublytic C5b-9 can protect oligodendrocytes from programmed cell death.

4.6. Complement Activation in MS CSF

Early studies reported a highly significant reduction of the terminal component of complement (C9) concentration in MS CSF [59]. Reduced level of CSF C9 in MS implies complement activation resulting in C9 consumption for the formation of membrane attack complexes, leading to myelin damage and neuronal cell membrane damage and causing a more widespread but reversible loss of function. More recently, it was shown that complement activation in the CSF in clinically isolated syndrome and early stages of relapsing-remitting MS [60]. Significantly, they found that CSF C1q and CSF C3a correlated with neuroinflammatory markers and neurofilament light chain levels; CSF C3a correlated with disease activity and progression. A very recent study reported that complement activation is associated with disease severity in MS [61]. They measured complement components and complement activation products in the CSF and plasma in 112 patients with clinically isolated syndrome (CIS), 127 patients with MS, and 75 CNS controls. They showed that C3a and C4a in MS CSF were associated with increased EDSS scores. The CNS compartmentalized activation of the classical and alternative pathways of complement support that complement activation contributes to MS pathology and disease severity.

5. Complement and IgG in Other Autoimmune and Neurodegenerative Diseases

5.1. Complement and IgG in Other Autoimmune Diseases

Complement activation is also a feature of other autoimmune diseases. For example, in systemic lupus erythematosus (SLE) with vasculitis, the levels of serum complement (C3, C4) were reduced [62], with complement levels being known to be lowered in SLE [63]. Complement activation was also reported in patients with nephrotic glomerular diseases [64]. In Myasthenia gravis, which is a prototype autoimmune disease, complement activation can be detected on the muscle membrane surface [63]. Given the known involvement of complement pathways in autoimmune diseases, anti-complement therapies have been considered and developed as therapeutic measures. Whether such therapies may be given to MS patients is a more controversial issue.

Myelin oligodendrocyte glycoprotein (MOG) antibody immunoglobulin G (IgG)-associated disease (MOGAD) has clinical and pathophysiological features-like but distinct from those of aquaporin-4 antibody (AQP4-IgG)-positive neuromyelitis optica spectrum disorders (AQP4-NMOSD). Both MOG-IgG and AQP4-IgG can activate the complement system [65,66]. A recent study showed higher complement activation and C3 degradation in MOGAD sera compared to NMOSD; in contrast, higher levels of soluble C5b-9 (sC5b-9) in NMOSD [67]. These studies further support the complement and IgG in the pathological roles of these disorders.

5.2. Complement and IgG in Other Neurodegenerative Diseases

The complement system is integral for protection against pathogens and for normal development of the CNS [68]. Several cell types in the brain have been implicated in the complement system, such as microglia and astrocytes [69]. Astrocytes are critical for many functions in the brain, such as clearance of cellular debris and aberrant proteins, maintaining the blood-brain barrier, and recycling of neurotransmitters [70]. Microglia are the resident immune cells in the brain. They continuously monitor for damaged cells and pathogenic invaders. When they locate something potentially harmful, they become activated, breaking down and engulfing foreign substances. They are also critical in many homeostatic functions, such as neurogenesis, astrogenesis, and maintaining synapses [71]. Astrocytes and microglia work in a concerted effort to maintain homeostasis in the brain and mediate inflammation [69]. Both microglia and astrocytes are known to play a role in cytokine production as well as secretion of both cytokines and complement proteins [72].

The complement pathway has been implicated in several neurodegenerative diseases such as Alzheimer's disease (AD), Parkinson's disease (PD), and Amyotrophic lateral sclerosis (ALS). Complement has a vital role in eliminating pathogens, clearing debris, and maintaining synaptic function in the central nervous system (CNS) [73]. However, when this highly regulated system begins to fail, it may exacerbate the progression of neurodegenerative diseases [69]. Several glial cell types have been implicated in neurodegeneration. Astrocytes and microglia are all necessary for normal functioning of the CNS, but when they become overactive in a diseased or aging state this may result in an increase of neuropathology in a degenerative state [73].

When astrocytes become reactive, they release C3. This results in the cleavage of C3 into its subunits, C3a and C3b. Microglia expresses the C3a receptor (C3aR). When bound by this ligand, microglia become activated, initiating the inflammatory response and phagocytosis, which is a potential mechanism of early synaptic loss in AD, resulting in memory impairments in mouse models [74]. Complement proteins in the CSF of AD patients have shown increased levels of C1q, C3d, and C4d [75,76]. Similarly, it has been shown in PD there is an increase of complement factors C3d, C4d, C7, and C9 in proximity to Lewy Bodies [77]. In mice, it was also shown that a C3 receptor (C3R) knockout was protected from the loss of dopaminergic neurons and motor impairments, which suggests that complement may contribute to the worsening of symptoms in PD [78].

5.3. The Involvement of Complement in Astrocytes and Microglia in MS

Complement C3 on microglial clusters in MS was found in the chronic stage of MS [79]. MS microglia nodules are associated with IgG transcription, complement activation, and MAC formation [80]. These studies provide evidence that complement activation in microglia may provide novel mechanisms of MS pathology.

6. Virus Infection and Complement Activation

Since a virus is thought by many to be involved in MS pathogenesis [27,81], it is relevant to consider whether complement activation occurs during viral infection. The role of complement in various viral infections has been reviewed extensively by Stoermer and Morrison [82]. Some viruses, such as Human Immunodeficiency virus (HIV) and cytomegalovirus (CMV), have developed strategies to evade the complement system by recruiting host complement regulatory proteins into their virions [82], thereby indicating that complement plays a role in virus-induced pathology. On the other hand, complement pathways can also protect the host against viral infection by mediating a degree of protection from damage and pathology induced by the virus. This protection against virus-induced disease can also occur via complement-mediated enhancement of B lymphocyte responses. This complement can either produce or protect against viral infection.

7. Hypothesis: IgG and Complement as Critical Players in Neurodegeneration in MS

7.1. Demyelination by Autoantibodies Requires Complement Activation without Fc Receptor Activation

Complement activation was found in both the early and later progressive phases of MS [83,84]. The ubiquitous occurrence of IgG and C1q staining in MS lesions indicates the dominant role of IgG antibodies and the classical complement pathway [56,85]. Antibody-antigen immunocomplexes were detected in foamy macrophages in the active lesion areas of MS [86]; demyelination by autoantibodies was found to be dependent on complement activation without Fc receptor activation [87]. A recent study demonstrated a strong association between neurodegeneration and local complement activation in progressive MS [88]. These studies provide further support for IgG-mediated complement activation in MS disease pathophysiology.

7.2. MS IgG and IgG Complexes Bind to the Surface of Neurons and Myelin

The specific antigens of the IgG within OCB in MS have remained to be identified [27]. Prominent antigen candidates include myelin proteins and viruses [81]. OCB antibodies were shown to be directed against ubiquitous intracellular antigens [89]. We previously reported MS OCB target patient-specific peptide antigens [81], and EBV epitopes were shown to be reactive to OCB [90]. It was reported that most MS serum antibodies are reactive to brain components [91].

Our study that MS plasma IgG aggregates/immune complexes induced apoptosis in neurons [41] suggests that the IgG aggregates may bind to the surfaces of neurons, which is consistent with the finding that MS had significantly higher levels of serum IgG bindings to the cell surfaces of neurons and oligodendrocytes [92]. Furthermore, IgG aggregates and immune complexes have been shown to bind myelin in MS [93]. Recent studies showed that MS CSF IgG binds to the human oligodendroglioma cell line, discriminating MS from controls with 96% specificity [92].

7.3. C1q Has a Low Affinity to Monomeric IgG but Binds to IgG Hexamers with High Avidity for Efficient Complement Activation

Complement activation is one of the most important biological functions of IgG. IgG immune complexes can activate all three pathways of the complement system, resulting in the generation of C3 and C5 cleavage products, activating a panel of different complement receptors on innate and adaptive immune cells [47]. The interaction of antibodies with specific antigens forms immune complexes/IgG aggregates. C1q has a very low affinity for monomeric IgG but avidity for IgG hexamers, which promotes complement activation [94], resulting in tissue damage and dynamic systemic activation of complement [57]. Indeed, our recent study demonstrated that MS IgG aggregate-induced neuronal cytotoxicity is complement-dependent; a co-localization of IgG and C3b was detected in neurons after incubation with MS plasma IgG aggregates [41]. We showed the staining of C5b-9 in the neurons, but we do not know whether C5b-9 is lytic or sublytic. Further investigations are needed to elucidate the mechanism.

7.4. Evidence Supporting the Presence of IgG Aggregates in MS CNS

Both IgG1 and IgG3 subclasses were reported to be present in the same OCB band, suggesting OCB consisted of heterogeneous antibodies [95]. We reported previously that there is no correlation between CSF IgG concentrations and the number of OCBs, suggesting the presence of IgG aggregates in the CSF [96]. In addition, significantly higher amounts of bound IgG were eluted from the MS brain using high- or low-pH buffers [51]. The antibody-antigen immune complexes were detected in foamy macrophages in the active lesion areas [97]. Furthermore, Mehta et al. [51] reported the presence of IgG immune complexes in MS brain tissues of white and gray matter. These studies support the presence of IgG aggregates in MS.

7.5. Potential Mechanism of IgG-Mediated, Complement-Dependent Neurodegeneration in MS

Neuron-specific activation of necroptosis was found in the cortical gray matter of MS [98]. Neuronal and axonal degeneration in MS was thought to be initiated by acute inflammation and subsequently driven by chronically smoldering, diffuse parenchymal myeloid, and meningeal lymphocytic inflammation. Oxidative stress, mitochondrial injury, and subsequent ion channel dysfunction secondary to chronic inflammation seem to have a constant impact on neurons and axons, leading to their demise during progressive MS [99].

It should be appreciated, however, that several distinct factors, in addition to complement and IgGs, may lead to the neurodegeneration seen in MS, and they are certainly not mutually exclusive. This issue has recently been discussed in detail by Morgan et al. [100] who pointed out that complement-induced MS pathology may be different in white versus gray matter in MS, that it is not known how important serum-derived as opposed to CNS-derived complement is, and that the specific complement pathways of importance remain unknown. However, both C1q and C3 may play a role, and different pathogenic mechanisms may operate at the various stages of MS, e.g., RRMS and SPMS. Morgan et al. [100] provided evidence for two fundamental neurodegenerations induced by complement pathways in MS brains. The first was called "outside-in", and the other was called the "inside-out" paradigm. In the former case, complement components enter the CNS via a breached Blood-brain barrier (BBB). They are activated either by the local presence of antigen/antibody complexes or else by "altered self" myelin epitopes. Local phagocytes and microglia then engulf myelin, which is attached to complement, resulting in demyelination and neurodegeneration. There is abundant experimental evidence to support this mechanism. By contrast, in the "inside-out" model, a so-far unknown key cellular degenerative event triggers myelin alterations and degeneration, which then allows very antigenic myelin epitopes to be exposed, with the complement pathways then acting as a secondary mechanism whereby they contribute to inflammatory responses followed by the tagging and phagocytosis of myelin [100]. Both scenarios are highly imaginative and creative, but it is yet unknown which, if either, operates in MS neurodegenerative processes. We hypothesize that both IgG and complement are key drivers of the neurodegeneration seen in MS. However, the actual mechanism(s) involved remains speculative if this is the case.

MS CNS is the target of the pathological immune responses affecting every part of the brain [101,102], characterized by the loss of neurons and oligodendrocytes, resulting in demyelination and neurodegeneration [103,104]. Gray matter pathology plays a central role in disease progression [105]. Meningeal B-cell follicles in SPMS are associated with the early onset of disease and severe cortical pathology [106], suggesting that soluble factors (IgG aggregates?) diffusing from these structures has a pathogenic role in neurons. We hypothesize that in MS CNS, the compartmentalized IgG antibodies form aggregates and produce complement-dependent cytotoxicity in neural cells.

8. Conclusions

The pathogenesis of demyelination and neurodegeneration in MS is not well understood. It is established that MS is characterized by the presence of oligoclonal bands and raised IgG. In addition, there is good evidence that complement activation is present in MS, both in histopathology of the MS brain lesions, in CSF, and in experimental studies in animal models of the disease. A key pathological feature of the MS lesion is axonal loss, which not only correlates with demyelination but is also likely to be the cause of the neurodegeneration seen as the disease progresses. It is hypothesized here that a combination of IgG and complement plays a significant role in the neurodegenerative process of MS. If this proves to be the case, then consideration should be given to developing effective therapies based on antagonists or antibodies to complement and its biochemical pathways. It is pertinent to consider future perspectives and directions. Further research is required to confirm and delineate a pathogenic role for IgG and complement in demyelination and neuronal cytotoxicity rather than just being epiphenomena. At this stage, it would be premature to consider the therapeutic options for blocking the actions of these two components in some way; nevertheless, this remains a potential treatment possibility in the future. It is also possible that clinical trials with other drugs or agents may add weight, or otherwise, to this notion.

Author Contributions: Conceptualization, X.Y.; Writings, P.G.E.K., M.F., J.P. and X.Y. All authors have read and agreed to the published version of the manuscript.

Funding: This research was funded by NIH NIMH (4R33MH118174).

Conflicts of Interest: The authors declare no conflict of interest.

References

1. Wallin, M.T.; Culpepper, W.J.; Campbell, J.D.; Nelson, L.M.; Langer-Gould, A.; Marrie, R.A.; Cutter, G.R.; Kaye, W.E.; Wagner, L.; Tremlett, H.; et al. The prevalence of MS in the United States: A population-based estimate using health claims data. *Neurology* **2019**, *92*, e1029–e1040. [CrossRef] [PubMed]
2. Trapp, B.D.; Peterson, J.; Ransohoff, R.M.; Rudick, R.; Mork, S.; Bo, L. Axonal transection in the lesions of multiple sclerosis. *N. Engl. J. Med.* **1998**, *338*, 278–285. [CrossRef]
3. Filippi, M.; Bozzali, M.; Rovaris, M.; Gonen, O.; Kesavadas, C.; Ghezzi, A.; Martinelli, V.; Grossman, R.I.; Scotti, G.; Comi, G.; et al. Evidence for widespread axonal damage at the earliest clinical stage of multiple sclerosis. *Brain* **2003**, *126*, 433–437. [CrossRef] [PubMed]
4. Calabrese, M.; Atzori, M.; Bernardi, V.; Morra, A.; Romualdi, C.; Rinaldi, L.; McAuliffe, M.J.; Barachino, L.; Perini, P.; Fischl, B.; et al. Cortical atrophy is relevant in multiple sclerosis at clinical onset. *J. Neurol.* **2007**, *254*, 1212–1220. [CrossRef]
5. Kennedy, P.G.E.; George, W.; Yu, X. The Possible Role of Neural Cell Apoptosis in Multiple Sclerosis. *Int. J. Mol. Sci.* **2022**, *23*, 7584. [CrossRef]
6. Yu, X.; Graner, M.; Kennedy, P.G.E.; Liu, Y. The Role of Antibodies in the Pathogenesis of Multiple Sclerosis. *Front. Neurol.* **2020**, *11*, 533388. [CrossRef] [PubMed]
7. Mahad, D.H.; Trapp, B.D.; Lassmann, H. Pathological mechanisms in progressive multiple sclerosis. *Lancet Neurol.* **2015**, *14*, 183–193. [CrossRef]
8. Mead, R.J.; Singhrao, S.K.; Neal, J.W.; Lassmann, H.; Morgan, B.P. The membrane attack complex of complement causes severe demyelination associated with acute axonal injury. *J. Immunol.* **2002**, *168*, 458–465. [CrossRef]
9. Damjanovic, D.; Valsasina, P.; Rocca, M.A.; Stromillo, M.L.; Gallo, A.; Enzinger, C.; Hulst, H.E.; Rovira, A.; Muhlert, N.; De Stefano, N.; et al. Hippocampal and Deep Gray Matter Nuclei Atrophy Is Relevant for Explaining Cognitive Impairment in MS: A Multicenter Study. *AJNR Am. J. Neuroradiol.* **2017**, *38*, 18–24. [CrossRef]
10. Eshaghi, A.; Marinescu, R.V.; Young, A.L.; Firth, N.C.; Prados, F.; Jorge Cardoso, M.; Tur, C.; De Angelis, F.; Cawley, N.; Brownlee, W.J.; et al. Progression of regional grey matter atrophy in multiple sclerosis. *Brain* **2018**, *141*, 1665–1677. [CrossRef]
11. Scalfari, A.; Romualdi, C.; Nicholas, R.S.; Mattoscio, M.; Magliozzi, R.; Morra, A.; Monaco, S.; Muraro, P.A.; Calabrese, M. The cortical damage, early relapses, and onset of the progressive phase in multiple sclerosis. *Neurology* **2018**, *90*, e2107–e2118. [CrossRef]
12. Andica, C.; Hagiwara, A.; Kamagata, K.; Yokoyama, K.; Shimoji, K.; Saito, A.; Takenaka, Y.; Nakazawa, M.; Hori, M.; Cohen-Adad, J. Gray matter alterations in early and late relapsing-remitting multiple sclerosis evaluated with synthetic quantitative magnetic resonance imaging. *Sci. Rep.* **2019**, *9*, 8147. [CrossRef] [PubMed]
13. Zhou, W.; Graner, M.; Beseler, C.; Domashevich, T.; Selva, S.; Webster, G.; Ledreux, A.; Zizzo, Z.; Lundt, M.; Alvarez, E.; et al. Plasma IgG aggregates as biomarkers for multiple sclerosis. *Clin. Immunol.* **2023**, *256*, 109801. [CrossRef] [PubMed]
14. Lubetzki, C.; Stankoff, B. Demyelination in multiple sclerosis. *Handb. Clin. Neurol.* **2014**, *122*, 89–99. [CrossRef] [PubMed]
15. Stassart, R.M.; Mobius, W.; Nave, K.A.; Edgar, J.M. The Axon-Myelin Unit in Development and Degenerative Disease. *Front. Neurosci.* **2018**, *12*, 467. [CrossRef]
16. Prineas, J.W.; Parratt, J.D. Oligodendrocytes and the early multiple sclerosis lesion. *Ann. Neurol.* **2012**, *72*, 18–31. [CrossRef]
17. Lei, Z.; Lin, W. Mechanisms Governing Oligodendrocyte Viability in Multiple Sclerosis and Its Animal Models. *Cells* **2024**, *13*, 116. [CrossRef]

18. Trapp, B.D.; Vignos, M.; Dudman, J.; Chang, A.; Fisher, E.; Staugaitis, S.M.; Battapady, H.; Mork, S.; Ontaneda, D.; Jones, S.E.; et al. Cortical neuronal densities and cerebral white matter demyelination in multiple sclerosis: A retrospective study. *Lancet Neurol.* **2018**, *17*, 870–884. [CrossRef]
19. Groh, J.; Abdelwahab, T.; Kattimani, Y.; Horner, M.; Loserth, S.; Gudi, V.; Adalbert, R.; Imdahl, F.; Saliba, A.E.; Coleman, M.; et al. Microglia-mediated demyelination protects against CD8$^+$ T cell-driven axon degeneration in mice carrying PLP defects. *Nat. Commun.* **2023**, *14*, 6911. [CrossRef]
20. Grundke Iqbal, I.; Raine, C.S.; Johnson, A.B.; Brosnan, C.F.; Bornstein, M.B. Experimental allergic encephalomyelitis. Characterization of serum factors causing demyelination and swelling of myelin. *J. Neurol. Sci.* **1981**, *50*, 63–79. [CrossRef]
21. Challa, D.K.; Bussmeyer, U.; Khan, T.; Montoyo, H.P.; Bansal, P.; Ober, R.J.; Ward, E.S. Autoantibody depletion ameliorates disease in murine experimental autoimmune encephalomyelitis. *MAbs* **2013**, *5*, 655–659. [CrossRef] [PubMed]
22. Linzey, M.; DiSano, K.; Welsh, N.; Pachner, A.; Gilli, F. Divergent complement system activation in two clinically distinct murine models of multiple sclerosis. *Front. Immunol.* **2022**, *13*, 924734. [CrossRef] [PubMed]
23. Hollen, C.W.; Paz Soldan, M.M.; Rinker, J.R., 2nd; Spain, R.I. The Future of Progressive Multiple Sclerosis Therapies. *Fed. Pract.* **2020**, *37*, S43–S49. [PubMed]
24. Wei, W.; Ma, D.; Li, L.; Zhang, L. Progress in the Application of Drugs for the Treatment of Multiple Sclerosis. *Front. Pharmacol.* **2021**, *12*, 724718. [CrossRef]
25. Hauser, S.L.; Bar-Or, A.; Comi, G.; Giovannoni, G.; Hartung, H.P.; Hemmer, B.; Lublin, F.; Montalban, X.; Rammohan, K.W.; Selmaj, K.; et al. Ocrelizumab versus Interferon Beta-1a in Relapsing Multiple Sclerosis. *N. Engl. J. Med.* **2017**, *376*, 221–234. [CrossRef]
26. Burman, J. Delaying the inevitable: Are disease modifying drugs for progressive MS worthwhile? *Mult. Scler. Relat. Disord.* **2021**, *54*, 103134. [CrossRef]
27. Kennedy, P.G.E.; George, W.; Yu, X. The elusive nature of the oligoclonal bands in multiple sclerosis. *J. Neurol.* **2024**, *271*, 116–124. [CrossRef]
28. Caroscio, J.T.; Kochwa, S.; Sacks, H.; Makuku, S.; Cohen, J.A.; Yahr, M.D. Quantitative cerebrospinal fluid IgG measurements as a marker of disease activity in multiple sclerosis. *Arch. Neurol.* **1986**, *43*, 1129–1131. [CrossRef]
29. Joseph, F.G.; Hirst, C.L.; Pickersgill, T.P.; Ben-Shlomo, Y.; Robertson, N.P.; Scolding, N.J. CSF oligoclonal band status informs prognosis in multiple sclerosis: A case control study of 100 patients. *J. Neurol. Neurosurg. Psychiatry* **2009**, *80*, 292–296. [CrossRef]
30. Avasarala, J.R.; Cross, A.H.; Trotter, J.L. Oligoclonal band number as a marker for prognosis in multiple sclerosis. *Arch. Neurol.* **2001**, *58*, 2044–2045. [CrossRef]
31. Calabrese, M.; Poretto, V.; Favaretto, A.; Alessio, S.; Bernardi, V.; Romualdi, C.; Rinaldi, F.; Perini, P.; Gallo, P. Cortical lesion load associates with progression of disability in multiple sclerosis. *Brain* **2012**, *135*, 2952–2961. [CrossRef] [PubMed]
32. Ferreira, D.; Voevodskaya, O.; Imrell, K.; Stawiarz, L.; Spulber, G.; Wahlund, L.O.; Hillert, J.; Westman, E.; Karrenbauer, V.D. Multiple sclerosis patients lacking oligoclonal bands in the cerebrospinal fluid have less global and regional brain atrophy. *J. Neuroimmunol.* **2014**, *274*, 149–154. [CrossRef] [PubMed]
33. Heussinger, N.; Kontopantelis, E.; Gburek-Augustat, J.; Jenke, A.; Vollrath, G.; Korinthenberg, R.; Hofstetter, P.; Meyer, S.; Brecht, I.; Kornek, B.; et al. Oligoclonal bands predict multiple sclerosis in children with optic neuritis. *Ann. Neurol.* **2015**, *77*, 1076–1082. [CrossRef] [PubMed]
34. Farina, G.; Magliozzi, R.; Pitteri, M.; Reynolds, R.; Rossi, S.; Gajofatto, A.; Benedetti, M.D.; Facchiano, F.; Monaco, S.; Calabrese, M. Increased cortical lesion load and intrathecal inflammation is associated with oligoclonal bands in multiple sclerosis patients: A combined CSF and MRI study. *J. Neuroinflamm.* **2017**, *14*, 40. [CrossRef]
35. Seraji-Bozorgzad, N.; Khan, O.; Cree, B.A.C.; Bao, F.; Caon, C.; Zak, I.; Razmjou, S.; Tselis, A.; Millis, S.; Bernitsas, E. Cerebral Gray Matter Atrophy Is Associated with the CSF IgG index in African American with Multiple Sclerosis. *J. Neuroimaging* **2017**, *27*, 476–480. [CrossRef]
36. Cristofanilli, M.; Rosenthal, H.; Cymring, B.; Gratch, D.; Pagano, B.; Xie, B.; Sadiq, S.A. Progressive multiple sclerosis cerebrospinal fluid induces inflammatory demyelination, axonal loss, and astrogliosis in mice. *Exp. Neurol.* **2014**, *261*, 620–632. [CrossRef]
37. Alcázar, A.; Regidor, I.; Masjuan, J.; Salinas, M.; Alvarez-Cermeño, J.C. Axonal damage induced by cerebrospinal fluid from patients with relapsing-remitting multiple sclerosis. *J. Neuroimmunol.* **2000**, *104*, 58–67. [CrossRef]
38. Cid, C.; Alvarez-Cermeño, J.C.; Regidor, I.; Plaza, J.; Salinas, M.; Alcázar, A. Caspase inhibitors protect against neuronal apoptosis induced by cerebrospinal fluid from multiple sclerosis patients. *J. Neuroimmunol.* **2003**, *136*, 119–124. [CrossRef]
39. Wong, J.K.; Lin, J.; Kung, N.J.; Tse, A.L.; Shimshak, S.J.E.; Roselle, A.K.; Cali, F.M.; Huang, J.; Beaty, J.M.; Shue, T.M.; et al. Cerebrospinal fluid immunoglobulins in primary progressive multiple sclerosis are pathogenic. *Brain* **2023**, *146*, 1979–1992. [CrossRef]
40. Lisak, R.P.; Nedelkoska, L.; Benjamins, J.A.; Schalk, D.; Bealmear, B.; Touil, H.; Li, R.; Muirhead, G.; Bar-Or, A. B cells from patients with multiple sclerosis induce cell death via apoptosis in neurons in vitro. *J. Neuroimmunol.* **2017**, *309*, 88–99. [CrossRef]
41. Zhou, W.; Graner, M.; Paucek, P.; Beseler, C.; Boisen, M.; Bubak, A.; Asturias, F.; George, W.; Graner, A.; Ormond, D.; et al. Multiple sclerosis plasma IgG aggregates induce complement-dependent neuronal apoptosis. *Cell Death Dis.* **2023**, *14*, 254. [CrossRef] [PubMed]
42. Saez-Calveras, N.; Stuve, O. The role of the complement system in Multiple Sclerosis: A review. *Front. Immunol.* **2022**, *13*, 970486. [CrossRef] [PubMed]

43. Lumsden, C.E. The immunogenesis of the multiple sclerosis plaque. *Brain Res.* **1971**, *28*, 365–390. [CrossRef] [PubMed]
44. Yu, X.; Zizzo, Z.; Kennedy, P.G. An appraisal of antigen identification and IgG effector functions driving host immune responses in multiple sclerosis. *Mult. Scler. Relat. Disord.* **2021**, *56*, 103328. [CrossRef]
45. Schartz, N.D.; Tenner, A.J. The good, the bad, and the opportunities of the complement system in neurodegenerative disease. *J. Neuroinflamm.* **2020**, *17*, 354. [CrossRef]
46. Westacott, L.J.; Wilkinson, L.S. Complement Dependent Synaptic Reorganisation During Critical Periods of Brain Development and Risk for Psychiatric Disorder. *Front. Neurosci.* **2022**, *16*, 840266. [CrossRef]
47. Dunkelberger, J.R.; Song, W.C. Complement and its role in innate and adaptive immune responses. *Cell Res.* **2010**, *20*, 34–50. [CrossRef]
48. Tatomir, A.; Talpos-Caia, A.; Anselmo, F.; Kruszewski, A.M.; Boodhoo, D.; Rus, V.; Rus, H. The complement system as a biomarker of disease activity and response to treatment in multiple sclerosis. *Immunol. Res.* **2017**, *65*, 1103–1109. [CrossRef]
49. Mastellos, D.C.; Hajishengallis, G.; Lambris, J.D. A guide to complement biology, pathology and therapeutic opportunity. *Nat. Rev. Immunol.* **2024**, *24*, 118–141. [CrossRef]
50. Ninomiya, H.; Sims, P.J. The human complement regulatory protein CD59 binds to the alpha-chain of C8 and to the "b" domain of C9. *J. Biol. Chem.* **1992**, *267*, 13675–13680. [CrossRef]
51. Mehta, P.D.; Frisch, S.; Thormar, H.; Tourtellotte, W.W.; Wisniewski, H.M. Bound antibody in multiple sclerosis brains. *J. Neurol. Sci.* **1981**, *49*, 91–98. [CrossRef] [PubMed]
52. Glynn, P.; Gilbert, H.; Newcombe, J.; Cuzner, M.L. Rapid analysis of immunoglobulin isoelectric focusing patterns with cellulose nitrate sheets and immunoperoxidase staining. *J. Immunol. Methods* **1982**, *51*, 251–257. [CrossRef] [PubMed]
53. Link, H.; Huang, Y.M. Oligoclonal bands in multiple sclerosis cerebrospinal fluid: An update on methodology and clinical usefulness. *J. Neuroimmunol.* **2006**, *180*, 17–28. [CrossRef] [PubMed]
54. Glynn, P.; Gilbert, H.M.; Newcombe, J.; Cuzner, M.L. Analysis of immunoglobulin G in multiple sclerosis brain: Quantitative and isoelectric focusing studies. *Clin. Exp. Immunol.* **1982**, *48*, 102–110. [PubMed]
55. Lucchinetti, C.; Bruck, W.; Parisi, J.; Scheithauer, B.; Rodriguez, M.; Lassmann, H. Heterogeneity of multiple sclerosis lesions: Implications for the pathogenesis of demyelination. *Ann. Neurol.* **2000**, *47*, 707–717. [CrossRef]
56. Breij, E.C.; Brink, B.P.; Veerhuis, R.; van den Berg, C.; Vloet, R.; Yan, R.; Dijkstra, C.D.; van der Valk, P.; Bö, L. Homogeneity of active demyelinating lesions in established multiple sclerosis. *Ann. Neurol.* **2008**, *63*, 16–25. [CrossRef]
57. Watkins, L.M.; Neal, J.W.; Loveless, S.; Michailidou, I.; Ramaglia, V.; Rees, M.I.; Reynolds, R.; Robertson, N.P.; Morgan, B.P.; Howell, O.W. Complement is activated in progressive multiple sclerosis cortical grey matter lesions. *J. Neuroinflamm.* **2016**, *13*, 161. [CrossRef]
58. Rus, H.; Cudrici, C.; Niculescu, F. The role of the complement system in innate immunity. *Immunol. Res.* **2005**, *33*, 103–112. [CrossRef]
59. Morgan, B.P.; Campbell, A.K.; Compston, D.A. Terminal component of complement (C9) in cerebrospinal fluid of patients with multiple sclerosis. *Lancet* **1984**, *2*, 251–254. [CrossRef]
60. Hakansson, I.; Ernerudh, J.; Vrethem, M.; Dahle, C.; Ekdahl, K.N. Complement activation in cerebrospinal fluid in clinically isolated syndrome and early stages of relapsing remitting multiple sclerosis. *J. Neuroimmunol.* **2020**, *340*, 577147. [CrossRef]
61. Oechtering, J.; Stein, K.; Schaedelin, S.A.; Maceski, A.M.; Orleth, A.; Meier, S.; Willemse, E.; Qureshi, F.; Heijnen, I.; Regeniter, A.; et al. Complement Activation Is Associated with Disease Severity in Multiple Sclerosis. *Neurol. Neuroimmunol. Neuroinflamm.* **2024**, *11*, e200212. [CrossRef] [PubMed]
62. Ayano, M.; Horiuchi, T. Complement as a Biomarker for Systemic Lupus Erythematosus. *Biomolecules* **2023**, *13*, 367. [CrossRef] [PubMed]
63. Thurman, J.M.; Yapa, R. Complement Therapeutics in Autoimmune Disease. *Front. Immunol.* **2019**, *10*, 672. [CrossRef]
64. Nell, D.; Wolf, R.; Podgorny, P.M.; Kuschnereit, T.; Kuschnereit, R.; Dabers, T.; Stracke, S.; Schmidt, T. Complement Activation in Nephrotic Glomerular Diseases. *Biomedicines* **2024**, *12*, 455. [CrossRef] [PubMed]
65. Keller, C.W.; Lopez, J.A.; Wendel, E.M.; Ramanathan, S.; Gross, C.C.; Klotz, L.; Reindl, M.; Dale, R.C.; Wiendl, H.; Rostasy, K.; et al. Complement Activation Is a Prominent Feature of MOGAD. *Ann. Neurol.* **2021**, *90*, 976–982. [CrossRef] [PubMed]
66. Asavapanumas, N.; Tradtrantip, L.; Verkman, A.S. Targeting the complement system in neuromyelitis optica spectrum disorder. *Expert Opin. Biol. Ther.* **2021**, *21*, 1073–1086. [CrossRef]
67. Cho, E.B.; Min, J.H.; Waters, P.; Jeon, M.; Ju, E.S.; Kim, H.J.; Kim, S.H.; Shin, H.Y.; Kang, S.Y.; Lim, Y.M.; et al. Differentiated pattern of complement system activation between MOG-IgG-associated disease and AQP4-IgG-positive neuromyelitis optica spectrum disorder. *Front. Immunol.* **2024**, *15*, 1320094. [CrossRef]
68. Lambris, J.D.; Reid, K.B.; Volanakis, J.E. The evolution, structure, biology and pathophysiology of complement. *Immunol. Today* **1999**, *20*, 207–211. [CrossRef]
69. Propson, N.E.; Gedam, M.; Zheng, H. Complement in Neurologic Disease. *Annu. Rev. Pathol.* **2021**, *16*, 277–298. [CrossRef]
70. Lee, H.G.; Wheeler, M.A.; Quintana, F.J. Function and therapeutic value of astrocytes in neurological diseases. *Nat. Rev. Drug Discov.* **2022**, *21*, 339–358. [CrossRef]
71. Lannes, N.; Eppler, E.; Etemad, S.; Yotovski, P.; Filgueira, L. Microglia at center stage: A comprehensive review about the versatile and unique residential macrophages of the central nervous system. *Oncotarget* **2017**, *8*, 114393–114413. [CrossRef] [PubMed]

72. Lian, H.; Litvinchuk, A.; Chiang, A.C.; Aithmitti, N.; Jankowsky, J.L.; Zheng, H. Astrocyte-Microglia Cross Talk through Complement Activation Modulates Amyloid Pathology in Mouse Models of Alzheimer's Disease. *J. Neurosci.* **2016**, *36*, 577–589. [CrossRef] [PubMed]
73. Lee, J.D.; Coulthard, L.G.; Woodruff, T.M. Complement dysregulation in the central nervous system during development and disease. *Semin. Immunol.* **2019**, *45*, 101340. [CrossRef] [PubMed]
74. Hong, S.; Beja-Glasser, V.F.; Nfonoyim, B.M.; Frouin, A.; Li, S.; Ramakrishnan, S.; Merry, K.M.; Shi, Q.; Rosenthal, A.; Barres, B.A.; et al. Complement and microglia mediate early synapse loss in Alzheimer mouse models. *Science* **2016**, *352*, 712–716. [CrossRef] [PubMed]
75. Daborg, J.; Andreasson, U.; Pekna, M.; Lautner, R.; Hanse, E.; Minthon, L.; Blennow, K.; Hansson, O.; Zetterberg, H. Cerebrospinal fluid levels of complement proteins C3, C4 and CR1 in Alzheimer's disease. *J. Neural Transm.* **2012**, *119*, 789–797. [CrossRef]
76. Wang, Y.; Hancock, A.M.; Bradner, J.; Chung, K.A.; Quinn, J.F.; Peskind, E.R.; Galasko, D.; Jankovic, J.; Zabetian, C.P.; Kim, H.M.; et al. Complement 3 and factor h in human cerebrospinal fluid in Parkinson's disease, Alzheimer's disease, and multiple-system atrophy. *Am. J. Pathol.* **2011**, *178*, 1509–1516. [CrossRef]
77. Yamada, T.; McGeer, P.L.; McGeer, E.G. Lewy bodies in Parkinson's disease are recognized by antibodies to complement proteins. *Acta Neuropathol.* **1992**, *84*, 100–104. [CrossRef]
78. Hou, L.; Wang, K.; Zhang, C.; Sun, F.; Che, Y.; Zhao, X.; Zhang, D.; Li, H.; Wang, Q. Complement receptor 3 mediates NADPH oxidase activation and dopaminergic neurodegeneration through a Src-Erk-dependent pathway. *Redox Biol.* **2018**, *14*, 250–260. [CrossRef]
79. Michailidou, I.; Naessens, D.M.; Hametner, S.; Guldenaar, W.; Kooi, E.J.; Geurts, J.J.; Baas, F.; Lassmann, H.; Ramaglia, V. Complement C3 on microglial clusters in multiple sclerosis occur in chronic but not acute disease: Implication for disease pathogenesis. *Glia* **2017**, *65*, 264–277. [CrossRef]
80. van den Bosch, A.M.R.; van der Poel, M.; Fransen, N.L.; Vincenten, M.C.J.; Bobeldijk, A.M.; Jongejan, A.; Engelenburg, H.J.; Moerland, P.D.; Smolders, J.; Huitinga, I.; et al. Profiling of microglia nodules in multiple sclerosis reveals propensity for lesion formation. *Nat. Commun.* **2024**, *15*, 1667. [CrossRef]
81. Graner, M.; Pointon, T.; Manton, S.; Green, M.; Dennison, K.; Davis, M.; Braiotta, G.; Craft, J.; Edwards, T.; Polonsky, B.; et al. Oligoclonal IgG antibodies in multiple sclerosis target patient-specific peptides. *PLoS ONE* **2020**, *15*, e0228883. [CrossRef] [PubMed]
82. Stoermer, K.A.; Morrison, T.E. Complement and viral pathogenesis. *Virology* **2011**, *411*, 362–373. [CrossRef] [PubMed]
83. Prineas, J.W.; Kwon, E.E.; Cho, E.S.; Sharer, L.R.; Barnett, M.H.; Oleszak, E.L.; Hoffman, B.; Morgan, B.P. Immunopathology of secondary-progressive multiple sclerosis. *Ann. Neurol.* **2001**, *50*, 646–657. [CrossRef] [PubMed]
84. Barnett, M.H.; Prineas, J.W. Relapsing and remitting multiple sclerosis: Pathology of the newly forming lesion. *Ann. Neurol.* **2004**, *55*, 458–468. [CrossRef] [PubMed]
85. Ingram, G.; Loveless, S.; Howell, O.W.; Hakobyan, S.; Dancey, B.; Harris, C.L.; Robertson, N.P.; Neal, J.W.; Morgan, B.P. Complement activation in multiple sclerosis plaques: An immunohistochemical analysis. *Acta Neuropathol. Commun.* **2014**, *2*, 53. [CrossRef]
86. Sadaba, M.C.; Tzartos, J.; Paino, C.; Garcia-Villanueva, M.; Alvarez-Cermeno, J.C.; Villar, L.M.; Esiri, M.M. Axonal and oligodendrocyte-localized IgM and IgG deposits in MS lesions. *J. Neuroimmunol.* **2012**, *247*, 86–94. [CrossRef]
87. Urich, E.; Gutcher, I.; Prinz, M.; Becher, B. Autoantibody-mediated demyelination depends on complement activation but not activatory Fc-receptors. *Proc. Natl. Acad. Sci. USA* **2006**, *103*, 18697–18702. [CrossRef]
88. Cooze, B.J.; Dickerson, M.; Loganathan, R.; Watkins, L.M.; Grounds, E.; Pearson, B.R.; Bevan, R.J.; Morgan, B.P.; Magliozzi, R.; Reynolds, R.; et al. The association between neurodegeneration and local complement activation in the thalamus to progressive multiple sclerosis outcome. *Brain Pathol.* **2022**, *32*, e13054. [CrossRef]
89. Brandle, S.M.; Obermeier, B.; Senel, M.; Bruder, J.; Mentele, R.; Khademi, M.; Olsson, T.; Tumani, H.; Kristoferitsch, W.; Lottspeich, F.; et al. Distinct oligoclonal band antibodies in multiple sclerosis recognize ubiquitous self-proteins. *Proc. Natl. Acad. Sci. USA* **2016**, *113*, 7864–7869. [CrossRef]
90. Wang, Z.; Kennedy, P.G.; Dupree, C.; Wang, M.; Lee, C.; Pointon, T.; Langford, T.D.; Graner, M.W.; Yu, X. Antibodies from Multiple Sclerosis Brain Identified Epstein-Barr Virus Nuclear Antigen 1 & 2 Epitopes which Are Recognized by Oligoclonal Bands. *J. Neuroimmune Pharmacol.* **2021**, *16*, 567–580. [CrossRef]
91. Prineas, J.W.; Parratt, J.D.E. Multiple sclerosis: Serum anti-CNS autoantibodies. *Mult. Scler.* **2018**, *24*, 610–622. [CrossRef] [PubMed]
92. Nazir, F.H.; Wiberg, A.; Muller, M.; Mangsbo, S.; Burman, J. Antibodies from serum and CSF of multiple sclerosis patients bind to oligodendroglial and neuronal cell-lines. *Brain Commun.* **2023**, *5*, fcad164. [CrossRef] [PubMed]
93. van der Poel, M.; Hoepel, W.; Hamann, J.; Huitinga, I.; Dunnen, J.D. IgG Immune Complexes Break Immune Tolerance of Human Microglia. *J. Immunol.* **2020**, *205*, 2511–2518. [CrossRef] [PubMed]
94. Burton, D.R.; Woof, J.M. Human antibody effector function. *Adv. Immunol.* **1992**, *51*, 1–84. [CrossRef] [PubMed]
95. Losy, J.; Mehta, P.D.; Wisniewski, H.M. Identification of IgG subclasses' oligoclonal bands in multiple sclerosis CSF. *Acta Neurol. Scand.* **1990**, *82*, 4–8. [CrossRef]

96. Beseler, C.; Vollmer, T.; Graner, M.; Yu, X. The complex relationship between oligoclonal bands, lymphocytes in the cerebrospinal fluid, and immunoglobulin G antibodies in multiple sclerosis: Indication of serum contribution. *PLoS ONE* **2017**, *12*, e0186842. [CrossRef]
97. Prineas, J.W.; Graham, J.S. Multiple sclerosis: Capping of surface immunoglobulin G on macrophages engaged in myelin breakdown. *Ann. Neurol.* **1981**, *10*, 149–158. [CrossRef]
98. Picon, C.; Jayaraman, A.; James, R.; Beck, C.; Gallego, P.; Witte, M.E.; van Horssen, J.; Mazarakis, N.D.; Reynolds, R. Neuron-specific activation of necroptosis signaling in multiple sclerosis cortical grey matter. *Acta Neuropathol.* **2021**, *141*, 585–604. [CrossRef]
99. Friese, M.A.; Schattling, B.; Fugger, L. Mechanisms of neurodegeneration and axonal dysfunction in multiple sclerosis. *Nat. Rev. Neurol.* **2014**, *10*, 225–238. [CrossRef]
100. Morgan, B.P.; Gommerman, J.L.; Ramaglia, V. An "Outside-In" and "Inside-Out" Consideration of Complement in the Multiple Sclerosis Brain: Lessons From Development and Neurodegenerative Diseases. *Front. Cell Neurosci.* **2020**, *14*, 600656. [CrossRef]
101. Meinl, E.; Krumbholz, M.; Derfuss, T.; Junker, A.; Hohlfeld, R. Compartmentalization of inflammation in the CNS: A major mechanism driving progressive multiple sclerosis. *J. Neurol. Sci.* **2008**, *274*, 42–44. [CrossRef] [PubMed]
102. Machado-Santos, J.; Saji, E.; Troscher, A.R.; Paunovic, M.; Liblau, R.; Gabriely, G.; Bien, C.G.; Bauer, J.; Lassmann, H. The compartmentalized inflammatory response in the multiple sclerosis brain is composed of tissue-resident CD8+ T lymphocytes and B cells. *Brain* **2018**, *141*, 2066–2082. [CrossRef]
103. Carassiti, D.; Altmann, D.R.; Petrova, N.; Pakkenberg, B.; Scaravilli, F.; Schmierer, K. Neuronal loss, demyelination and volume change in the multiple sclerosis neocortex. *Neuropathol. Appl. Neurobiol.* **2018**, *44*, 377–390. [CrossRef] [PubMed]
104. Dulamea, A.O. The contribution of oligodendrocytes and oligodendrocyte progenitor cells to central nervous system repair in multiple sclerosis: Perspectives for remyelination therapeutic strategies. *Neural Regen. Res.* **2017**, *12*, 1939–1944. [CrossRef]
105. Möck, E.E.A.; Honkonen, E.; Airas, L. Synaptic Loss in Multiple Sclerosis: A Systematic Review of Human Post-mortem Studies. *Front. Neurol.* **2021**, *12*, 782599. [CrossRef] [PubMed]
106. Magliozzi, R.; Howell, O.; Vora, A.; Serafini, B.; Nicholas, R.; Puopolo, M.; Reynolds, R.; Aloisi, F. Meningeal B-cell follicles in secondary progressive multiple sclerosis associate with early onset of disease and severe cortical pathology. *Brain* **2007**, *130*, 1089–1104. [CrossRef]

biomolecules

MDPI

Review

Therapeutic Strategies for Spinocerebellar Ataxia Type 1

Laurie M. C. Kerkhof [1,2], Bart P. C. van de Warrenburg [3], Willeke M. C. van Roon-Mom [1,2] and Ronald A. M. Buijsen [1,*]

1 Department of Human Genetics, Leiden University Medical Center, 2333 ZA Leiden, The Netherlands
2 Dutch Center for RNA Therapeutics, Leiden University Medical Center, 2333 ZA Leiden, The Netherlands
3 Department of Neurology, Donders Institute for Brain, Cognition, and Behaviour, Radboud University Medical Center, 6525 GA Nijmegen, The Netherlands
* Correspondence: r.a.m.buijsen@lumc.nl

Abstract: Spinocerebellar ataxia type 1 (SCA1) is an autosomal dominant neurodegenerative disorder that affects one or two individuals per 100,000. The disease is caused by an extended CAG repeat in exon 8 of the *ATXN1* gene and is characterized mostly by a profound loss of cerebellar Purkinje cells, leading to disturbances in coordination, balance, and gait. At present, no curative treatment is available for SCA1. However, increasing knowledge on the cellular and molecular mechanisms of SCA1 has led the way towards several therapeutic strategies that can potentially slow disease progression. SCA1 therapeutics can be classified as genetic, pharmacological, and cell replacement therapies. These different therapeutic strategies target either the (mutant) *ATXN1* RNA or the ataxin-1 protein, pathways that play an important role in downstream SCA1 disease mechanisms or which help restore cells that are lost due to SCA1 pathology. In this review, we will provide a summary of the different therapeutic strategies that are currently being investigated for SCA1.

Keywords: Spinocerebellar ataxia type 1; neurodegenerative disorder; therapeutic strategies; preclinical research; clinical trials; gain-of-function mechanism; ataxin1; *ATXN1*; polyQ disorders

1. Introduction

Spinocerebellar ataxia type 1 (SCA1) is a progressive, autosomal dominant neurodegenerative disorder characterized by ataxia, speech, and swallowing difficulties, spasticity, and abnormal control of eye movements. At later stages of the disease, patients show signs of muscular atrophy and cognitive defects [1–4]. Eventually, patients show bulbar dysfunction that leads to respiratory failure, which is the main cause of death [5]. In SCA1, the first clinical signs may develop between 4 and 74 years of age, but they typically manifest in the third or fourth decade of life [6]. The time span from onset of the disease to death varies from 10 to 30 years, with an average of 15 years. Individuals with a more juvenile onset show a more rapid disease progression [7]. Epidemiological information about the prevalence of SCA1 is limited to only a few studies but is thought to be one to two cases per 100,000 people [6,8–10].

SCA1 is caused by an expanded CAG repeat in the coding region of the human *ATXN1* gene [2]. Normal alleles range from 6 to 38 CAG repeats, with 1 to 3 CAT interruptions that are thought to be involved in the stability of the trinucleotide stretch during DNA replication. Disease-causing alleles have 39 to 44 CAG repeats without stabilizing CAT interruptions or larger expansions with CAT interruptions [11–14]. There is an inverse correlation between the number of uninterrupted CAG repeats and the age of disease onset [13,15]. The *ATXN1* gene is organized in nine exons of which the first seven fall in the 5′ untranslated region (UTR). The 5′UTR exons can undergo alternative splicing, suggesting that transcriptional and translational regulation of the SCA1 encoded protein, ataxin-1, may be complex [16,17]. The protein is expressed throughout the brain and is present mainly in the nucleus of neuronal cells, where it is thought to play a role in transcriptional regulation.

However, in Purkinje Cells (PC) in the cerebellum, the ataxin-1 protein is localized in both the cytoplasm and the nucleus [16]. Expansion of the *ATXN1* gene results in an expanded polyglutamine tract in the ataxin-1 protein, thereby inducing a polyglutamine-induced, toxic gain-of-function mechanism leading to SCA1 disease pathology [18].

Currently, there is no disease-modifying treatment available for SCA1 patients, but several therapies, including physiotherapy, occupational therapy, and speech therapy are offered which provide some symptomatic relief and help in improving the quality of life [19]. Multiple disease-modifying treatments are currently being developed. In this review, we will discuss SCA1 pathophysiology, followed by an outline of the therapeutic strategies that are being investigated, limiting our review to studies that have been tested in animal models of SCA1 or have already progressed to clinical trials. We have divided the strategies into the following three categories: (1) modulation of ataxin-1 levels, (2) pharmacological targets, and (3) cell replacement therapies. Lastly, we discuss some of the limitations that hamper drug development for SCA1 and outline our future perspectives on SCA1 drug development.

2. SCA1 Pathophysiology

SCA1 pathophysiology is characterized mainly by degeneration of the cerebellar cortex, deep cerebellar nuclei, and brainstem [3]. Loss of PCs from the cerebellar cortex and the inferior olivary nucleus is most prominent, but a loss of neurons in cortical, subcortical, and spinal structures is also observed [20]. SCA1 patients also show degeneration of the peripheral nervous system (PNS), with a loss of motor neurons in the lumbosacral spinal cord and the thin anterior spinal roots [21]. Additionally, a loss of myelin as well as an early activation of astrocytes and microglia have been observed [22,23].

In the last decades, research has been conducted to study the molecular mechanisms behind SCA1 pathogenesis. However, the disease mechanisms are still not fully understood [24,25]. Genetic studies have shown that *Sca1* knock-out mice do not display any signs of ataxia, motor incoordination, or cerebellar degeneration [26]. However, *Sca1* knock-out mice do demonstrate deficits in several learning and memory behavioral tests and neurophysiological studies. These results show that SCA1 is not caused by a loss-of-function mechanism and indicate a role for ataxin-1 in learning and memory [26]. On the other hand, expression of mutant ataxin-1 in mouse and *Drosophila* models leads to relevant disease phenotypes and pathology, suggesting that SCA1 is caused by a toxic gain-of-function mechanism [27–29].

The polyglutamine repeat expansion is thought to exert its toxicity by altering the interactions of several factors with key domains, including the C-terminus and AXH domain in ataxin-1 (Figure 1, [30,31]). The ataxin-1 C-terminus contains a highly conserved region, with a nuclear localization signal (NLS) and domains for binding of the 14-3-3 protein, which is a multifunctional regulatory molecule, as well as RBM17, which is a transcription factor (Figure 1, [30–32]). The NLS shuttles the ataxin-1 protein from the cytoplasm to the nucleus. Mice expressing mutant ataxin-1 without NLS did not show any SCA1 disease signatures, while mice expressing mutant ataxin-1 containing a deletion within the self-association region spanning amino acids 495–605 developed ataxia and PC pathology in the absence of nuclear ataxin-1 aggregates [33–35]. This demonstrates that nuclear ataxin-1 localization is critical to develop pathology [30]. Other research suggests that aggregation is important in SCA1 pathology, as several genetic modifiers identified in a *Drosophila* screen highlight the role of protein folding and clearance in SCA1 [29]. Furthermore, it is hypothesized that phosphorylation of ataxin-1 at S776 facilitates 14-3-3 protein binding, thereby stabilizing the ataxin-1 protein, causing accumulation and subsequent neurodegeneration in SCA1 [31,36–38]. The expanded CAG repeat in ataxin-1 has also been shown to enhance the formation of the protein-complex-containing transcription factor RBM17 [39].

Ataxin-1 additionally contains an AXH domain of 120 amino acids, which mediates many protein–protein interactions and RNA-binding activity [24,25]. The AXH domain forms a complex with the transcriptional repressor capicua (CIC), and mutant ataxin-1 reduces complex formation and repressor function of CIC in both *Drosophila* and cell models. This contributes to SCA1 pathogenesis through a partial loss-of-function mechanism [39,40]. By inhibiting this repressor activity, gene expression of glutamatergic receptor genes and synaptic long-term depression in the PCs is altered, which contributes to disease pathophysiology [40]. Indeed, alterations in PC firing output are observed in SCA1, which are thought to be caused by alterations in the PC synaptic input as well as changes in the intrinsic PC firing [41]. Mutations in the binding site of CIC with the AXH domain in SCA1$^{154Q/2Q}$ mice normalized genome-wide CIC binding. However, transcriptional and behavioral phenotypes were only partially rescued, suggesting the involvement of additional factors in SCA1 disease pathogenesis [34,42]. The AXH domain is also known to bind to several other transcriptional regulators, including RORα and Tip60, the RoRα co-activator, thereby mediating the expression of a group of genes that play a role in PC development and function [43,44]. Additionally, AXH acts as a dimerization domain for ataxin-1-like (*ATXN1L*). *ATXN1L*, or Boat (brother-of-ataxin-1), shares 33% homology with *ATXN1*, including the conserved AXH domain, where both *ATXN1* and *ATXN1L* interact with the transcriptional repressor CIC [24,25,45]. The AXH domain also binds to RNA, and binding is dependent on the size of the CAG repeat expansion in such a way that the ability of *ATXN1* binding to RNA decreases when the repeat expansion increases, possibly altering its role in RNA metabolism [46].

Figure 1. Ataxin-1 protein and its functional domains. The ataxin-1 protein with the polyglutamine (polyQ) region, the AXH domain, and the C-terminal domain containing the NLS and the S776 phosphorylation site. Binding partners of these domains are depicted under their respective domains.

3. Therapeutic Strategies

SCA1 therapeutics can be tested in several disease model systems, including cell culture and animal models. Here, we provide a short overview of the most important mouse models used for the development of SCA1 therapeutics. Subsequently, a summary of treatment strategies targeting ataxin-1 levels, SCA1 relevant pathways, or cell replacement studies for SCA1, which have been tested in preclinical animal models or are in clinical studies, will be discussed. SCA1 is modeled in different animals, including fly, zebrafish, and mouse models [27–29,47]. The most frequently used SCA1 mouse models are the SCA1 transgenic (B05) and the knock-in SCA1 mouse model, SCA1$^{154Q/2Q}$ (Table 1, [27,28]).

The B05 transgenic mouse model uses the Purkinje cell protein 2 (Pcp2) promoter to ensure overexpression of human *ATXN1* cDNA in the cerebellar PCs specifically [27]. The human transgene was modified to contain 82 uninterrupted CAG repeats. Overexpression of this transgene in mice resulted in an ataxic phenotype as well as neuropathological changes [27,48]. Mice showed a significant loss of PCs in the cerebellum and accumulation of nuclear inclusions in the PCs [27,48]. However, the transgenic mouse model does not reflect the full disease pathology, which also involves a dysfunction in a variety of other neurons next to PCs, leading to, for instance, the cognitive dysfunction observed in humans [3,49].

The SCA1$^{154Q/2Q}$ knock-in mice expresses mutant *Atxn1*, with 154 CAG repeats under the control of endogenous regulators, thereby expressing mutant *Atxn1* throughout the brain and spinal cord. SCA1$^{154Q/2Q}$ mice display motor incoordination and muscle wasting, as well as premature lethality, reduced adult hippocampal neurogenesis, and kyphosis (exaggerated, forward rounding of the upper back due to weakness in the spinal bones) [28]. Furthermore, the mice display cognitive deficits, such as reduced memory and cognitive flexibility [50,51].

In addition to mouse models to study SCA1, a fly model was developed [29,47]. The *Drosophila melanogaster* fly model expresses human 82Q *ATXN1* in the eye retina using the GAL4/UAS system. This leads to nuclear inclusion formation in the eye photoreceptor cells and neurons of the CNS, subsequently resulting in degeneration of the retina and neurodegeneration [29]. However, the fly model lacks a cerebellum, making it difficult to model the cerebellar neurodegeneration of SCA1 [47].

Table 1. Animal models used in the pre-clinical development of SCA1 therapeutics.

SCA1 Animal Model	Genetic Background	Advantages	Disadvantages
SCA1 transgenic B05 mice	Human *ATXN1* gene containing 82 CAG repeats under control of a pcp2 promoter	Mimics cerebellar component of SCA1, including ataxia and classical neuropathological changes. Contains human *ATXN1*.	Expression only in the PCs of the cerebellum. Has 50 to 100 times overexpression of *ATXN1* compared with endogenous levels [27,48].
SCA1 knock-in mice (SCA1$^{154Q/2Q}$)	154 CAG repeats in the mouse *Sca1* locus under control of endogenous promoters	Expresses endogenous *Atxn1* levels. Shows development of slow progressive neurodegeneration, ataxia, and deficits in memory (reflecting extra-cerebellar pathology).	Nuclear inclusions were found in more brain regions in the mice than were found in humans. Model contains very long CAG repeats, not in the range of SCA1 patients [28].
SCA1 *Drosophila melanogaster* (fly) model	Expresses human *ATXN1* gene containing 82Q in the eye retina using the GAL4/UAS system	Shows nuclear inclusion formation in the eye photoreceptor cells and CNS neurons as well as degeneration of the retina and neurodegeneration.	The model lacks a cerebellum [29].

3.1. Modulating Ataxin-1 Levels

As discussed before, the cellular pathological mechanisms underlying SCA1 are driven mainly by the presence of the mutant ataxin-1 protein. Presently, many intervention strategies focus on the development of therapies that can reduce the levels of mutant ataxin-1, which will have an effect on all downstream pathological processes. The ataxin-1-lowering strategies that are currently in development target *ATXN1* RNA act by either modulating RNA levels or by blocking translation. Here, we will provide an overview of the different ataxin-1 modulating therapies that are currently being explored for SCA1 (Figure 2).

3.1.1. RNA Interference

RNA interference (RNAi), or post-transcriptional gene silencing, is a conserved biological process in which noncoding double-stranded RNA molecules are involved in suppression of gene expression through translational or transcriptional repression. [52,53]. The process can be induced by tailor-made genetic sequences, thus providing a method for deliberately silencing a gene of interest in model systems [53]. The RNAi process is mediated by three functionally different noncoding dsRNA molecules, namely microRNA (miRNA), short hairpin RNA (shRNA), and small interfering RNA (siRNA), which converge into the same RNAi pathway. In general, RNAi strategies cleave the ds-RNA with an endonuclease called Dicer into single RNA fragments. The guide strands of these RNA fragments are then subsequently loaded into an RNA-induced silencing complex (RISC) to promote endonucleolytic cleavage of the homologous mRNA by the RNase Argonaute 2 (Figure 2, [54]).

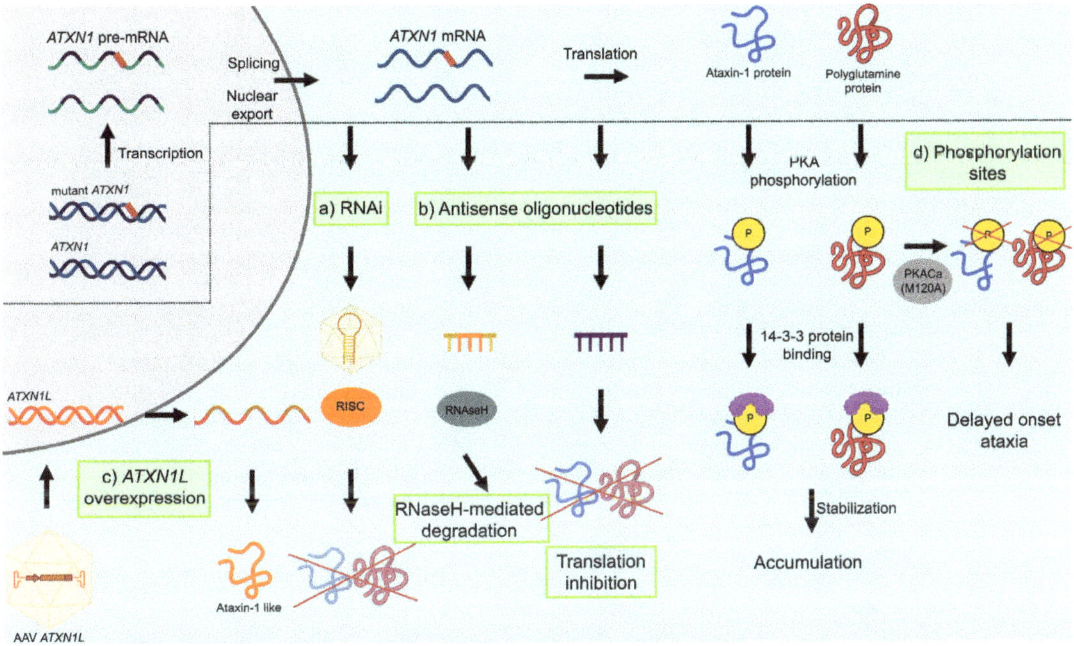

Figure 2. Ataxin-1 modulating therapies. Ataxin-1 modulating therapies that are currently being investigated as possible therapeutics for SCA1. These strategies include (**a**) RNAi strategies, which involve binding to the RISC, (**b**) ASOs, including ASOs that mediate RNaseH-mediated degradation and translation inhibition, (**c**) ATXN1L, which can be overexpressed to preserve wildtype protein function, or (**d**) the S776 phosphorylation site, which can be targeted to reduce ataxin-1 levels.

A first proof-of-concept study for SCA1 in 2004 showed the ability of RNAi to inhibit neurodegeneration in the SCA1 B05 transgenic mouse model. Upon intracerebellar injection, recombinant adeno-associated virus (AAV) vectors expressing shRNAs targeting human *ATXN1* sequences improved motor coordination, restored cerebellar morphology, and resolved ataxin-1 inclusions in PCs of these mice (Table 2, [55]). RNAi therapies can be delivered using lentiviruses and AAVs, as they are minimally immunogenic and result in the stable expression in the brain regions of interest [56]. Keiser, Davidson, and colleagues delivered improved viral vectors expressing miRNAs targeting human ataxin-1 (miS1 vectors) to the cerebellum of pre-symptomatic SCA1 B05 transgenic mice. This resulted in widespread cerebellar PC transduction and improved behavioral and histological phenotypes [57]. Next, to assess the extra-cerebellar therapeutic effects, the miS1 vector was injected into the deep cerebellar nuclei of pre-symptomatic SCA1$^{154Q/2Q}$ knock-in mice, where the mutant ataxin-1 is expressed throughout the brain. This approach reduced ataxin-1 expression in the cerebellar cortex and brainstem. Cerebellar lobule integrity, rotarod performance, and histological phenotypes were preserved, and disease-related transcriptional changes were prevented for over a year, suggesting that delivery to deep cerebellar nuclei is sufficient for widespread therapeutic benefit [58]. Importantly, as treatment in humans is started after symptom onset, a follow-up study in symptomatic transgenic B05 mice was performed. Results of this study showed that the SCA1 phenotypes could be reversed when the miS1 vector was delivered after symptom onset [59]. To study biodistribution and tolerability, the compound was subsequently administered into the deep cerebellar nuclei of adult rhesus macaques. Eight weeks after injection, transduction was seen in SCA1 relevant brain regions, including the deep cerebellar nuclei, cerebellar PCs, and the brainstem, where a significant reduction of endogenous *ATXN1* mRNA levels of more than 30% compared with the uninjected hemisphere was found [60]. There were no clinical complications as assessed by behavioral abnormalities or neuropathological findings [60]. Altogether, these data are supportive of a clinical application of an AAV-based RNAi therapy for SCA1 [57–60]. To advance this technology to patients, investigational new drug (IND)-enabling studies were performed. However, 3 months after AAV-based delivery into the deep cerebellar nuclei of rhesus macaques, cerebellar toxicity was observed [61]. RNA-sequencing studies showed that, despite limited amounts of vector, there was substantial 3′ inverted terminal repeat promoter activity showing neurotoxic effects in the rhesus macaques [61]. By altering the miS1 expression context, the promoter activity was reduced. These findings stress the importance of extended safety studies in multiple species when assessing new therapeutics for human application [61].

3.1.2. Antisense Oligonucleotides

A second RNA-based therapy approach to reduce gene or protein expression is the use of antisense oligonucleotides (ASOs). ASOs are chemically modified, single-stranded oligonucleotides of generally 12 to 30 bases, which bind to RNA through Watson–Crick–Franklin base pairing. Depending on the sequence and modifications, ASOs can alter RNA functions through several mechanisms. They can be used to reduce expression of a toxic protein, to modify mutant proteins, to reduce their toxicity, or to restore protein expression (Figure 2, [66]).

ASO therapies can be delivered through intravenous administration or directly into the CSF via intrathecal (IT) or intracerebroventricular (ICV) injections [54,56,67]. Although systemic administration is less invasive and, therefore, favorable for clinical applications, it requires doses that are approximately 100 times higher compared with ICV, which increases the risk of toxicity [68]. To minimize toxicity, transporters or drug carriers can be used, which allow the ASOs to cross the blood–brain barrier (BBB) [68–72]. To avoid the BBB, ASOs can also be delivered directly in the CSF through IT or ICV injections [54,56,67,68]. Local delivery into the CSF is preferred, as this allows for direct and local targeting, thereby reducing the required doses and reducing systemic exposure and toxicity as well as renal and hepatic elimination [67,68]. It has been shown that after IT and ICV injections, ASOs are widely distributed in the brain, although the delivery into deeper cerebellar brain structures seems to be less efficient [68]. IT injections have already been successfully used in human clinical trials for amyotrophic lateral sclerosis and spinal muscular atrophy, without major side effects [73,74]. However, as these injections are quite invasive and treatment with ASOs are transient, re-administration of the ASOs every 3 to 4 months limits its clinical use [66–68].

Gapmers are chimeric ASOs that contain a central block of DNA nucleotides flanked by modified RNA sequences, usually containing 2′-O-modified chemistries. The modified sequences improve target affinity and stability, while the central DNA sequence forms a DNA/RNA hybrid, which stimulates RNA cleavage through the recruitment of RNase H (Figure 2). Following a single ICV injection of an *Atxn1* gapmer ASO353 in pre-symptomatic $SCA1^{154Q/2Q}$ mice, a reduction of *Atxn1* levels at 6 and 24 weeks after ASO administration was shown. Furthermore, a rescue of disease-associated phenotypes was demonstrated, including motor performance, survival, analysis of neurochemicals, and transcriptional disease signatures (Table 2, [62]).

Table 2. Summary of the different therapeutic strategies aimed at modulating ataxin-1 levels.

Mechanism	Model System	Molecular Findings		Pathology	Behavioral/Functional Tests
		RNA Level	Protein Level		
shRNA targeting human *ATXN1*	SCA1 transgenic B05 mice	-	-	Amelioration of PC pathology and rescue of nuclear inclusions in PCs.	Improvement in motor performance as assessed by the rotarod assay [55].
AAV expressing an artificial miRNA targeting human ataxin-1 (miS1)	SCA1 knock-in mice (SCA1$^{154Q/2Q}$); pre-symptomatic	Up to 30% reduction	58 to 72% reduction of ataxin-1 levels	Amelioration of PC pathology and no glial activation.	Improvement in motor performance as assessed by the rotarod assay and improvement in gait analysis [58].
	SCA1 transgenic B05 mice; pre-symptomatic	Up to 70% reduction		Amelioration of PC pathology and no glial activation.	Improvement of motor performance as assessed by the rotarod assay [57].
	SCA1 transgenic B05 mice; symptomatic	Up to 67% reduction		Amelioration of PC pathology. Mice treated with the highest dose showed no *ATXN1*-positive PCs but showed enhanced immunoreactivity.	Rescue of motor performance to wildtype levels as assessed by the rotarod assay in a dose-dependent manner; rescue in NAA/inositol ratios [59].
	Rhesus monkey	Up to 30% reduction in the left cerebellar hemisphere		Eight weeks post-injection, microglial activation and astrogliosis were enhanced in the left cerebellar cortex and deep cerebellar nuclei [60].	
	Rhesus monkey	Significant reduction of *ATXN1* in medial and lateral cerebellar cortex		Toxicity in deep cerebellar nuclei was observed; necrosis, demyelination, perivascular/leptomeningeal lymphoid infiltrates, increase in lesion severity, PC loss, and enhanced immunoreactivity.	Monkeys developed ataxia, tremor, head-tilt, and dysmetria 3 months after treatment [61].
ASO353	SCA1 knock-in mice (SCA1$^{154Q/2Q}$)	Significant reduction of total mouse *Atxn1* in brainstem and cerebellum		Rescue of analysis of neurochemicals and transcriptional disease signatures. No unwanted side effects involving BACE1, CIC activity or reduction in neuronal precursor cells.	Improved motor performance at 28 weeks, as assessed by the balance beam test. Pre-symptomatic treatment improved motor performance on both the rotarod and balance beam as well as prolonged survival [62,63].
(CUG)^7 VO659	SCA1 knock-in mice (SCA1$^{154Q/2Q}$)	An exon 8 skip product of the *Atxn1* gene was detected	A significant dose-dependent reduction in polyglutamine ataxin-1 in several brain regions. Reduction up to 45% in cerebellum and up to 56% in brainstem [64]		
Combination of miS1 and vectors expressing human *ATXN1L*	SCA1 transgenic B05 mice	Significant increase in human *ATXN1L* expression and a significant reduction in human *ATXN1* levels		Improvement in autonomous transcriptional changes and partial restoration of glial activation.	Significant improvement in latency to fall on the rotarod assay [65]

Steric or RNA blocking ASOs can inhibit (or activate) protein translation through binding to regulatory elements, including the upstream open reading frame (Figure 2, [66]). Using a 2′-O-methyl-modified $(CUG)^7$ ASO with a phosphorothioate backbone that specifically targets expanded CAG stretches and does not activate RNase H-dependent RNA degradation, a lowering of polyglutamine protein levels was seen in polyglutamine patient cell lines and mouse models [64,75]. By administering 75 or 150 µg of $(CUG)^7$ by ICV infusion into the right lateral ventricle of $SCA1^{154Q/2Q}$ mice weekly for a total of 8 weeks, a reduction of *Atxn1* levels was seen in all SCA1-relevant brain regions, including cerebellum, brainstem, and spinal cord. Quantification of ASO levels showed a widespread distribution of $(CUG)^7$ in all brain regions. Most importantly, ASO levels in the different brain regions and the spinal cord were similar, despite the differences in distance from the injection site [64].

In addition to delivery of the ASO to the brain, targeting *ATXN1* using RNA therapy faces another challenge, as a complete loss of *ATXN1* may contribute to Alzheimer's disease by increasing the transcription of BACE1, which encodes for the ß-secretase enzyme [25,76]. In addition to these possible adverse effects after a loss of ataxin-1, another study investigated the change in BACE1 levels after downregulating ataxin-1 with the gapmer ASO353 [63]. Results from this study showed that ASO-mediated reduction of *Atxn1* did not lead to an unwanted increase in BACE1 levels in $SCA1^{154Q/2Q}$ mice after 3 ASO353 ICV injections [63]. This suggests that a partial loss of ataxin-1 function later in life due to an ASO or other RNA therapy is safe and can be used in clinical applications [63]. However, to preserve the expression and function of ataxin-1, specific silencing of the expanded CAG repeat allele can also be achieved. This can be accomplished by either targeting single nucleotide polymorphisms (SNPs) or by specifically restoring the unexpanded protein [77–79].

3.1.3. *ATXN1L* Overexpression

A different therapeutic strategy to treat SCA1 is overexpression of the *ATXN1* paralog, *ATXN1L* (Figure 2). Gene duplication of *Atxn1l* in $SCA1^{154Q/2Q}$ mice reduced neuropathology and behavioral deficits through displacement of mutant ataxin-1 from its native complex with CIC, an interaction that is involved in SCA1 cerebellar pathology [45]. Two studies have shown that treatment of AAVs expressing human *ATXN1L* improves motor coordination and pathology in pre-symptomatic SCA1 B05 mice [57,65]. Given the fact that the toxic gain-of-function mechanism can be treated with the previously described RNAi or ASO-mediated knockdown of ataxin-1, overexpression of *ATXN1L* might help by also treating the loss-of-function caused by these ataxin-1 lowering therapies. When combining upregulation of *ATXN1L* with RNAi-induced downregulation of *ATXN1*, the improvement of gene expression changes and motor behavior impairments for the combined treatment were greater than for the individual treatments [65]. However, as overexpression of wild-type ataxin-1 (30Q) in *Drosophila* results in neurodegenerative phenotypes similar to those caused by the expanded protein, it is essential to monitor possible side effects of increased *ATXN1L* [29].

3.1.4. S776 Phosphorylation

Since phosphorylation of ataxin-1 at the serine 776 residue (*ATXN1*-pS776) was shown to play a significant role in protein toxicity (Figure 2, [38], inhibition of S776 phosphorylation could be a promising therapeutic target for SCA1. In SCA1 cell models, pharmacological inhibition of S776 phosphorylation led to a decrease in ataxin-1 [38]. Depending on the brain region, phosphorylation of ataxin-1 at S776 is regulated by different factors. In the brainstem, phosphorylation is modulated mainly by Rsk3, whereas in the cerebellum, it is modulated via Msk1 [80]. Furthermore, protein kinase A (PKA) can mediate phosphorylation of S776. B05 mice with a mutation in PKA, generated via a Cre-lox system, showed a reduced PKA-mediated phosphorylation of *ATXN1*-S776 in PCs as well as enhanced degradation of ataxin-1 and improved cerebellar-dependent motor performance [38]. In the

SCA1$^{154Q/2Q}$ mouse model, where the S776 phosphorylation site was mutated to an Ala776 (S776A) on either the wildtype allele or the mutant allele, it was shown that targeting S776 phosphorylation could ameliorate SCA1 pathology [36,81]. Only disruption of the S776 phosphorylation site on the mutant allele, but not the wildtype allele, reduced ataxin-1 protein levels and ameliorated pathological disease hallmarks, including the reduced thickness of the cerebellar molecular layer and nuclear inclusion formation. Furthermore, there was a partial rescue of motor function and an expanded life span, but learning and memory deficits were not rescued [36]. As therapeutic approaches are likely to decrease S776 phosphorylation on both alleles, these studies show the importance of developing allele-specific intervention strategies for SCA1. Furthermore, the region-specific modulation of ataxin-1 S776 phosphorylation, by Rsk3 and Msk1 in the brainstem and cerebellum, respectively, underlines the need for combinatorial treatment targeting both kinases, which could improve therapeutic efficacy for SCA1 [80].

3.2. Pharmacological Targets

There are several mediators that are known to play a role in SCA1 pathology that could be potential pharmacological targets. For instance, activity of the PCs can be influenced by several drugs that target synaptic and non-synaptic receptors (Figure 3, [82–84]). PCs receive excitatory synaptic input from climbing fibers (CF) and parallel fibers (PF), whereas it receives inhibitory inputs from the molecular interneurons, the basket cells and stellate cells (Figure 4, [41]). Alterations in the glutamatergic input of climbing fibers onto PCs have shown to be altered in SCA1 mice [85,86]. Deficits in CF-PC synaptic transmission have been observed in B05 transgenic mice at 6 weeks of age [87]. Furthermore, mutant ataxin-1 markedly increases inhibitory inputs from basket cells onto the PCs during cerebellar development, thereby disrupting cerebellar PC function [88]. In addition to alterations in synaptic input, the intrinsic activity of PCs can determine the PC output, which has been shown to be altered in several SCAs [41]. PCs are able to produce spontaneous electrical activity in the absence of synaptic input. This autonomous spiking is dependent on the proper function of a number of potassium channels that are highly expressed in the dendritic membrane of PCs [83,84,89]. For instance, voltage-gated potassium channels can modulate intrinsic firing, and the calcium-dependent potassium channels are involved in maintaining the autonomous spiking of PCs [41]. Alterations in PC intrinsic firing have been observed in several SCAs, including SCA1, where a loss of spontaneous cerebellar PC spiking is observed [41]. This is caused by a disruption of calcium homeostasis, leading to reduced function of the large conductance calcium-activated potassium (BK) channels. Additionally, a dysfunction in other potassium channels, such as voltage-gated potassium channels, can lead to higher A-type potassium currents (Ika) [82–84,89,90]. This dysfunction may contribute to dendritic hyperexcitability, thereby eventually contributing to neurodegeneration [84,90].

Next to the potassium channel dysfunction, contributing to alterations in the PC intrinsic activity, altered excitatory glutamatergic inputs, linked to the type-1 metabotropic glutamate receptors (mGluR1), also contribute to the altered PC activity observed in SCA1 (Figure 3, [91–93]). Reductions in mGluR1 mRNA levels as well protein levels have been observed in symptomatic elderly SCA1$^{154Q/2Q}$ mice [91,94]. Furthermore, in pre-symptomatic B05 transgenic mice, reduced mGluR-excitatory postsynaptic current (EPSC) amplitudes were observed, leading to altered PC activity and eventually to PC death [93]. These results suggest that a loss of mGluR1 in PCs is associated with the pathological phenotype of SCA1 [91,93]. Contradictorily, another study showed prolonged mGluR1 currents, despite reduced EPSC amplitudes, in cerebellar parallel fiber synapses in symptomatic B05 mice, which are thought to be caused by a loss of glutamate transporter activity [95]. However, the majority of studies indicate a reduction in mGluR1 activity. Additionally, it has been shown that genes encoding for mGluR1 signaling proteins, such as Homer-3, are downregulated in B05 mice cerebellar PCs, which can lead to neuronal dysfunction in SCA1 [86,91,96]. As PC firing dysfunction is a common pathologic phenomenon in multiple SCAs, therapies restoring this dysfunction by targeting changes in the synaptic input as well as the intrinsic firing of PCs might be applicable to treat multiple ataxias [41]. The SCA-induced alterations in both potassium and glutamate receptors could be targeted by several drugs [97].

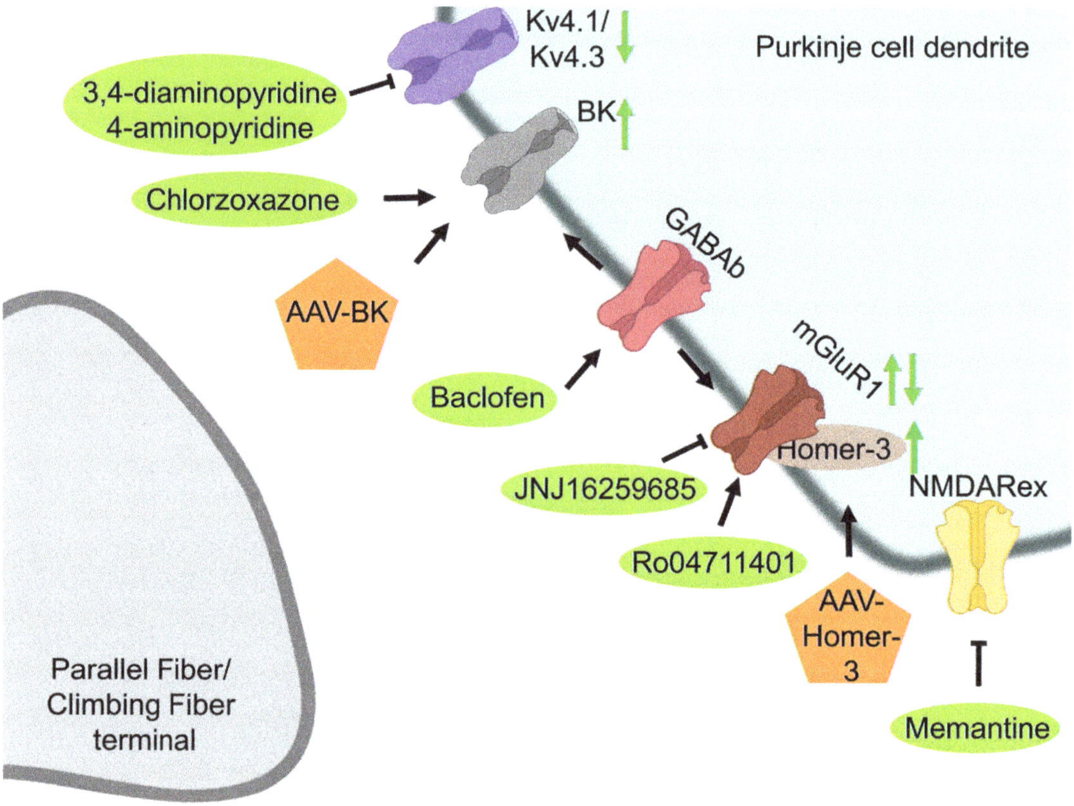

Figure 3. Receptor-mediated therapies. Cerebellar PC activity is mediated through several receptors and ion channels. In SCA1, cerebellar PC activity is dysregulated and can be restored through several drugs targeting either the potassium channels (Kv4.1/Kv4.3 and BK channels) or the mGluR1, NMDARex, or GABAb receptors. Green arrows indicate the effect of the therapies on the receptor and ion channel function.

3.2.1. Potassium Channels

To restore PC spiking in several SCAs, BK channel function could be restored, and IKa currents could be reduced. BK channel function can be restored either by enhancing BK channel expression or by activating the BK channels that are present (Figure 3, [83,84,90]). Dual administration of both stereotactically injected BK-AAV and intraperitoneally injected baclofen, a GABAb receptor agonist that potentiates a sub-threshold-activated potassium channel current in PCs, was performed. Treatment effectively normalized the dendritic excitability in the PCs, rescued neuropathology, and, moreover, significantly improved motor coordination two weeks after initiation of therapy in B05 transgenic mice (Table 3, [83,84,90]). BK channels can also be activated through pharmacological targeting with chlorzoxazone, which is a known activator of calcium-activated potassium channels. Dual treatment with chlorzoxazone and baclofen, dissolved in drinking water, normalized dendritic excitability in the PCs and significantly improved motor performance of B05 as well as in SCA1$^{154Q/2Q}$ mice in a stage where motor impairment is cerebellar in origin [84,90]. On the other hand, at 20 weeks of age, when motor dysfunction is caused by motor neuron rather than cerebellar dysfunction, dual treatment with chlorzoxazone and baclofen did not rescue motor dysfunction in the SCA1$^{154Q/2Q}$ mice [90]. As chlorzoxazone and baclofen are both muscle relaxants, they have tolerability concerns in patients with neurological disorders and older adults [90,98]. However, retrospective SCA patient data show that co-administration is tolerated and may improve SCA symptoms, suggesting that this treatment might be promising for use in clinical trials [90].

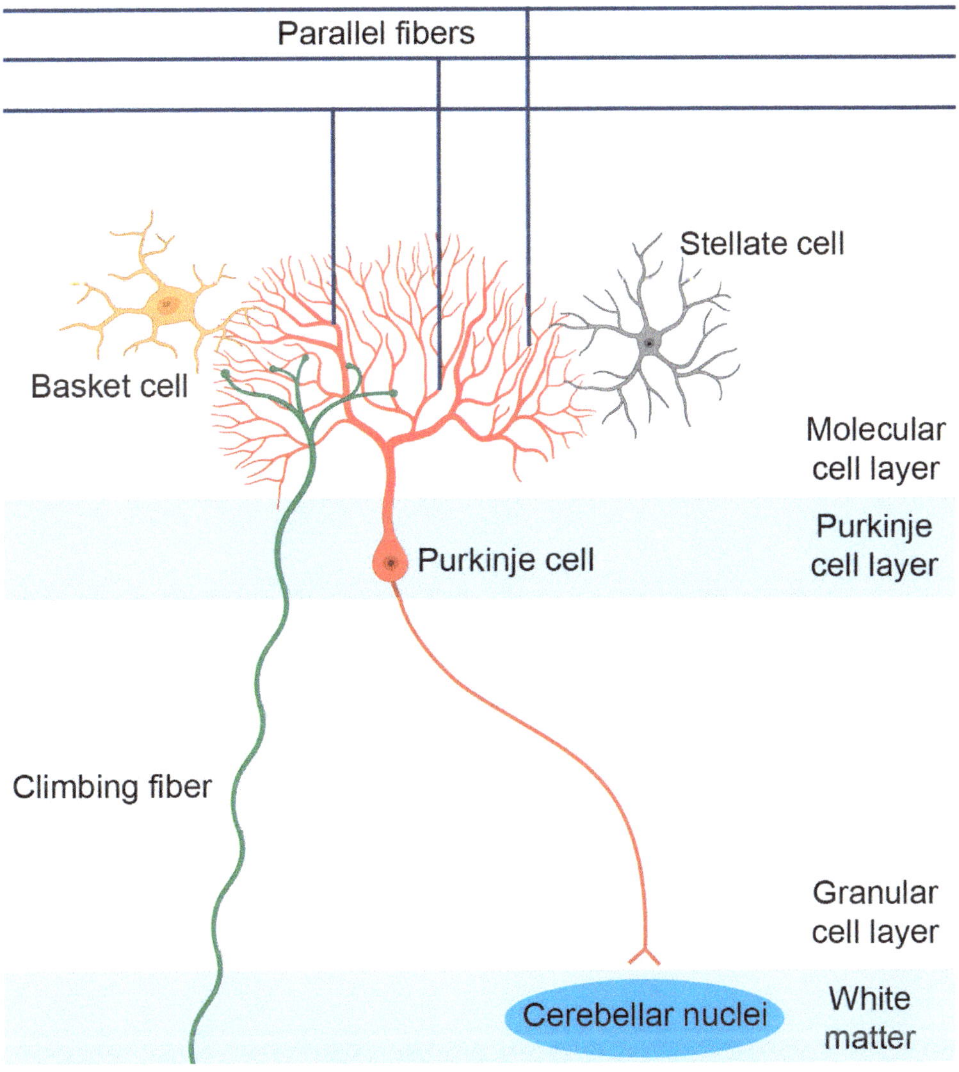

Figure 4. Overview of the cerebellar structure. The cerebellum consists of three layers, namely the granular cell layer, the purkinje cell layer, and the molecular cell layer. The PCs receive excitatory synaptic input from the CF and PF as well as inhibitory input from the basket and stellate cells.

Table 3. Overview of the different pharmacological therapeutic strategies for SCA1.

Mechanism	Model System	Pathology	Behavioral/Functional Test
BK-AAV and baclofen (intracerebellar injection)	SCA1 B05 transgenic mice	Mitigation of dendritic degeneration	Improves motor performance [83]
Chlorzoxazone and baclofen (oral administration)	SCA1 B05 transgenic mice	Rescue of dendritic excitability in PCs	Improves motor performance [84]
	SCA1 knock-in mice (SCA1$^{154Q/2Q}$)	Rescue of dendritic excitability in PCs	Improves motor performance without affecting muscle strength [90]
4-aminopyridine and 3,4-diaminopyridine (subcutaneous injection)	SCA1 B05 transgenic mice	Normalization of firing rate of PCs; partial protection against cell atrophy	Improves motor performance [82]
mGlu1 receptor PAM (Ro0711401; subcutaneous injection)	SCA1 knock-in mice (SCA1$^{154Q/2Q}$)	Increase in the number of dendritic spines	Significantly improves motor performance; effect lasted for 6 days [91]
Baclofen (intrathecal injection)	SCA1 B05 transgenic mice	Enhancement of mGluR1 signaling	Improves motor performance [93]
AAV-Homer-3 (ICV)	SCA1 B05 transgenic mice	Restoration of mTORC1 signaling and neuronal activation; improvement in PC morphology	Significantly improves motor performance [85]
Memantine (oral administration)	SCA1 B05 transgenic mice	Rescue of PC density; significant reduction in PC nuclear inclusions	Attenuates body weight loss and extends the life span [99]
IGF-1 (intranasal administration)	SCA1 B05 transgenic mice	Restoration of PC pathology	Improves motor performance [100]
Recombinant VEGF (ICV)	SCA1 knock-in mice (SCA1$^{154Q/2Q}$)	Restoration of PC dendrite pathology and firing rate; improved microvascular health.	Improves motor performance [101]
Nano-VEGF (ICV)	SCA1 knock-in mice (SCA1$^{154Q/2Q}$)	Restoration of PC dendrite pathology and firing rate; improved microvascular health.	Improves motor performance [102]
BDNF	SCA1 knock-in mice (SCA1$^{154Q/2Q}$)	Improvement in cerebellar and hippocampal pathology	Ameliorates motor and cognitive deficits [103]
Lithium carbonate (oral administration)	SCA1 knock-in mice (SCA1$^{154Q/2Q}$)	Partial rescue of dendrite pathology, restoration of isoprenylcystein carboxyl methyltransferase levels; reduction of GSK3ß levels; restoration of purine, oxidative stress, and energy production metabolic pathways	Improves motor performance, spatial learning, and memory [104,105]
MitoQ (oral administration)	SCA1 knock-in mice (SCA1$^{154Q/2Q}$)	Amelioration of mitochondrial morphology and function; improvement in PC numbers and function	Improves motor performance [106]

Table 3. *Cont.*

Mechanism	Model System	Pathology	Behavioral/Functional Test
Succinic acid (oral administration)	SCA1 B05 transgenic mice	Rescue of complex I OXPHOS dysfunction; amelioration of molecular layer and PC layer degeneration	Improves motor performance [107]
AAV1-HMGB1 (intrathecal injection)	SCA1 knock-in mice (SCA1$^{154Q/2Q}$)	Rescue of PC loss; improvement in PC morphology and molecular layer thickness as well as nuclear and mitochondrial DNA damage	Improves motor performance and increased survival ratio [108]
AAV-RpA1 (intrathecal injection)	SCA1 knock-in mice (SCA1$^{154Q/2Q}$)	Rescue of γH2AX and 53BP1 levels; recovery of mitochondrial DNA damage, dendritic pathology; rescue of molecular layer thickness, impaired splicing, transcription, and abnormal cell cycle	Improves motor performance [109]
panPAK inhibitor (intraperitoneal injection)	SCA1 knock-in mice (SCA1$^{154Q/2Q}$)	Significant reduction in both expanded and wildtype *ATXN1* RNA levels in the cerebellum [110]	
PLX3397 (oral administration)	SCA1 B05 transgenic mice	Reduction in microglial density in the cerebellum; reduction in TNF-α and increase in PSD-95 expression; increase in WT ataxin-1 protein levels	Improves motor performance [111]

Another therapeutic strategy that targets voltage-gated potassium channels reduces the observed enhanced IKa currents in SCA1 pathology. Aminopyridines (APs) are potassium channel blockers with a high affinity for A-type potassium channels. Subcutaneous injection of APs has been shown to normalize the firing frequency of PCs and the motor dysfunction of early symptomatic B05 mice [82]. However, no effects were observed when treating B05 mice with an advanced SCA1 phenotype, suggesting that motor dysfunction at later stages is caused primarily by PC atrophy and death and not by electrophysiological dysfunction [82]. Surprisingly, starting subcutaneous injections with 3,4-diaminopyridine in an early stage of SCA1 delay the onset of motor dysfunction and partially prevent neurodegeneration in B05 mice [82]. This emphasizes the importance of starting treatment early in the disease process or even before the onset of disease symptoms.

3.2.2. Glutamatergic Signaling

The observed reduction in mGluR1-mediated EPSC amplitude can be modulated by so-called positive allosteric modulators (PAMs), which amplify the mGluR1 receptor function [91,94,95]. Subcutaneous injection of SCA1$^{154Q/2Q}$ mice with the Ro0711401 PAM led to a significant increase in motor performance after 30 min [91]. Surprisingly, this effect on motor performance lasted for 6 days, which was well beyond the time of drug clearance from the cerebellar tissue. This could be explained by the induction of cerebellar long-term depression during the repeated execution of the rotarod assay, which could possibly induce a form of procedural memory [91]. Additionally, defects in learning and memory were improved [94]. On the other hand, intraperitoneal injection of the JNJ16259685 negative allosteric modulator (NAM) in SCA1$^{154Q/2Q}$ mice resulted in markedly reduced motor function [91]. Contradictorily, one study found an improvement in motor performance after subcutaneous injections in conditional B05 transgenic mice with the same NAM [95].

Furthermore, the modulatory effects of baclofen on mGluR1 have been tested [93]. Subarachnoid injections of baclofen enhanced cerebellar mGluR1 signaling and improved the motor performance in B05 mice for approximately 1 week [93]. It is suggested that baclofen can improve motor functions in the transgenic mice either through interacting with the mGluR1 and enhancing mGluR1 signaling through the secretion of brain-derived neurotrophic factor (BDNF) or through activation of insulin growth factor 1 (IGF-1) receptors [93]. However, as previously mentioned, baclofen could additionally exert its effects through potentiation of sub-threshold-activated potassium channel currents in PCs [90].

In addition to targeting mGluR1 directly, mGluR1-related signaling molecules can be targeted to ameliorate SCA1 pathology. For instance, a reduction in Homer-3, which is an adaptor protein of mGluR1, has been observed in early symptomatic SCA1$^{154Q/2Q}$ mice [85,93]. This reduction is likely mediated through reduced RORα-mediated transcriptional activity caused by mutant *ATXN1* [93]. Homer-3 binds to mGluRs, and its reduction leads to altered PC activation responses [85,93]. To improve these altered PC activation responses, Ruegsegger and colleagues enhanced Homer-3 expression using an AAV [85]. AAV-Homer-3 treatment in SCA1$^{154Q/2Q}$ mice ameliorated PC climbing fiber deficits, rescued dendritic spine loss, and improved behavioral outcomes [85].

3.2.3. NMDA Receptor

Although there is no convincing evidence that the extra-synaptic *N*-methyl-D-aspartate receptor (NMDAR) is involved in SCA1, administration of memantine, which is a NMDAR antagonist that preferentially inhibits the extra-synaptic NMDAR, had beneficial effects (Figure 3, Table 3, [99]). Oral treatment with memantine attenuated body weight loss and extended the life span of treated SCA1$^{154Q/2Q}$, possibly though preventing neuronal cell death in the brainstem or PC death in the cerebellum [99]. This suggests that extra-synaptic NMDAR contributes to the pathogenesis of SCA1 and that treatment with memantine could ameliorate this [99].

3.2.4. Protection of PC Survival and Function

Next to restoration of the PC electrical activity, several other pharmacological therapeutics have been developed that aim to protect PC survival and PC function. For SCA1, IGF-1, vascular endothelial growth factor (VEGF), BDNF, and lithium carbonate have been tested in mouse models (Figure 5, Table 3). IGF-1 is a trophic factor for both glia and neurons in the cerebellum, and it promotes PC survival and dendritic growth. IGF-1 and its receptors are also present in the climbing afferents of the inferior olive nucleus, where it is involved in motor learning processes, thereby influencing PC synaptic activity [100]. It has been found that impairments in the IGF-1 pathway are involved in neurodegenerative processes [112]. Intranasal administration of IGF-1 in B05 mice resulted in a dose-dependent improvement in motor performance as well as the restoration of the calcium buffer protein calbindin-D28K and protein kinase C-γ (PKC-γ) expression, which are both reduced in SCA1 [100]. It is suggested that restoration of these proteins could lead to improved calcium homeostasis and PKC-γ-mediated signaling in SCA1 PCs [100].

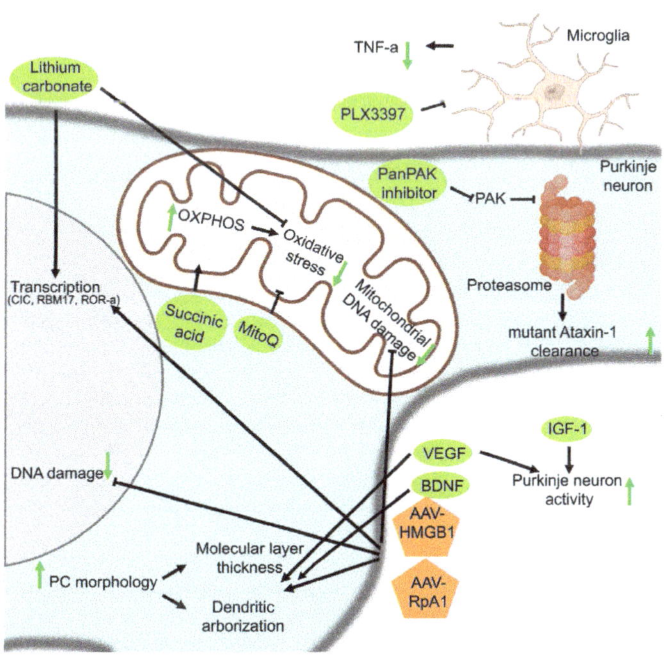

Figure 5. Other pharmacological targets for SCA1: These aim to restore PC survival and function, improve mitochondrial function, increase DNA damage repair, and enhance mutant ataxin-1 clearance by the proteasome or lower inflammation. Green arrows indicate the effect of the therapies on the different cell functions that are targeted.

VEGF is another trophic factor, which additionally functions as an angiogenic factor. VEGF is widely expressed in PCs in the cerebellum and is downregulated in SCA1$^{154Q/2Q}$ mice as well as human patients, possibly through a repressive effect of the mutant ataxin-1 [101,102]. Reduced levels of cerebellar VEGF in SCA1 lead to a decrease in microvessel density and hypoxia as well as a decrease in the growth and survival of PCs [101]. To increase VEGF expression, pharmacological delivery of recombinant VEGF (rVEGF) and a synthetic VEGF mimetic peptide (nano-VEGF) were tested in SCA1$^{154Q/2Q}$ mice models [101,102]. ICV administration of rVEGF and nano-VEGF improved motor performance and resulted in a reduction in neuropathology in SCA1$^{154Q/2Q}$ mice [101,102]. However, nano-VEGF administration outperformed rVEGF therapy in improving the levels of capillary proteins and amelioration of microvascular health [102]. These improved outcomes of nano-VEGF delivery could possibly be assigned to its increased biological stability compared with rVEGF amongst others [102].

Brain-derived neurotrophic factor (BDNF) is an important trophic factor for cognitive and motor function and plays key roles in both survival signaling and neuroplasticity. BDNF levels are reduced in the cerebellum and medulla of SCA1 patients [103]. Furthermore, BDNF levels have shown to be decreased in the cerebellum of transgenic B05 mice, and treatment with BDNF improved cerebellar pathology and motor impairments in these mice [103,113,114]. Subcutaneous BDNF treatment in $SCA1^{154Q/2Q}$ mice additionally showed amelioration of motor and cognitive deficits as well as improved cerebellar and hippocampal pathology [103].

Lithium carbonate has an effect on several cellular functions, and it has shown to exert neuroprotective effects, possibly by affecting gene transcription or by reducing oxidative stress (Figure 5, [104,105,115,116]). Furthermore, lithium carbonate has been shown to suppress neurodegeneration by partially rescuing cell death in $SCA1^{154Q/2Q}$ mice [104]. Administration of 0.2% lithium carbonate in the chow of $SCA1^{154Q/2Q}$ mice resulted in a rescue of motor function as well as spatial learning and memory impairment [104]. Treatment also partially rescued dendritic pathology and reduced the GSK3-ß activity, thereby possibly inhibiting apoptosis [104]. Moreover, lithium carbonate was shown to regulate purine, oxidative stress, and energy production metabolic pathways [105]. Although is not clear how lithium carbonate exerts neuroprotective effects, this could be mediated through apoptosis inhibition, restored purine metabolite levels, improvement in transcriptional dysregulation, or by affecting neurotransmission or neurogenesis [104,105].

3.2.5. Mitochondrial Functioning

PCs have a high metabolic demand due to their synaptic trafficking of several proteins and vesicles as well as for the maintenance of the resting membrane potential [106,107]. Because neurons have limited glycolytic capacity, they are highly dependent on oxidative phosphorylation (OXPHOS) in the mitochondria, which is a major producer of reactive oxygen species (ROS) [106]. OXPHOS dysfunction and oxidative damage has been observed in the PCs in $SCA1^{154Q/2Q}$ mice [106,107]. Therefore, targeting the OXPHOS dysfunction or ROS could ameliorate SCA1 disease pathology (Figure 5, Table 3).

Treatment with MitoQ, a mitochondria-targeted anti-oxidant, restored mitochondrial functioning and ameliorated disease symptoms by reducing oxidative stress in $SCA1^{154Q/2Q}$ mice [106]. Furthermore, in B05 transgenic mice, OXPHOS complex I deficits were observed [107]. Targeting this dysfunction with succinic acid boosts cerebellar function in these animals. Succinic acid is thought to bypass the dysfunctional complex I without disrupting complex II or downstream oxygen consumption in the OXPHOS [107]. It additionally restores complex III inhibition and prevents damage of the outer mitochondrial membranes. In addition to amelioration of mitochondrial function, succinic acid treatment reduced PC dendritic atrophy and the loss of PCs accompanied by a less severe cerebellar ataxia phenotype [107].

3.2.6. DNA Damage Repair, Transcription, and Replication

Through the altered interaction of mutant *ATXN1* with several transcription and splicing factors, transcriptional dysregulation occurs [39,40]. Furthermore, it is thought that DNA damage repair might also be impaired in SCA1 pathology [108,109]. Several studies have tried to target the transcriptional dysregulation as well as the DNA damage repair pathway, which are involved in SCA1 pathology (Figure 5).

In SCA1, the expression of mutant *ATXN1* was shown to reduce high-mobility group box 1 (HMGB1), thereby inhibiting damage repair [108]. HMGB1 is a chromatin protein which regulates the higher structure of genomic DNA, thereby facilitating the binding of other proteins and influencing transcription and DNA damage repair [106,108]. To rescue the imbalance of gene expression observed in SCA1, HMGB1 was used as a therapeutic strategy [108]. Restoring HMGB1 expression in SCA1$^{154Q/2Q}$ mice with an AAV-based method (AAV1-HMGB1) showed an increase in lifespan and improvement in motor activity [108]. However, the SCA1 phenotype could not be completely normalized after HMGB1 treatment, suggesting that there was insufficient upregulation of HMGB1 or that other mechanisms also contribute to SCA1 pathology [108]. Furthermore, there are several concerns for the therapeutic usage of HMGB1, including inflammation induction and the broad effect of HMGB1 on general gene expression, which might induce unwanted side effects in cells [108].

RpA1 is another molecule that is involved in multiple DNA damage repair pathways, including homologous recombination and non-homologous end joining (Figure 5). Genetic screens in *Drosophila* as well as in silico models have shown that RpA1 might be involved in DNA damage repair in SCA1 pathology [109,117]. Therefore, a gene therapy with AAV-RpA1 was tested in SCA1$^{154Q/2Q}$ mice. Treatment with AAV-RpA1 showed a recovery of motor function, DNA damage, dendritic shrinkage, and spine abnormalities [109]. It is suggested that these abnormalities are resolved through decreasing the DNA damage in the affected neurons and PCs [109]. However, RpA1 might also ameliorate disease symptoms by affecting RNA splicing, transcription, and the cell cycle [109].

3.2.7. The Proteostatic Machinery

As SCA1 is a protein aggregation disease, therapeutics that target the proteostatic machinery to promote protein breakdown and prevent aggregation have also been tested. P21-activated kinases (PAKs) are a family of serine/threonine kinases (Figure 5, Table 3, [110]). A genetic screen in a fly SCA1 model has shown that PAK1 might be a potential modulator of *ATXN1* [110]. It was suggested that PAK1 can regulate *ATXN1* clearance through interfering with the proteasome-mediated degradation, possibly by inhibiting a ubiquitin E3 ligase or other factors in the proteasome pathway that normally promote ubiquitination and degradation of *ATXN1*. Indeed, pharmacological inhibition of all PAKs resulted in a reduction in *Atxn1* levels in SCA1$^{154Q/2Q}$ mice, suggesting that PAK1 regulates *ATXN1* stability and toxicity [110].

3.2.8. Inflammation

As gliosis correlates with disease severity in SCA1 patients, and microglia are activated in several mouse models of SCA1, therapeutic strategies targeting microglia are also studied for SCA1 [23,118]. Colony-stimulating factor 1 receptor (CSF1R) signaling is essential for microglial survival [111]. Therefore, pharmacological inhibition of the CSF1R signaling through administration of PLX3397 can be used to cause microglial depletion after several days of administration (Figure 5, Table 3, [111]). Administering PLX3397 in in pre-symptomatic B05 mice resulted in a 69% decrease in microglial density and improved motor function, although no significant alterations in PC atrophy, synaptic loss, or altered astrogliosis were observed [111]. It is suggested that the improved motor function after treatment could be caused by an increased expression of post-synaptic density 95 (PSD95) and wildtype *ATXN1* as well as the reduction of the pro-inflammatory cytokine TNF-α [111]. Increased PSD95 expression may preserve the functions of the remaining synapses, whereas increased levels of wild-type *ATXN1* promote the function of PCs, thereby possibly being protective in SCA1.

3.3. Stem Cell Replacement Therapies

Instead of targeting either *ATXN1* or other molecules involved in disease pathology, another therapeutic strategy aims to replace the cells that undergo neurodegeneration in SCA1 using stem cell replacement [119]. Several stem cell types can be used for replacement therapies, including pluripotent stem cells (PSCs) and multipotent stem cells, such as neural progenitor cells (NPCs) and mesenchymal stem cells (MSCs) (Table 4, [21,119–125]).

Table 4. Overview of the different stem cell replacement therapies investigated for SCA1.

Cell Type	Delivery Method	Delivery Location	Model	Treatment Age	Pathology Outcome
MSCs	Single IT injection	Subarachnoid space	SCA1 B05 transgenic mice	5 weeks, pre-symptomatic stage	Amelioration of PC and PC dendritic spine pathology as well as improved motor coordination [121]
Human fetal MSCs	Single intracranial injection	Cerebellar cortex	SCA1 B05 transgenic mice	4 weeks, pre-symptomatic stage as well as 6–8-month old symptomatic mice	Fusion of MSCs with cerebellar neurons in the symptomatic mice only [120]
Human umbilical MSCs	Single bilaterial intracranial injection	Left and right cerebellar lobules IV	SCA1 B05 transgenic mice	4 weeks, pre-symptomatic stage	Improvement in molecular layer thickness, PC loss, and PC dendritic morphology; improvement in motor coordination [122]
MSCs	Single IT injection	Subarachnoid space	SCA1 knock-in mice ($SCA1^{154Q/2Q}$)	5 weeks, pre-symptomatic stage	Larger axon and myelin sizes [21]
MSC conditioned medium	Single IT injection and/or multiple intravenous injections	Subarachnoid space and/or systemic circulation	SCA1 knock-in mice ($SCA1^{154Q/2Q}$)	4 weeks, pre-symptomatic stage	Improvement in axon and myelin degeneration in the spinal neurons as well as improvement in motor coordination in all administration routes tested [123]
E13-E15 primordium mouse NPCs	Intracranial injection	Deep cerebellar nuclei	SCA1 B05 transgenic mice	6–9 weeks, early symptomatic stage	Improved motor performance and amelioration of PC dendritic morphology [125]
Mouse NPCs from subventricular zone	Single, bilateral intracranial injection	Cerebellar white matter	SCA1 B05 transgenic mice	5 or 13 weeks, pre-symptomatic; 24 weeks, symptomatic stage	Improvement in motor coordination, PC survival, PC function, and PC morphology in the symptomatic mice [124]

MSCs are multipotent and reside in several tissues but mainly in the bone marrow and fat, where they support hematopoiesis and produce cells of the mesodermal lineage. Additionally, they also have immunomodulatory and neurotrophic properties [119]. The underlying mechanism of the therapeutic effects of MSCs is not yet fully understood, but it is thought to occur through cellular replacement, fusion with degenerating neurons, or neuronal trophic support [119,120]. Pre-clinical studies using MSCs have been tested in both the B05 SCA1 transgenic as well as the SCA1$^{154Q/2Q}$ mouse model [119]. Results from different pre-clinical studies using B05 transgenic mice show similar results [120–122]. IT administration of MSCs in pre-symptomatic B05 mice results in a thicker granular layer, rescue of dendritic pathology, and improvements in motor coordination [121]. Another study found that transplantation of MSCs isolated from the Wharton's jelly of the umbilical cord directly into the left and right cerebellar cortex of pre-symptomatic transgenic B05 mice improved cerebellar atrophy as well as motor and behavioral deficits [122]. These ameliorating effects of MSC therapy were not caused by a direct differentiation of the MSCs into neurons or astrocytes after transplantation [122]. This suggests that the neuroprotective effects observed are possibly attributed through neurotrophic factors [121,122]. On the other hand, a study injecting human fetal MSCs into the cerebellum of symptomatic transgenic B05 mice showed fusion of the MSCs with PCs and molecular layer interneurons, thereby rescuing cerebellar degeneration [120]. However, this fusion was not observed in non-symptomatic SCA1 mice or wildtype mice, suggesting that fusion occurs only when degeneration is ongoing [120].

In addition to transgenic B05 mice, MSC therapies were also tested in SCA1$^{154Q/2Q}$ mice to assess the effects of these therapies on the PNS, as motor neuron degeneration in this area is observed in SCA1 patients. IT administration of GFP-labeled MSCs into the subarachnoid space in pre-symptomatic SCA1$^{154Q/2Q}$ mice showed distribution of the MSCs into the ventral root of the spinal cord, which resulted in larger axons and bigger myelin sheaths compared with non-treated SCA1$^{154Q/2Q}$ mice [21]. In this study, administration of MSCs was already efficient in a dose 10 to 100 times lower than the dose used in human studies, suggesting that lower doses might be sufficient to reach therapeutic effects in SCA1 patients, although this will need to be confirmed in human clinical trials [21]. These results might also suggest that the neuronal degeneration is more likely rescued due to trophic factors released from MSCs and not due to the direct differentiation of MSCs into neurons [21]. In another study, administration of MSC-conditioned medium, was tested in SCA1$^{154Q/2Q}$ mice [123]. Mice either received a single IT injection at 4 weeks of age or they received intravenous injections every 2 weeks from 4-12 weeks of age. Both IT and intravenous injections from the pre-symptomatic stage onwards improved the progressive degeneration of both axon and myelin in the spinal neurons, supporting the notion that therapy is effective mainly through the secretion of neurotrophic factors [123]. Both administration routes resulted in improved motor coordination, although repetitive intravenous injections resulted in reduced rotarod performance, suggesting that a single IT injection might be safer than multiple intravenous injections [123].

Next to MSCs, NPCs can also be used as a cell source for stem cell therapy [124,125]. NPCs derive from the ectoderm and can give rise to neuronal and glial cell populations of the CNS. NPCs either can be directly isolated from primary CNS tissue or can be differentiated in vitro from ESCs or iPSCs using specific growth factors [119]. The therapeutic effect of NPCs could be due to cell replacement or differentiation into oligodendrocyte progenitor cells, thereby allowing remyelination or through trophic support [119]. As the transplantation of NPCs has proven to be safe and effective in Parkinson's disease patients, its translation into the clinic seems feasible [119,126]. In a pre-clinical study, embryonic 13–15 day primordium NPCs were isolated from mice and subsequently injected into the deep cerebellar nuclei of 6- to 9-week-old, early symptomatic, transgenic B05 SCA1 mice [125]. Transplantation led to improved motor performance and amelioration of PC dendritic morphology [125]. Despite the ongoing neurodegeneration in the SCA1 mice brains, the donor cells survived and elicited a beneficial effect in the mice up to 20 weeks [125]. Another pre-clinical study transplanted mouse NPC derived from the adult subventricular zone into the cerebellar white matter of transgenic symptomatic B05 SCA1 mice [124]. Transplantation in the symptomatic SCA1 mice improved motor coordination and improved PC survival, function, and morphology [124]. However, treatment in pre-symptomatic SCA1 mice, when PC degeneration was minimal, did not show any improvements, suggesting that the time of treatment is crucial [124].

Pre-clinical studies have shown that replacement therapies using MSCs and NPCs can improve neuropathology and motor coordination in SCA1 mice [121,122,124,125]. Because MSCs can be more rapidly expanded ex vivo compared with NPCs, it is easier to obtain sufficient MSC numbers to treat a large number of patients [119,120,122]. Additionally, the self-renewal capacity and differentiation of MSCs give them great therapeutic potential, although caution should be taken, as passaging the MSCs for 20 times or more might acquire a similar profile to Ewing's sarcoma [120,127]. Furthermore, through the immunomodulatory actions of MSCs, they have low immunogenicity and can avoid host immune responses, thereby allowing for allogenic engraftments [119,127]. Therefore, several phase I/II clinical trials using MSCs have been performed on SCA1 patients, which are discussed below [127,128].

4. Clinical Trials

As outlined in this review, several therapeutics for SCA1 have been tested in pre-clinical animal models. Therapies focus either on modulating the levels of ataxin-1, restoring the normal function of several targets involved in SCA1 pathophysiology, or replacing the cells that have been lost due to neurodegeneration. Pre-clinical studies have shown promising results for these strategies, and some therapies were additionally tested in clinical trials [127,128].

Most clinical trials in SCA1 have focused on restoring the PC activity by targeting the potassium channels and the glutamatergic channels. A clinical, open label trial with the slow-release form of 4-AP (Dalfampiridine) was performed in 16 patients with chronic cerebellar ataxia, including 3 patients with SCA1 [129,130]. Treatment induced modest, short-term improvements in ataxia [129]. A long-term clinical, placebo-controlled cross-over trial (NCT01811706) with Dalfampiridine was additionally performed in 20 patients with SCA types 1, 2, 3, or 6 [131]. However, results showed no significant improvement of ataxia or other measures of gait [131].

Other clinical studies have been performed to improve the electrical function of PCs by using Riluzole, Troriluzole, or CAD-1883, which all act by promoting the opening of small-conductance calcium-activated potassium channels and by enhancing glutamate transporters [132]. A randomized, double-blind, placebo-controlled pilot trial with Riluzole was performed in 40 patients with cerebellar ataxias, including two SCA1 patients [133]. Results showed that treatment with Riluzole improved the ICARS score in patients [133]. These results were confirmed in a 1-year randomized, double-blind, placebo-controlled trial with Riluzole in patients with several types of genetic ataxia, including SCAs, showing a higher proportion of improved SARA scores in the Riluzole-treated group [134]. However, a recent similar trial in SCA2 patients showed neutral results (NCT03347344) [135]. Troriluzole, a pro-drug of Riluzole, was studied in a clinical trial in a cohort of SCA patients, including 35 SCA1 patients (NCT02960893). Subsequently, a phase 3 clinical trial using this drug in SCA has been performed, although results have not yet been published (NCT03701399). On the basis of these trial results, it is not clear whether Riluzole, and variants thereof, could possibly be a treatment option for patients with SCA [136]. Furthermore, CAD-1883, an allosteric modulator of the small-conductance calcium-activated potassium channels, has been investigated [132]. An open-label phase 2 clinical trial with CAD-1883 has begun for patients with SCA, although results still need to be published (NCT03688685).

Additionally, oral lithium has been investigated in an open-label, non-randomized phase I trial (NCT00683943), but results have never been published [7]. Moreover, in collaboration with VICO therapeutics, the $(CUG)^7$ ASO (VO659) has recently entered a phase I/II open-label clinical trial, including 95 SCA1, SCA3, and Huntington's disease patients to investigate the safety, tolerability, pharmacokinetics, and pharmacodynamics of ascending doses of intrathecally administered V0659 (Clinical Trials Register).

In one open label clinical study, 14 SCA patients were IT injected with umbilical cord MSCs for 4 weeks [127]. Treatment resulted in significantly improved ataxia severity and quality of life, as assessed by the ICARS and ADL scores, respectively, without any serious adverse effects [127]. In another open-label, uncontrolled phase I/II clinical study, a combination of umbilical cord MSC intravenous injections and IT injections were tested [128]. Treatment resulted in significantly better improvements in the Berg Balance Scale compared with baseline [128]. Additionally, ICARS scores improved in the 3rd and 6th month after treatment, suggesting that the MSC therapy had an effect on ataxia severity.

Although results from these clinical studies on MSC transplants seem efficient and safe, a meta-analysis of these studies showed no significant differences in the ICARS scores before and after treatment, low certainty in estimates of quality, and a high risk of bias, as both studies were uncontrolled clinical trials [137]. Furthermore, in general, a lack of standardization in the used protocols, dosage, and type of administration in pre-clinical studies might slow down the translation of SCA1 therapies into the clinic [119]. To be able to assess the efficiency of the different therapies in SCA patients, more clinical studies with a larger sample size and robust design should be performed [137].

5. Future Perspectives on SCA1 Therapeutics

Although several human clinical trials have been performed, no therapies for SCA1 have reached market authorization yet. For future SCA1 therapeutics to reach the clinic, several challenges, mainly in the translation of preclinical findings into clinical trials and in clinical trial design, need to be overcome [138,139]. Mouse models are essential to study biodistribution and bioavailability as well as the effect of a potential treatment on the ataxia phenotype. However, they either overexpress the mutant ataxin-1 protein, or the mutant protein is expressed only in a subset of brain cells. Furthermore, there are considerable differences between the mouse and human genomes, making translation of findings from animal models to patients often difficult [27,28,48]. Including models from patient-derived induced pluripotent stem cells (iPSCs) into the pre-clinical drug-development pipeline might improve translation into clinical research, as these reflect endogenous protein expression levels as well as the patient's genetic background [138,140].

Because of the BBB, delivery of drugs to the brain is a major challenge. Finding the right administration route and dose of a drug in order to reach a therapeutic dose in the brain without inducing toxicity, remains challenging and should be assessed for each therapeutic strategy [54]. Another issue is determining the stage of the disease process at which treatment should commence. As SCA1 is a monogenetic disorder, the cause of the disease is present from birth, whereas symptom onset usually occurs later in life, when neurodegeneration already occurred. It will be critical to determine at what stage of the disease progression pathology is irreversible and at what stage it can still be slowed down or maybe even halted [141–143]. Currently, the presence of symptoms is an inclusion criteria for clinical trials, and maybe neurodegeneration at this stage of the disease is already too far advanced, reducing the chances of showing a treatment effect within the relatively short duration of a trial. However, the inclusion of non-symptomatic patients in clinical trials might raise ethical concerns. Future research into biomarkers that can accurately detect changes in SCA1 patients well before onset of clinical symptoms will be crucial to starting clinical trials at earlier disease stages when neurodegeneration is less advanced and the likelihood of beneficial effects of the treatment are higher [141,144].

Drug development is a long and expensive process, and as SCA1 is a relatively rare disease, it is less attractive for industry to invest in SCA1 drug development [54]. Therefore, it is important to evaluate the costs and benefits of SCA1 therapeutics [145]. Therapy costs can be reduced by slowing progression of SCA1 in early disease stages and avoiding treating disease in later, more severe stages [145]. Moreover, it is difficult to recruit sufficient patients into clinical trials and to show significant treatment effects with small group sizes [144,146]. To overcome this problem, improvements in clinical trial designs could allow for the inclusion of smaller numbers of patients. For instance, multiple drugs could be tested against a single placebo arm, or a cross-over design could be used, thereby reducing the number of patients that need to be included in the clinical trial [144]. Moreover, drugs that target common pathogenic mechanisms observed in multiple SCAs, such as the altered PC activity, could be beneficial [41]. There is a world-wide increase in awareness that new and innovative solutions are needed for the development of treatments for rare diseases [147].

In SCA1, disease progression is slow, and clinical rating scales might not be sensitive to measure effects in trials that run for one year [4,146]. Improving and unifying these clinical rating scales will reduce variability in clinical trial outcome measures. The variability in disease severity, age of onset, and symptom progression among SCA1 patients further complicates the accurate measurement of the response to potential therapies [4]. An integrated staging system, such as recently proposed for Huntington's disease, would reduce variability in clinical data acquisition but could also facilitate the inclusion of patients before the onset of clinical signs and symptoms of ataxia [148]. Additionally, natural history data are crucial to better understand SCA1 disease progression and to identify biomarkers and factors that influence the disease course [144,149,150]. Several natural history consortia have been set up, including the READISCA and the Dutch SCA1 study [151].

SCA1 is a devastating disease, and to date only symptomatic treatment is available. With the advancements made in understanding SCA1 pathology and the progression of different intervention strategies, there is great promise that a therapy to treat SCA1 patients will become available in the coming years. There are some pharmacological treatments targeting dysregulated pathways in SCA1 as well as promising strategies modulating ataxin-1 levels, with several RNA targeting therapies for SCA1 in advanced stages of preclinical development. Another promising strategy for SCA1 is gene editing using CRISPR/Cas9, which has already successfully been tested in a pre-clinical study in hiPSCs [62,152]. However, before CRISPR/Cas9-based gene editing can be evaluated in human clinical trials, limitations regarding off-target effects and the use of viral vectors for delivery to the brain will first have to be resolved [152,153]. Additionally, future treatment strategies might involve a combination of treatments both targeting the polyglutamine ataxin-1 as well as other dysregulated pathways that are further downstream in the SCA1 disease pathology. Additionally, combinatorial treatments might be necessary to target region-specific regulators of ataxin-1. For a rare disease such as SCA1, worldwide collaborations are essential to move forward to gather, standardize, and share data towards successful clinical development of a therapy for SCA1.

Funding: This research is funded by the Dutch brain foundation project DR-2018-00253 (R.A.M.B.), ZonMW PSIDER project 10250022110002 (W.M.C.v.R.-M. and L.M.C.K.), and ZonMw Pto2 grant 446002002 (R.A.M.B., W.M.C.v.R.-M. and B.P.C.v.d.W.).

Acknowledgments: We would like to thank the Dutch SCA1 patients united in the Dutch SCA1 Families Fund for their contribution in the project. Figures were created with the help of BioRender.com.

Conflicts of Interest: All authors are supported by research grants from ZonMw. B.P.C.v.d.W., W.M.C.v.R.-M. and R.A.M.B. are supported by research grants from the Dutch Brain Foundation. B.P.C.v.d.W. is supported by research grants form NWO, Radboud university medical center, and the Christina Foundation. He has served on scientific advisory boards for UniQure and Servier, and receives royalities from BSL-Springer Nature.

References

1. Zoghbi, H.Y.; Orr, H.T. Spinocerebellar ataxia type 1. *Semin Cell Biol.* **1995**, *6*, 29–35. [CrossRef] [PubMed]
2. Banfi, S.; Servadio, A.; Chung, M.-Y.; Kwiatkowski, T.J.; McCall, A.E.; Duvick, L.A.; Shen, Y.; Roth, E.J.; Orr, H.T.; Zoghbi, H. Identification and characterization of the gene causing type 1 spinocerebellar ataxia. *Nat. Genet.* **1994**, *7*, 513–520. [CrossRef] [PubMed]
3. Buijsen, R.A.; Toonen, L.J.; Gardiner, S.L.; van Roon-Mom, W.M. Genetics, Mechanisms, and Therapeutic Progress in Polyglutamine Spinocerebellar Ataxias. *Neurotherapeutics* **2019**, *16*, 263–286. [CrossRef] [PubMed]
4. Jacobi, H.; Bauer, P.; Giunti, P.; Labrum, R.; Sweeney, M.G.; Charles, P.; Durr, A.; Marelli, C.; Globas, C.; Linnemann, C.; et al. The natural history of spinocerebellar ataxia type 1, 2, 3, and 6: A 2-year follow-up study. *Neurology* **2011**, *77*, 1035–1041. [CrossRef]
5. Orengo, J.P.; van der Heijden, M.E.; Hao, S.; Tang, J.; Orr, H.T.; Zoghbi, H.Y. Motor neuron degeneration correlates with respiratory dysfunction in SCA1. *Dis. Model. Mech.* **2018**, *11*, dmm032623. [CrossRef]
6. Schöls, L.; Bauer, P.; Schmidt, T.; Schulte, T.; Riess, O. Autosomal dominant cerebellar ataxias: Clinical features, genetics, and pathogenesis. *Lancet Neurol.* **2004**, *3*, 291–304. [CrossRef]
7. Opal, P.; Ashizawa, T. Spinocerebellar Ataxia Type 1. In *GeneReviews®*; Adam, M.P., Everman, D.B., Mirzaa, G.M., Pagon, R.A., Wallace, S.E., Bean, L.J.H., Gripp, K.W., Amemiya, A., Eds.; University of Washington Seattle: Seattle, WA, USA, 1993.
8. van de Warrenburg, B.; Sinke, R.; Verschuuren-Bemelmans, C.; Scheffer, H.; Brunt, E.; Ippel, P.; Maat-Kievit, J.; Dooijes, D.; Notermans, N.; Lindhout, D.; et al. Spinocerebellar ataxias in the Netherlands: Prevalence and age at onset variance analysis. *Neurology* **2002**, *58*, 702–708. [CrossRef]
9. Tang, B.; Liu, C.; Shen, L.; Dai, H.; Pan, Q.; Jing, L.; Ouyang, S.; Xia, J. Frequency of SCA1, SCA2, SCA3/MJD, SCA6, SCA7, and DRPLA CAG trinucleotide repeat expansion in patients with hereditary spinocerebellar ataxia from Chinese kindreds. *Arch. Neurol.* **2000**, *57*, 540–544. [CrossRef]
10. Bryer, A.; Krause, A.; Bill, P.; Davids, V.; Bryant, D.; Butler, J.; Heckmann, J.; Ramesar, R.; Greenberg, J. The hereditary adult-onset ataxias in South Africa. *J. Neurol. Sci.* **2003**, *216*, 47–54. [CrossRef]
11. Orr, H.; Chung, M.-Y.; Banfi, S.; Kwiatkowski, T.J.; Servadio, A.; Beaudet, A.L.; McCall, A.E.; Duvick, L.A.; Ranum, L.P.W.; Zoghbi, H. Expansion of an unstable trinucleotide CAG repeat in spinocerebellar ataxia type 1. *Nat. Genet.* **1993**, *4*, 221–226. [CrossRef]
12. Zühlke, C.; Dalski, A.; Hellenbroich, Y.; Bubel, S.; Schwinger, E.; Bürk, K. Spinocerebellar ataxia type 1 (SCA1): Phenotype-genotype correlation studies in intermediate alleles. *Eur. J. Hum. Genet.* **2002**, *10*, 204–209. [CrossRef]
13. Menon, R.P.; Nethisinghe, S.; Faggiano, S.; Vannocci, T.; Rezaei, H.; Pemble, S.; Sweeney, M.G.; Wood, N.W.; Davis, M.B.; Pastore, A.; et al. The Role of Interruptions in polyQ in the Pathology of SCA1. *PLoS Genet.* **2013**, *9*, e1003648. [CrossRef]
14. Gardiner, S.L.; Boogaard, M.W.; Trompet, S.; De Mutsert, R.; Rosendaal, F.R.; Gussekloo, J.; Jukema, J.W.; Roos, R.A.C.; Aziz, N.A. Prevalence of Carriers of Intermediate and Pathological Polyglutamine Disease–Associated Alleles Among Large Population-Based Cohorts. *JAMA Neurol.* **2019**, *76*, 650–656. [CrossRef]
15. Kraus-Perrotta, C.; Lagalwar, S. Expansion, mosaicism and interruption: Mechanisms of the CAG repeat mutation in spinocerebellar ataxia type 1. *Cerebellum Ataxias* **2016**, *3*, 20. [CrossRef]
16. Servadio, A.; Koshy, B.; Armstrong, D.D.; Antalffy, B.; Orr, H.; Zoghbi, H. Expression analysis of the ataxin–1 protein in tissues from normal and spinocerebellar ataxia type 1 individuals. *Nat. Genet.* **1995**, *10*, 94–98. [CrossRef]
17. Manek, R.; Nelson, T.; Tseng, E.; Rodriguez-Lebron, E. 5′UTR-mediated regulation of Ataxin-1 expression. *Neurobiol. Dis.* **2020**, *134*, 104564. [CrossRef]
18. Paulson, H.L.; Shakkottai, V.G.; Clark, H.B.; Orr, H. Polyglutamine spinocerebellar ataxias—From genes to potential treatments. *Nat. Rev. Neurosci.* **2017**, *18*, 613–626. [CrossRef]
19. Ilg, W.; Bastian, A.J.; Boesch, S.; Burciu, R.G.; Celnik, P.; Claaßen, J.; Feil, K.; Kalla, R.; Miyai, I.; Nachbauer, W.; et al. Consensus Paper: Management of Degenerative Cerebellar Disorders. *Cerebellum* **2013**, *13*, 248–268. [CrossRef]
20. Rüb, U.; Bürk, K.; Timmann, D.; Dunnen, W.D.; Seidel, K.; Farrag, K.; Brunt, E.; Heinsen, H.; Egensperger, R.; Bornemann, A.; et al. Spinocerebellar ataxia type 1 (SCA1): New pathoanatomical and clinico-pathological insights. *Neuropathol. Appl. Neurobiol.* **2012**, *38*, 665–680. [CrossRef]
21. Mieda, T.; Suto, N.; Iizuka, A.; Matsuura, S.; Iizuka, H.; Takagishi, K.; Nakamura, K.; Hirai, H. Mesenchymal stem cells attenuate peripheral neuronal degeneration in spinocerebellar ataxia type 1 knockin mice. *J. Neurosci. Res.* **2015**, *94*, 246–252. [CrossRef]
22. Rüb, U.; Schöls, L.; Paulson, H.; Auburger, G.; Kermer, P.; Jen, J.C.; Seidel, K.; Korf, H.; Deller, T. Clinical features, neurogenetics and neuropathology of the polyglutamine spinocerebellar ataxias type 1, 2, 3, 6 and 7. *Prog. Neurobiol.* **2013**, *104*, 38–66. [CrossRef] [PubMed]
23. Cvetanovic, M.; Ingram, M.; Orr, H.; Opal, P. Early activation of microglia and astrocytes in mouse models of spinocerebellar ataxia type 1. *Neuroscience* **2015**, *289*, 289–299. [CrossRef] [PubMed]
24. Zoghbi, H.; Orr, H. Pathogenic Mechanisms of a Polyglutamine-mediated Neurodegenerative Disease, Spinocerebellar Ataxia Type 1. *J. Biol. Chem.* **2009**, *284*, 7425–7429. [CrossRef] [PubMed]
25. Tejwani, L.; Lim, J. Pathogenic mechanisms underlying spinocerebellar ataxia type 1. *Cell. Mol. Life Sci.* **2020**, *77*, 4015–4029. [CrossRef]
26. Matilla, A.; Roberson, E.D.; Banfi, S.; Morales, J.; Armstrong, D.L.; Burright, E.N.; Orr, H.T.; Sweatt, J.D.; Zoghbi, H.Y.; Matzuk, M.M. Mice Lacking Ataxin-1 Display Learning Deficits and Decreased Hippocampal Paired-Pulse Facilitation. *J. Neurosci.* **1998**, *18*, 5508–5516. [CrossRef]

27. Burright, E.N.; Clark, H.B.; Servadio, A.; Matilla, T.; Feddersen, R.M.; Yunis, W.S.; Duvick, L.A.; Zoghbi, H.Y.; Orr, H.T. SCA1 transgenic mice: A model for neurodegeneration caused by an expanded CAG trinucleotide repeat. *Cell* **1995**, *82*, 937–948. [CrossRef]
28. Watase, K.; Weeber, E.J.; Xu, B.; Antalffy, B.; Yuva-Paylor, L.; Hashimoto, K.; Kano, M.; Atkinson, R.; Sun, Y.; Armstrong, D.L.; et al. A Long CAG Repeat in the Mouse Sca1 Locus Replicates SCA1 Features and Reveals the Impact of Protein Solubility on Selective Neurodegeneration. *Neuron* **2002**, *34*, 905–919. [CrossRef]
29. Fernandez-Funez, P.; Nino-Rosales, M.L.; De Gouyon, B.; She, W.-C.; Luchak, J.M.; Martinez, P.; Turiegano, E.; Benito, J.; Capovilla, M.; Skinner, P.J.; et al. Identification of genes that modify ataxin-1-induced neurodegeneration. *Nature* **2000**, *408*, 101–106. [CrossRef]
30. Klement, I.A.; Skinner, P.J.; Kaytor, M.D.; Yi, H.; Hersch, S.M.; Clark, H.; Zoghbi, H.Y.; Orr, H.T. Ataxin-1 Nuclear Localization and Aggregation: Role in Polyglutamine-Induced Disease in SCA1 Transgenic Mice. *Cell* **1998**, *95*, 41–53. [CrossRef]
31. Chen, H.-K.; Fernandez-Funez, P.; Acevedo, S.F.; Lam, Y.C.; Kaytor, M.D.; Fernandez, M.H.; Aitken, A.; Skoulakis, E.M.; Orr, H.T.; Botas, J.; et al. Interaction of Akt-Phosphorylated Ataxin-1 with 14-3-3 Mediates Neurodegeneration in Spinocerebellar Ataxia Type 1. *Cell* **2003**, *113*, 457–468. [CrossRef]
32. Cummings, C.J.; Mancini, M.A.; Antalffy, B.; DeFranco, D.B.; Orr, H.; Zoghbi, H. Chaperone suppression of aggregation and altered subcellular proteasome localization imply protein misfolding in SCA1. *Nat. Genet.* **1998**, *19*, 148–154. [CrossRef]
33. Handler, H.P.; Duvick, L.; Mitchell, J.S.; Cvetanovic, M.; Reighard, M.; Soles, A.; Mather, K.B.; Rainwater, O.; Serres, S.; Nichols-Meade, T.; et al. Decreasing mutant ATXN1 nuclear localization improves a spectrum of SCA1-like phenotypes and brain region transcriptomic profiles. *Neuron* **2022**, *111*, 493–507.e6. [CrossRef]
34. Putka, A.F.; McLoughlin, H.S. Diverse regional mechanisms drive spinocerebellar ataxia type 1 phenotypes. *Neuron* **2023**, *111*, 447–449. [CrossRef]
35. Burright, E.N.; Davidson, J.D.; Duvick, L.A.; Koshy, B.; Zoghbi, H.Y.; Orr, H.T. Identification of a self-association region within the SCA1 gene product, ataxin-1. *Hum. Mol. Genet.* **1997**, *6*, 513–518. [CrossRef]
36. Nitschke, L.; Coffin, S.L.; Xhako, E.; El-Najjar, D.B.; Orengo, J.P.; Alcala, E.; Dai, Y.; Wan, Y.-W.; Liu, Z.; Orr, H.T.; et al. Modulation of ATXN1 S776 phosphorylation reveals the importance of allele-specific targeting in SCA1. *J. Clin. Investig.* **2021**, *6*, e144955. [CrossRef]
37. Jorgensen, N.D.; Andresen, J.M.; Lagalwar, S.; Armstrong, B.; Stevens, S.; Byam, C.E.; Duvick, L.A.; Lai, S.; Jafar-Nejad, P.; Zoghbi, H.Y.; et al. Phosphorylation of ATXN1 at Ser776 in the cerebellum. *J. Neurochem.* **2009**, *110*, 675–686. [CrossRef]
38. Ortiz, J.M.P.; Mollema, N.; Toker, N.; Adamski, C.J.; O'Callaghan, B.; Duvick, L.; Friedrich, J.; Walters, M.A.; Strasser, J.; Hawkinson, J.E.; et al. Reduction of protein kinase A-mediated phosphorylation of ATXN1-S776 in Purkinje cells delays onset of Ataxia in a SCA1 mouse model. *Neurobiol. Dis.* **2018**, *116*, 93–105. [CrossRef]
39. Lim, J.; Crespo-Barreto, J.; Jafar-Nejad, P.; Bowman, A.B.; Richman, R.; Hill, D.E.; Orr, H.T.; Zoghbi, H.Y. Opposing effects of polyglutamine expansion on native protein complexes contribute to SCA1. *Nature* **2008**, *452*, 713–718. [CrossRef]
40. Lam, Y.C.; Bowman, A.B.; Jafar-Nejad, P.; Lim, J.; Richman, R.; Fryer, J.D.; Hyun, E.D.; Duvick, L.A.; Orr, H.T.; Botas, J.; et al. ATAXIN-1 Interacts with the Repressor Capicua in Its Native Complex to Cause SCA1 Neuropathology. *Cell* **2006**, *127*, 1335–1347. [CrossRef]
41. Cook, A.A.; Fields, E.; Watt, A.J. Losing the Beat: Contribution of Purkinje Cell Firing Dysfunction to Disease, and Its Reversal. *Neuroscience* **2021**, *462*, 247–261. [CrossRef]
42. Coffin, S.L.; Durham, M.A.; Nitschke, L.; Xhako, E.; Brown, A.M.; Revelli, J.-P.; Gonzalez, E.V.; Lin, T.; Handler, H.P.; Dai, Y.; et al. Disruption of the ATXN1-CIC complex reveals the role of additional nuclear ATXN1 interactors in spinocerebellar ataxia type 1. *Neuron* **2022**, *111*, 481–492.e8. [CrossRef] [PubMed]
43. Serra, H.G.; Duvick, L.; Zu, T.; Carlson, K.; Stevens, S.; Jorgensen, N.; Lysholm, A.; Burright, E.; Zoghbi, H.; Clark, H.B.; et al. RORα-Mediated Purkinje Cell Development Determines Disease Severity in Adult SCA1 Mice. *Cell* **2006**, *127*, 697–708. [CrossRef] [PubMed]
44. Gehrking, K.M.; Andresen, J.M.; Duvick, L.; Lough, J.; Zoghbi, H.Y.; Orr, H.T. Partial loss of Tip60 slows mid-stage neurodegeneration in a spinocerebellar ataxia type 1 (SCA1) mouse model. *Hum. Mol. Genet.* **2011**, *20*, 2204–2212. [CrossRef] [PubMed]
45. Bowman, A.; Lam, Y.C.; Jafar-Nejad, P.; Chen, H.-K.; Richman, R.; Samaco, R.; Fryer, J.D.; Kahle, J.J.; Orr, H.; Zoghbi, H. Duplication of Atxn1l suppresses SCA1 neuropathology by decreasing incorporation of polyglutamine-expanded ataxin-1 into native complexes. *Nat. Genet.* **2007**, *39*, 373–379. [CrossRef] [PubMed]
46. Yue, S.; Serra, H.G.; Zoghbi, H.Y.; Orr, H.T. The spinocerebellar ataxia type 1 protein, ataxin-1, has RNA-binding activity that is inversely affected by the length of its polyglutamine tract. *Hum. Mol. Genet.* **2001**, *10*, 25–30. [CrossRef]
47. Elsaey, M.A.; Namikawa, K.; Köster, R.W. Genetic Modeling of the Neurodegenerative Disease Spinocerebellar Ataxia Type 1 in Zebrafish. *Int. J. Mol. Sci.* **2021**, *22*, 7351. [CrossRef]
48. Clark, H.B.; Burright, E.N.; Yunis, W.S.; Larson, S.; Wilcox, C.; Hartman, B.; Matilla, A.; Zoghbi, H.Y.; Orr, H.T. Purkinje Cell Expression of a Mutant Allele of *SCA1* in Transgenic Mice Leads to Disparate Effects on Motor Behaviors, Followed by a Progressive Cerebellar Dysfunction and Histological Alterations. *J. Neurosci.* **1997**, *17*, 7385–7395. [CrossRef]
49. Burk, K.; Globas, C.; Bosch, S.; Klockgether, T.; Zuhlke, C.; Daum, I.; Dichgans, J. Cognitive deficits in spinocerebellar ataxia type 1, 2, and 3. *J. Neurol.* **2003**, *250*, 207–211. [CrossRef]

50. Tichanek, F.; Salomova, M.; Jedlicka, J.; Kuncova, J.; Pitule, P.; Macanova, T.; Petrankova, Z.; Tuma, Z.; Cendelin, J. Hippocampal mitochondrial dysfunction and psychiatric-relevant behavioral deficits in spinocerebellar ataxia 1 mouse model. *Sci. Rep.* **2020**, *10*, 5418. [CrossRef]
51. Asher, M.; Rosa, J.-G.; Rainwater, O.; Duvick, L.; Bennyworth, M.; Lai, R.-Y.; Kuo, S.-H.; Cvetanovic, M.; Sca, C. Cerebellar contribution to the cognitive alterations in SCA1: Evidence from mouse models. *Hum. Mol. Genet.* **2019**, *29*, 117–131. [CrossRef]
52. Fire, A.; Xu, S.; Montgomery, M.K.; Kostas, S.A.; Driver, S.E.; Mello, C.C. Potent and specific genetic interference by double-stranded RNA in Caenorhabditis elegans. *Nature* **1998**, *391*, 806–811. [CrossRef]
53. Elbashir, S.M.; Harborth, J.; Lendeckel, W.; Yalcin, A.; Weber, K.; Tuschl, T. Duplexes of 21-nucleotide RNAs mediate RNA interference in cultured mammalian cells. *Nature* **2001**, *411*, 494–498. [CrossRef]
54. Vázquez-Mojena, Y.; León-Arcia, K.; González-Zaldivar, Y.; Rodríguez-Labrada, R.; Velázquez-Pérez, L. Gene Therapy for Polyglutamine Spinocerebellar Ataxias: Advances, Challenges, and Perspectives. *Mov. Disord.* **2021**, *36*, 2731–2744. [CrossRef]
55. Xia, H.; Mao, Q.; Eliason, S.L.; Harper, S.Q.; Martins, I.H.; Orr, H.T.; Paulson, H.L.; Yang, L.; Kotin, R.M.; Davidson, B.L. RNAi suppresses polyglutamine-induced neurodegeneration in a model of spinocerebellar ataxia. *Nat. Med.* **2004**, *10*, 816–820. [CrossRef]
56. Keiser, M.S.; Kordasiewicz, H.B.; McBride, J.L. Gene suppression strategies for dominantly inherited neurodegenerative diseases: Lessons from Huntington's disease and spinocerebellar ataxia. *Hum. Mol. Genet.* **2015**, *25*, R53–R64. [CrossRef]
57. Keiser, M.S.; Geoghegan, J.C.; Boudreau, R.L.; Lennox, K.A.; Davidson, B.L. RNAi or overexpression: Alternative therapies for Spinocerebellar Ataxia Type 1. *Neurobiol. Dis.* **2013**, *56*, 6–13. [CrossRef]
58. Keiser, M.S.; Boudreau, R.; Davidson, B.L. Broad Therapeutic Benefit After RNAi Expression Vector Delivery to Deep Cerebellar Nuclei: Implications for Spinocerebellar Ataxia Type 1 Therapy. *Mol. Ther.* **2014**, *22*, 588–595. [CrossRef]
59. Keiser, M.S.; Monteys, A.M.; Corbau, R.; Gonzalez-Alegre, P.; Davidson, B.L. RNAi prevents and reverses phenotypes induced by mutant human ataxin-1. *Ann. Neurol.* **2016**, *80*, 754–765. [CrossRef]
60. Keiser, M.S.; Kordower, J.H.; Gonzalez-Alegre, P.; Davidson, B.L. Broad distribution of ataxin 1 silencing in rhesus cerebella for spinocerebellar ataxia type 1 therapy. *Brain* **2015**, *138*, 3555–3566. [CrossRef]
61. Keiser, M.S.; Ranum, P.T.; Yrigollen, C.M.; Carrell, E.M.; Smith, G.R.; Muehlmatt, A.L.; Chen, Y.H.; Stein, J.M.; Wolf, R.L.; Radaelli, E.; et al. Toxicity after AAV delivery of RNAi expression constructs into nonhuman primate brain. *Nat. Med.* **2021**, *27*, 1982–1989. [CrossRef]
62. Friedrich, J.; Kordasiewicz, H.B.; O'Callaghan, B.L.; Handler, H.P.; Wagener, C.; Duvick, L.; Swayze, E.; Rainwater, O.; Hofstra, B.; Benneyworth, M.; et al. Antisense oligonucleotide–mediated ataxin-1 reduction prolongs survival in SCA1 mice and reveals disease-associated transcriptome profiles. *J. Clin. Investig.* **2018**, *3*, e123193. [CrossRef] [PubMed]
63. O'callaghan, B.; Hofstra, B.; Handler, H.P.; Kordasiewicz, H.B.; Cole, T.; Duvick, L.; Friedrich, J.; Rainwater, O.; Yang, P.; Benneyworth, M.; et al. Antisense Oligonucleotide Therapeutic Approach for Suppression of Ataxin-1 Expression: A Safety Assessment. *Mol. Ther.-Nucleic Acids* **2020**, *21*, 1006–1016. [CrossRef]
64. Kourkouta, E.; Weij, R.; González-Barriga, A.; Mulder, M.; Verheul, R.; Bosgra, S.; Groenendaal, B.; Puolivāli, J.; Toivanen, J.; van Deutekom, J.C.; et al. Suppression of Mutant Protein Expression in SCA3 and SCA1 Mice Using a CAG Repeat-Targeting Antisense Oligonucleotide. *Mol. Ther.-Nucleic Acids* **2019**, *17*, 601–614. [CrossRef] [PubMed]
65. Carrell, E.M.; Keiser, M.S.; Robbins, A.B.; Davidson, B.L. Combined overexpression of ATXN1L and mutant ATXN1 knockdown by AAV rescue motor phenotypes and gene signatures in SCA1 mice. *Mol. Ther.-Methods Clin. Dev.* **2022**, *25*, 333–343. [CrossRef] [PubMed]
66. Hammond, S.M.; Aartsma-Rus, A.; Alves, S.; Borgos, S.E.; Buijsen, R.A.M.; Collin, R.W.J.; Covello, G.; Denti, M.A.; Desviat, L.R.; Echevarría, L.; et al. Delivery of oligonucleotide-based therapeutics: Challenges and opportunities. *EMBO Mol. Med.* **2021**, *13*, e13243. [CrossRef]
67. Quemener, A.M.; Bachelot, L.; Forestier, A.; Donnou-Fournet, E.; Gilot, D.; Galibert, M.-D. The powerful world of antisense oligonucleotides: From bench to bedside. *Wiley Interdiscip. Rev. RNA* **2020**, *11*, e1594. [CrossRef]
68. Silva, A.C.; Lobo, D.D.; Martins, I.; Lopes, S.M.; Henriques, C.; Duarte, S.P.; Dodart, J.-C.; Nobre, R.J.; De Almeida, L.P. Antisense oligonucleotide therapeutics in neurodegenerative diseases: The case of polyglutamine disorders. *Brain* **2019**, *143*, 407–429. [CrossRef]
69. Ashizawa, A.T.; Holt, J.; Faust, K.; Liu, W.; Tiwari, A.; Zhang, N.; Ashizawa, T. Intravenously Administered Novel Liposomes, DCL64, Deliver Oligonucleotides to Cerebellar Purkinje Cells. *Cerebellum* **2018**, *18*, 99–108. [CrossRef]
70. Kozlu, S.; Caban, S.; Yerlikaya, F.; Fernandez-Megia, E.; Novoa-Carballal, R.; Riguera, R.; Yemisci, M.; Gürsoy-Ozdemir, Y.; Dalkara, T.; Couvreur, P.; et al. An aquaporin 4 antisense oligonucleotide loaded, brain targeted nanoparticulate system design. *Die Pharm.* **2014**, *69*, 340–345.
71. Du, L.; Kayali, R.; Bertoni, C.; Fike, F.; Hu, H.; Iversen, P.L.; Gatti, R.A. Arginine-rich cell-penetrating peptide dramatically enhances AMO-mediated ATM aberrant splicing correction and enables delivery to brain and cerebellum. *Hum. Mol. Genet.* **2011**, *20*, 3151–3160. [CrossRef]
72. Zeniya, S.; Kuwahara, H.; Daizo, K.; Watari, A.; Kondoh, M.; Yoshida-Tanaka, K.; Kaburagi, H.; Asada, K.; Nagata, T.; Nagahama, M.; et al. Angubindin-1 opens the blood–brain barrier in vivo for delivery of antisense oligonucleotide to the central nervous system. *J. Control. Release* **2018**, *283*, 126–134. [CrossRef]

73. Miller, T.M.; Cudkowicz, M.E.; Genge, A.; Shaw, P.J.; Sobue, G.; Bucelli, R.C.; Chiò, A.; Van Damme, P.; Ludolph, A.C.; Glass, J.D.; et al. Trial of Antisense Oligonucleotide Tofersen for *SOD1* ALS. *N. Engl. J. Med.* **2022**, *387*, 1099–1110. [CrossRef]
74. Acsadi, G.; Crawford, T.O.; Müller-Felber, W.; Shieh, P.B.; Richardson, R.; Natarajan, N.; Castro, D.; Ramirez-Schrempp, D.; Gambino, G.; Sun, P.; et al. Safety and efficacy of nusinersen in spinal muscular atrophy: The EMBRACE study. *Muscle Nerve* **2021**, *63*, 668–677. [CrossRef]
75. Evers, M.M.; Pepers, B.A.; Van Deutekom, J.C.T.; Mulders, S.A.M.; Dunnen, J.T.D.; Aartsma-Rus, A.; Van Ommen, G.-J.B.; Van Roon-Mom, W.M.C. Targeting Several CAG Expansion Diseases by a Single Antisense Oligonucleotide. *PLoS ONE* **2011**, *6*, e24308. [CrossRef]
76. Suh, J.; Romano, D.M.; Nitschke, L.; Herrick, S.P.; DiMarzio, B.A.; Dzhala, V.; Bae, J.-S.; Oram, M.K.; Zheng, Y.; Hooli, B.; et al. Loss of Ataxin-1 Potentiates Alzheimer's Pathogenesis by Elevating Cerebral BACE1 Transcription. *Cell* **2019**, *178*, 1159–1175.e17. [CrossRef]
77. Alves, S.; Nascimento-Ferreira, I.; Auregan, G.; Hassig, R.; Dufour, N.; Brouillet, E.; de Lima, M.P.; Hantraye, P.; de Almeida, L.P.; Déglon, N. Allele-Specific RNA Silencing of Mutant Ataxin-3 Mediates Neuroprotection in a Rat Model of Machado-Joseph Disease. *PLoS ONE* **2008**, *3*, e3341. [CrossRef]
78. Scholefield, J.; Greenberg, L.J.; Weinberg, M.; Arbuthnot, P.B.; Abdelgany, A.; Wood, M.J.A. Design of RNAi Hairpins for Mutation-Specific Silencing of Ataxin-7 and Correction of a SCA7 Phenotype. *PLoS ONE* **2009**, *4*, e7232. [CrossRef]
79. Kubodera, T.; Yokota, T.; Ishikawa, K.; Mizusawa, H. New RNAi Strategy for Selective Suppression of a Mutant Allele in Polyglutamine Disease. *Oligonucleotides* **2005**, *15*, 298–302. [CrossRef]
80. Lee, W.; Lavery, L.; Rousseaux, M.W.C.; Rutledge, E.B.; Jang, Y.; Wan, Y.; Wu, S.; Kim, W.; Al-Ramahi, I.; Rath, S.; et al. Dual targeting of brain region-specific kinases potentiates neurological rescue in Spinocerebellar ataxia type 1. *EMBO J.* **2021**, *40*, e106106. [CrossRef]
81. Emamian, E.S.; Kaytor, M.D.; Duvick, L.A.; Zu, T.; Tousey, S.K.; Zoghbi, H.; Clark, H.; Orr, H.T. Serine 776 of Ataxin-1 Is Critical for Polyglutamine-Induced Disease in SCA1 Transgenic Mice. *Neuron* **2003**, *38*, 375–387. [CrossRef]
82. Hourez, R.; Servais, L.; Orduz, D.; Gall, D.; Millard, I.; D'Exaerde, A.d.K.; Cheron, G.; Orr, H.T.; Pandolfo, M.; Schiffmann, S.N. Aminopyridines Correct Early Dysfunction and Delay Neurodegeneration in a Mouse Model of Spinocerebellar Ataxia Type 1. *J. Neurosci.* **2011**, *31*, 11795–11807. [CrossRef] [PubMed]
83. Chopra, R.; Bushart, D.D.; Shakkottai, V.G. Dendritic potassium channel dysfunction may contribute to dendrite degeneration in spinocerebellar ataxia type 1. *PLoS ONE* **2018**, *13*, e0198040. [CrossRef] [PubMed]
84. Bushart, D.; Chopra, R.; Singh, V.; Murphy, G.G.; Wulff, H.; Shakkottai, V.G. Targeting potassium channels to treat cerebellar ataxia. *Ann. Clin. Transl. Neurol.* **2018**, *5*, 297–314. [CrossRef] [PubMed]
85. Ruegsegger, C.; Stucki, D.M.; Steiner, S.; Angliker, N.; Radecke, J.; Keller, E.; Zuber, B.; Rüegg, M.A.; Saxena, S. Impaired mTORC1-Dependent Expression of Homer-3 Influences SCA1 Pathophysiology. *Neuron* **2016**, *89*, 129–146. [CrossRef] [PubMed]
86. Lin, X.; Antalffy, B.; Kang, D.; Orr, H.T.; Zoghbi, H.Y. Polyglutamine expansion down-regulates specific neuronal genes before pathologic changes in SCA1. *Nat. Neurosci.* **2000**, *3*, 157–163. [CrossRef]
87. Barnes, J.A.; Ebner, B.A.; Duvick, L.A.; Gao, W.; Chen, G.; Orr, H.T.; Ebner, T.J. Abnormalities in the Climbing Fiber-Purkinje Cell Circuitry Contribute to Neuronal Dysfunction in *ATXN1[82Q]* Mice. *J. Neurosci.* **2011**, *31*, 12778–12789. [CrossRef]
88. Edamakanti, C.R.; Do, J.; Didonna, A.; Martina, M.; Opal, P. Mutant ataxin1 disrupts cerebellar development in spinocerebellar ataxia type 1. *J. Clin. Investig.* **2018**, *128*, 2252–2265. [CrossRef]
89. Dell'Orco, J.M.; Wasserman, A.; Chopra, R.; Ingram, M.A.C.; Hu, Y.-S.; Singh, V.; Wulff, H.; Opal, P.; Orr, H.; Shakkottai, V.G. Neuronal Atrophy Early in Degenerative Ataxia Is a Compensatory Mechanism to Regulate Membrane Excitability. *J. Neurosci.* **2015**, *35*, 11292–11307. [CrossRef]
90. Bushart, D.D.; Huang, H.; Man, L.J.; Bs, L.M.M.; Shakkottai, V.G. A Chlorzoxazone-Baclofen Combination Improves Cerebellar Impairment in Spinocerebellar Ataxia Type 1. *Mov. Disord.* **2020**, *36*, 622–631. [CrossRef]
91. Notartomaso, S.; Zappulla, C.; Biagioni, F.; Cannella, M.; Bucci, D.; Mascio, G.; Scarselli, P.; Fazio, F.; Weisz, F.; Lionetto, L.; et al. Pharmacological enhancement of mGlu1 metabotropic glutamate receptors causes a prolonged symptomatic benefit in a mouse model of spinocerebellar ataxia type 1. *Mol. Brain* **2013**, *6*, 48. [CrossRef]
92. Mitsumura, K.; Hosoi, N.; Furuya, N.; Hirai, H. Disruption of metabotropic glutamate receptor signalling is a major defect at cerebellar parallel fibre-Purkinje cell synapses in *staggerer* mutant mice. *J. Physiol.* **2011**, *589*, 3191–3209. [CrossRef]
93. Shuvaev, A.N.; Hosoi, N.; Sato, Y.; Yanagihara, D.; Hirai, H. Progressive impairment of cerebellar mGluR signalling and its therapeutic potential for cerebellar ataxia in spinocerebellar ataxia type 1 model mice. *J. Physiol.* **2016**, *595*, 141–164. [CrossRef]
94. Liberatore, F.; Antenucci, N.; Tortolani, D.; Mascio, G.; Fanti, F.; Sergi, M.; Battaglia, G.; Bruno, V.; Nicoletti, F.; Maccarrone, M.; et al. Targeting mGlu1 Receptors in the Treatment of Motor and Cognitive Dysfunctions in Mice Modeling Type 1 Spinocerebellar Ataxia. *Cells* **2022**, *11*, 3916. [CrossRef]
95. Power, E.M.; Morales, A.; Empson, R.M. Prolonged Type 1 Metabotropic Glutamate Receptor Dependent Synaptic Signaling Contributes to Spino-Cerebellar Ataxia Type 1. *J. Neurosci.* **2016**, *36*, 4910–4916. [CrossRef]
96. Serra, H.G.; Byam, C.E.; Lande, J.D.; Tousey, S.K.; Zoghbi, H.Y.; Orr, H.T. Gene profiling links SCA1 pathophysiology to glutamate signaling in Purkinje cells of transgenic mice. *Hum. Mol. Genet.* **2004**, *13*, 2535–2543. [CrossRef]
97. Nag, N.; Tarlac, V.; Storey, E. Assessing the Efficacy of Specific Cerebellomodulatory Drugs for Use as Therapy for Spinocerebellar Ataxia Type 1. *Cerebellum* **2012**, *12*, 74–82. [CrossRef]

98. Fick, D.M.; Semla, T.P.; Steinman, M.; Beizer, J.; Brandt, N.; Dombrowski, R. American Geriatrics Society 2019 Updated AGS Beers Criteria®for Potentially Inappropriate Medication Use in Older Adults. *J. Am. Geriatr. Soc.* **2019**, *67*, 674–694. [CrossRef]
99. Iizuka, A.; Nakamura, K.; Hirai, H. Long-term oral administration of the NMDA receptor antagonist memantine extends life span in spinocerebellar ataxia type 1 knock-in mice. *Neurosci. Lett.* **2015**, *592*, 37–41. [CrossRef]
100. Vig, P.; Subramony, S.; D'Souza, D.; Wei, J.; Lopez, M. Intranasal administration of IGF-I improves behavior and Purkinje cell pathology in SCA1 mice. *Brain Res. Bull.* **2006**, *69*, 573–579. [CrossRef]
101. Cvetanovic, M.; Patel, J.; Marti, H.; Kini, A.R.; Opal, P. Vascular endothelial growth factor ameliorates the ataxic phenotype in a mouse model of spinocerebellar ataxia type 1. *Nat. Med.* **2011**, *17*, 1445–1447. [CrossRef]
102. Hu, Y.-S.; Do, J.; Edamakanti, C.R.; Kini, A.R.; Martina, M.; Stupp, S.I.; Opal, P. Self-assembling vascular endothelial growth factor nanoparticles improve function in spinocerebellar ataxia type 1. *Brain* **2019**, *142*, 312–321. [CrossRef] [PubMed]
103. Rosa, J.-G.; Hamel, K.; Soles, A.; Sheeler, C.; Borgenheimer, E.; Gilliat, S.; Sbrocco, K.; Ghanoum, F.; Handler, H.P.; Forster, C.; et al. BDNF is altered in a brain-region specific manner and rescues deficits in Spinocerebellar Ataxia Type 1. *Neurobiol. Dis.* **2023**, *178*, 106023. [CrossRef] [PubMed]
104. Watase, K.; Gatchel, J.; Sun, Y.; Emamian, E.; Atkinson, R.; Richman, R.; Mizusawa, H.; Orr, H.; Shaw, C.; Zoghbi, H.Y. Lithium Therapy Improves Neurological Function and Hippocampal Dendritic Arborization in a Spinocerebellar Ataxia Type 1 Mouse Model. *PLoS Med.* **2007**, *4*, e182. [CrossRef] [PubMed]
105. Perroud, B.; Jafar-Nejad, P.; Wikoff, W.R.; Gatchel, J.R.; Wang, L.; Barupal, D.K.; Crespo-Barreto, J.; Fiehn, O.; Zoghbi, H.Y.; Kaddurah-Daouk, R. Pharmacometabolomic Signature of Ataxia SCA1 Mouse Model and Lithium Effects. *PLoS ONE* **2013**, *8*, e70610. [CrossRef]
106. Stucki, D.M.; Ruegsegger, C.; Steiner, S.; Radecke, J.; Murphy, M.P.; Zuber, B.; Saxena, S. Mitochondrial impairments contribute to Spinocerebellar ataxia type 1 progression and can be ameliorated by the mitochondria-targeted antioxidant MitoQ. *Free. Radic. Biol. Med.* **2016**, *97*, 427–440. [CrossRef]
107. Ferro, A.; Carbone, E.; Zhang, J.; Marzouk, E.; Villegas, M.; Siegel, A.; Nguyen, D.; Possidente, T.; Hartman, J.; Polley, K.; et al. Short-term succinic acid treatment mitigates cerebellar mitochondrial OXPHOS dysfunction, neurodegeneration and ataxia in a Purkinje-specific spinocerebellar ataxia type 1 (SCA1) mouse model. *PLoS ONE* **2017**, *12*, e0188425. [CrossRef]
108. Ito, H.; Fujita, K.; Tagawa, K.; Chen, X.; Homma, H.; Sasabe, T.; Shimizu, J.; Shimizu, S.; Tamura, T.; Muramatsu, S.; et al. HMGB 1 facilitates repair of mitochondrial DNA damage and extends the lifespan of mutant ataxin-1 knock-in mice. *EMBO Mol. Med.* **2014**, *7*, 78–101. [CrossRef]
109. Taniguchi, J.B.; Kondo, K.; Fujita, K.; Chen, X.; Homma, H.; Sudo, T.; Mao, Y.; Watase, K.; Tanaka, T.; Tagawa, K.; et al. RpA1 ameliorates symptoms of mutant ataxin-1 knock-in mice and enhances DNA damage repair. *Hum. Mol. Genet.* **2016**, *25*, 4432–4447. [CrossRef]
110. Bondar, V.V.; Adamski, C.J.; Onur, T.S.; Tan, Q.; Wang, L.; Diaz-Garcia, J.; Park, J.; Orr, H.T.; Botas, J.; Zoghbi, H.Y. PAK1 regulates ATXN1 levels providing an opportunity to modify its toxicity in spinocerebellar ataxia type 1. *Hum. Mol. Genet.* **2018**, *27*, 2863–2873. [CrossRef]
111. Qu, W.; Johnson, A.; Kim, J.H.; Lukowicz, A.; Svedberg, D.; Cvetanovic, M. Inhibition of colony-stimulating factor 1 receptor early in disease ameliorates motor deficits in SCA1 mice. *J. Neuroinflamm.* **2017**, *14*, 107. [CrossRef]
112. Fernandez, A.; Carro, E.; Lopez-Lopez, C.; Torres-Aleman, I. Insulin-like growth factor I treatment for cerebellar ataxia: Addressing a common pathway in the pathological cascade? *Brain Res. Rev.* **2005**, *50*, 134–141. [CrossRef]
113. Sheeler, C.; Rosa, J.-G.; Borgenheimer, E.; Mellesmoen, A.; Rainwater, O.; Cvetanovic, M. Post-symptomatic Delivery of Brain-Derived Neurotrophic Factor (BDNF) Ameliorates Spinocerebellar Ataxia Type 1 (SCA1) Pathogenesis. *Cerebellum* **2021**, *20*, 420–429. [CrossRef]
114. Mellesmoen, A.; Sheeler, C.; Ferro, A.; Rainwater, O.; Cvetanovic, M. Brain Derived Neurotrophic Factor (BDNF) Delays Onset of Pathogenesis in Transgenic Mouse Model of Spinocerebellar Ataxia Type 1 (SCA1). *Front. Cell. Neurosci.* **2019**, *12*, 509. [CrossRef]
115. Chen, G.; Zeng, W.-Z.; Yuan, P.-X.; Huang, L.-D.; Jiang, Y.-M.; Zhao, Z.-H.; Manji, H.K. The Mood-Stabilizing Agents Lithium and Valproate RobustlIncrease the Levels of the Neuroprotective Protein bcl-2 in the CNS. *J. Neurochem.* **1999**, *72*, 879–882. [CrossRef]
116. Jope, R.S. Lithium and GSK-3: One inhibitor, two inhibitory actions, multiple outcomes. *Trends Pharmacol. Sci.* **2003**, *24*, 441–443. [CrossRef]
117. Barclay, S.S.; Tamura, T.; Ito, H.; Fujita, K.; Tagawa, K.; Shimamura, T.; Katsuta, A.; Shiwaku, H.; Sone, M.; Imoto, S.; et al. Systems biology analysis of Drosophila in vivo screen data elucidates core networks for DNA damage repair in SCA1. *Hum. Mol. Genet.* **2013**, *23*, 1345–1364. [CrossRef]
118. Öz, G.; Hutter, D.; Tkáč, I.; Clark, H.B.; Gross, M.D.; Jiang, H.; Eberly, L.E.; Bushara, K.O.; Gomez, C.M. Neurochemical alterations in spinocerebellar ataxia type 1 and their correlations with clinical status. *Mov. Disord.* **2010**, *25*, 1253–1261. [CrossRef]
119. Maciel, P.; Correia, J.; Duarte-Silva, S.; Salgado, A. Cell-based therapeutic strategies for treatment of spinocerebellar ataxias: An update. *Neural Regen. Res.* **2023**, *18*, 1203. [CrossRef]
120. Huda, F.; Fan, Y.; Suzuki, M.; Konno, A.; Matsuzaki, Y.; Takahashi, N.; Chan, J.K.Y.; Hirai, H. Fusion of Human Fetal Mesenchymal Stem Cells with "Degenerating" Cerebellar Neurons in Spinocerebellar Ataxia Type 1 Model Mice. *PLoS ONE* **2016**, *11*, e0164202. [CrossRef]
121. Matsuura, S.; Shuvaev, A.N.; Iizuka, A.; Nakamura, K.; Hirai, H. Mesenchymal Stem Cells Ameliorate Cerebellar Pathology in a Mouse Model of Spinocerebellar Ataxia Type 1. *Cerebellum* **2013**, *13*, 323–330. [CrossRef]

122. Tsai, P.-J.; Yeh, C.-C.; Huang, W.-J.; Min, M.-Y.; Huang, T.-H.; Ko, T.-L.; Huang, P.-Y.; Chen, T.-H.; Hsu, S.P.C.; Soong, B.-W.; et al. Xenografting of human umbilical mesenchymal stem cells from Wharton's jelly ameliorates mouse spinocerebellar ataxia type 1. *Transl. Neurodegener.* **2019**, *8*, 29. [CrossRef] [PubMed]
123. Suto, N.; Mieda, T.; Iizuka, A.; Nakamura, K.; Hirai, H. Morphological and Functional Attenuation of Degeneration of Peripheral Neurons by Mesenchymal Stem Cell-Conditioned Medium in Spinocerebellar Ataxia Type 1-Knock-in Mice. *CNS Neurosci. Ther.* **2016**, *22*, 670–676. [CrossRef] [PubMed]
124. Chintawar, S.; Hourez, R.; Ravella, A.; Gall, D.; Orduz, D.; Rai, M.; Bishop, D.P.; Geuna, S.; Schiffmann, S.N.; Pandolfo, M. Grafting Neural Precursor Cells Promotes Functional Recovery in an SCA1 Mouse Model. *J. Neurosci.* **2009**, *29*, 13126–13135. [CrossRef] [PubMed]
125. Kaemmerer, W.F.; Low, W.C. Cerebellar Allografts Survive and Transiently Alleviate Ataxia in a Transgenic Model of Spinocerebellar Ataxia Type-1. *Exp. Neurol.* **1999**, *158*, 301–311. [CrossRef] [PubMed]
126. Lige, L.; Zengmin, T. Transplantation of neural precursor cells in the treatment for parkinson disease: An efficacy and safety analysis. *Turk. Neurosurg.* **2015**, *26*, 378–383. [CrossRef]
127. Dongmei, H.; Jing, L.; Mei, X.; Ling, Z.; Hongmin, Y.; Zhidong, W.; Li, D.; Zikuan, G.; Hengxiang, W. Clinical analysis of the treatment of spinocerebellar ataxia and multiple system atrophy-cerebellar type with umbilical cord mesenchymal stromal cells. *Cytotherapy* **2011**, *13*, 913–917. [CrossRef]
128. Jin, J.-L.; Liu, Z.; Lu, Z.-J.; Guan, D.-N.; Wang, C.; Chen, Z.-B.; Zhang, J.; Zhang, W.-Y.; Wu, J.-Y.; Xu, Y. Safety and Efficacy of Umbilical Cord Mesenchymal Stem Cell Therapy in Hereditary Spinocerebellar Ataxia. *Curr. Neurovascular Res.* **2013**, *10*, 11–20. [CrossRef]
129. Giordano, I.; Bogdanow, M.; Jacobi, H.; Jahn, K.; Minnerop, M.; Schoels, L.; Synofzik, M.; Teufel, J.; Klockgether, T. Experience in a short-term trial with 4-Aminopyridine in cerebellar ataxia. *J. Neurol.* **2013**, *260*, 2175–2176. [CrossRef]
130. Schniepp, R.; Wuehr, M.; Neuhaeusser, M.; Benecke, A.K.; Adrion, C.; Brandt, T.; Strupp, M.; Jahn, K. 4-Aminopyridine and cerebellar gait: A retrospective case series. *J. Neurol.* **2012**, *259*, 2491–2493. [CrossRef]
131. Wagner, J.L.; O'Connor, D.M.; Donsante, A.; Boulis, N.M. Gene, Stem Cell, and Alternative Therapies for SCA 1. *Front. Mol. Neurosci.* **2016**, *9*, 67. [CrossRef]
132. Gala, D.; Scharf, S.; Kudlak, M.; Green, C.; Khowaja, F.; Shah, M.; Kumar, V.; Ullal, G. A Comprehensive Review of the Neurological Manifestations of Celiac Disease and Its Treatment. *Diseases* **2022**, *10*, 111. [CrossRef]
133. Ristori, G.; Romano, S.; Visconti, A.; Cannoni, S.; Spadaro, M.; Frontali, M.; Pontieri, F.E.; Vanacore, N.; Salvetti, M. Riluzole in cerebellar ataxia: A randomized, double-blind, placebo-controlled pilot trial. *Neurology* **2010**, *74*, 839–845. [CrossRef]
134. Romano, S.; Coarelli, G.; Marcotulli, C.; Leonardi, L.; Piccolo, F.; Spadaro, M.; Frontali, M.; Ferraldeschi, M.; Vulpiani, M.C.; Ponzelli, F.; et al. Riluzole in patients with hereditary cerebellar ataxia: A randomised, double-blind, placebo-controlled trial. *Lancet Neurol.* **2015**, *14*, 985–991. [CrossRef]
135. Coarelli, G.; Heinzmann, A.; Ewenczyk, C.; Fischer, C.; Chupin, M.; Monin, M.-L.; Hurmic, H.; Calvas, F.; Calvas, P.; Goizet, C.; et al. Safety and efficacy of riluzole in spinocerebellar ataxia type 2 in France (ATRIL): A multicentre, randomised, double-blind, placebo-controlled trial. *Lancet Neurol.* **2022**, *21*, 225–233. [CrossRef]
136. Schniepp, R.; Möhwald, K.; Wuehr, M. Gait ataxia in humans: Vestibular and cerebellar control of dynamic stability. *J. Neurol.* **2017**, *264*, 87–92. [CrossRef]
137. Appelt, P.A.; Comella, K.; de Souza, L.A.P.S.; Luvizutto, G.J. Effect of stem cell treatment on functional recovery of spinocerebellar ataxia: Systematic review and meta-analysis. *Cerebellum Ataxias* **2021**, *8*, 8. [CrossRef]
138. Hommersom, M.P.; Buijsen, R.A.M.; van Roon-Mom, W.M.C.; van de Warrenburg, B.P.C.; van Bokhoven, H. Human Induced Pluripotent Stem Cell-Based Modelling of Spinocerebellar Ataxias. *Stem Cell Rev. Rep.* **2021**, *18*, 441–456. [CrossRef]
139. Olmos, V.; Gogia, N.; Luttik, K.; Haidery, F.; Lim, J. The extra-cerebellar effects of spinocerebellar ataxia type 1 (SCA1): Looking beyond the cerebellum. *Cell. Mol. Life Sci.* **2022**, *79*, 404. [CrossRef]
140. Buijsen, R.A.; Gardiner, S.L.; Bouma, M.J.; van der Graaf, L.M.; Boogaard, M.W.; Pepers, B.A.; Eussen, B.; de Klein, A.; Freund, C.; van Roon-Mom, W.M. Generation of 3 spinocerebellar ataxia type 1 (SCA1) patient-derived induced pluripotent stem cell lines LUMCi002-A, B, and C and 2 unaffected sibling control induced pluripotent stem cell lines LUMCi003-A and B. *Stem Cell Res.* **2018**, *29*, 125–128. [CrossRef]
141. Rubinsztein, D.C.; Orr, H. Diminishing return for mechanistic therapeutics with neurodegenerative disease duration? There may be a point in the course of a neurodegenerative condition where therapeutics targeting disease-causing mechanisms are futile. *Bioessays* **2016**, *38*, 977–980. [CrossRef]
142. Zu, T.; Duvick, L.A.; Kaytor, M.D.; Berlinger, M.S.; Zoghbi, H.Y.; Clark, H.B.; Orr, H.T. Recovery from Polyglutamine-Induced Neurodegeneration in Conditional *SCA1* Transgenic Mice. *J. Neurosci.* **2004**, *24*, 8853–8861. [CrossRef] [PubMed]
143. Ibrahim, M.F.; Power, E.M.; Potapov, K.; Empson, R.M. Motor and Cerebellar Architectural Abnormalities during the Early Progression of Ataxia in a Mouse Model of SCA1 and How Early Prevention Leads to a Better Outcome Later in Life. *Front. Cell. Neurosci.* **2017**, *11*, 292. [CrossRef] [PubMed]
144. Brooker, S.M.; Edamakanti, C.R.; Akasha, S.M.; Kuo, S.; Opal, P. Spinocerebellar ataxia clinical trials: Opportunities and challenges. *Ann. Clin. Transl. Neurol.* **2021**, *8*, 1543–1556. [CrossRef] [PubMed]

145. van Prooije, T.; Ruigrok, S.; Berkmortel, N.v.D.; Maas, R.P.P.W.M.; Wijn, S.; van Roon-Mom, W.M.C.; van de Warrenburg, B.; Grutters, J.P.C. The potential value of disease-modifying therapy in patients with spinocerebellar ataxia type 1: An early health economic modeling study. *J. Neurol.* 2023, *Online ahead of print*. [CrossRef]
146. Aartsma-Rus, A.; van Roon-Mom, W.; Lauffer, M.; Siezen, C.; Duijndam, B.; Coenen-de Roo, T.; Schüle, R.; Synofzik, M.; Graessner, H. Development of tailored splice switching oligonucleotides for progressive brain disorders in Europe: Development, regulation and implementation considerations. *RNA* **2023**, *29*, 446–454. [CrossRef]
147. Pizzamiglio, C.; Vernon, H.J.; Hanna, M.G.; Pitceathly, R.D.S. Designing clinical trials for rare diseases: Unique challenges and opportunities. *Nat. Rev. Methods Prim.* **2022**, *2*, 13. [CrossRef]
148. Tabrizi, S.J.; Schobel, S.; Gantman, E.C.; Mansbach, A.; Borowsky, B.; Konstantinova, P.; Mestre, T.A.; Panagoulias, J.; Ross, C.A.; Zauderer, M.; et al. A biological classification of Huntington's disease: The Integrated Staging System. *Lancet Neurol.* **2022**, *21*, 632–644. [CrossRef]
149. Diallo, A.; Jacobi, H.; du Montcel, S.T.; Klockgether, T. Natural history of most common spinocerebellar ataxia: A systematic review and meta-analysis. *J. Neurol.* **2020**, *268*, 2749–2756. [CrossRef]
150. Wilke, C.; Mengel, D.; Schöls, L.; Hengel, H.; Rakowicz, M.; Klockgether, T.; Durr, A.; Filla, A.; Melegh, B.; Schüle, R.; et al. Levels of Neurofilament Light at the Preataxic and Ataxic Stages of Spinocerebellar Ataxia Type 1. *Neurology* **2022**, *98*, e1985–e1996. [CrossRef]
151. Lin, C.-C.; Ashizawa, T.; Kuo, S.-H. Collaborative Efforts for Spinocerebellar Ataxia Research in the United States: CRC-SCA and READISCA. *Front. Neurol.* **2020**, *11*, 902. [CrossRef]
152. Pappadà, M.; Bonuccelli, O.; Buratto, M.; Fontana, R.; Sicurella, M.; Caproni, A.; Fuselli, S.; Benazzo, A.; Bertorelli, R.; De Sanctis, V., et al. Suppressing gain of function proteins via CRISPR/Cas9 system in SCA1 cells. *Sci. Rep.* **2022**, *12*, 20285. [CrossRef]
153. Uddin, F.; Rudin, C.M.; Sen, T. CRISPR Gene Therapy: Applications, Limitations, and Implications for the Future. *Front. Oncol.* **2020**, *10*, 1387. [CrossRef]

biomolecules

MDPI

Systematic Review

A Systematic Review on Dementia and Translocator Protein (TSPO): When Nuclear Medicine Highlights an Underlying Expression

Miriam Conte [1], Maria Silvia De Feo [1], Ferdinando Corica [1], Joana Gorica [1], Marko Magdi Abdou Sidrak [1], Flaminia De Cristofaro [1], Luca Filippi [2], Maria Ricci [3], Giuseppe De Vincentis [1] and Viviana Frantellizzi [1,*]

1 Department of Radiological Sciences, Oncology and Anatomo-Pathology, Sapienza, University of Rome, 00161 Rome, Italy
2 Department of Nuclear Medicine, Santa Maria Goretti Hospital, 04100 Latina, Italy
3 Nuclear Medicine Unit, Cardarelli Hospital, 86100 Campobasso, Italy
* Correspondence: viviana.frantellizzi@uniroma1.it

Abstract: Background: Translocator protein (TSPO) is a neuroinflammation hallmark. Different TSPO affinity compounds have been produced and over time, the techniques of radiolabeling have been refined. The aim of this systematic review is to summarize the development of new radiotracers for dementia and neuroinflammation imaging. Methods: An online search of the literature was conducted in the PubMed, Scopus, Medline, Cochrane Library, and Web of Science databases, selecting published studies from January 2004 to December 2022. The accepted studies considered the synthesis of TSPO tracers for nuclear medicine imaging in dementia and neuroinflammation. Results: A total of 50 articles was identified. Twelve papers were selected from the included studies' bibliographies and 34 were excluded. Thus, 28 articles were ultimately selected for quality assessment. Conclusion: Huge efforts in developing specific and stable tracers for PET/SPECT imaging have been made. The long half-life of ^{18}F makes this isotope a preferable choice to ^{11}C. An emerging limitation to this however is that neuroinflammation involves all of the brain which inhibits the possibility of detecting a slight inflammation status change in patients. A partial solution to this is using the cerebellum as a reference region and developing higher TSPO affinity tracers. Moreover, it is necessary to consider the presence of distomers and racemic compounds interfering with pharmacological tracers' effects and increasing the noise ratio in images.

Keywords: TSPO; PBR; neuroinflammation; SPECT; PET; dementia

1. Introduction

Translocator protein (TSPO), previously designated as peripheral benzodiazepine receptor (PBR), is a protein on the outer mitochondrial membrane with five transmembrane helical domains combined with a voltage-dependent anion channel and a nucleoside carrier. It has been well conserved throughout evolution and is expressed both by prokaryotes and eukaryotes. TSPO binds benzodiazepines, namely RO5-4864 and derivates of isoquinoline carboxamide including PK11195 [1]. In fact, this protein belongs to the family of benzodiazepine receptors which can be classically distinguished into two types of receptors: the central benzodiazepine receptor (CBR) localized in the central nervous system, through which benzodiazepines exert their pharmacological effects [2], and the peripheral benzodiazepine receptor (PBR), or peripheral benzodiazepine binding site (PBBS) (Schoemaker et al. 1981), the object of this systematic review. TSPO can be found in steroid cells as well as in the central nervous system. It is ubiquitous in cerebral phagocytic cells (microglia), which are the primary immune cells in the central nervous system, but it can also be found in the heart, lungs, kidneys, and liver as well [3–6]. The primary function of TSPO is its involvement in cholesterol transportation through the inner mitochondrial membrane for steroid synthesis. TSPO seems to be involved in the regulation of mitochondrial membrane potential and metabolism, apoptosis and proliferation, inflammation, immunomodulation, the transport of porphyrin, heme synthesis, calcium signaling, and in the regulation of oxidative stress [1]. Regarding its regulation, in a normal brain, density expression is modest [7]. On the contrary, TSPO expression is upregulated in activated microglia, which occurs after an injury, in the choroid plexus and, though in lower amounts, in reactive astrocytes and the olfactory bulb's neurons. Upregulation can also be observed in two brain malignancies: neuroblastoma and glioblastoma [8]. TSPO is therefore considered a sensitive hallmark for microglial activation and glial response. Since inflammation is tightly linked to neuronal dysfunction and loss in Alzheimer's disease (AD), TSPO could represent a promising tool for neuroinflammation detection and in particular, in conditions such as AD [9]. In fact, inflammation seems to be a phenomenon that predates amyloid deposition while the microglial response holds off Aβ plaque formation [10]. As glial function lowers with aging, microglia become more sensitive to stimuli and less efficient in removing damaged cells which themselves are a potent inflammatory trigger. This phenomenon is a vicious circle that leads to ulterior neuronal damage [11]. A variety of radiolabeled ligands has been produced for PET and SPECT imaging [12,13]. Their low affinity, low binding specificity, and high lipophilicity which is reflected in high background noise [14] have led to the development of novel second- and third-generation radiotracers. To name a few, the most famous TSPO PET tracer is ^{11}C-PK11195, while the newest tracers are ^{11}C-SSR180575, ^{11}C-PBR28, ^{18}F-FEDAA1106, ^{18}F-PBR06, ^{18}F-FEPPA, ^{11}C-DPA713, ^{18}F-PBR111, ^{11}C-DAC, ^{11}C-AC-5216, ^{18}F-DPA-714, and ^{123}I-CLINDE [15]. However, newer generation tracers are affected by variability in binding affinity due to human genetic polymorphism (rs6971) in exon 4 of the TSPO gene, which determines the substitution of alanine for threonine at position 147 (A147T) [16]. To give an example, Owen et al. discovered that ^{11}C-PBR28 did not bind in a cohort of non-binding patients; however, this was not detected using ^{11}C-PK11195 [17]. Based on this premise, tracers can be divided into high-affinity binders (HABs), low-affinity binders (LABs), and mixed-affinity binders (MABs). The predominant polymorphism Ala/Ala is responsible for high affinity, while the forms Ala/Thr and Thr/Thr are involved in mixed and low affinity, respectively [15]. Since in low-affinity binders, the quality of the image is degraded, a feasible genetic leukocyte analysis could be conducted to determine the TSPO polymorphism. Difficulties in establishing a reference region remain in patients with huge widespread inflammation, in which a slight neuroinflammation status change could be challenging to detect. A partial solution has been found using the cerebellum as a reference region for ^{11}C-PBR28-TSPO binding in AD even if further studies for the other tracers are needed [18]. In addition, a pivotal aspect to contemplate in pharmacology is the presence of enantiomers of a considered molecule and thus the racemic form of a mixture. For instance, GE180 is a chiral molecule synthesized as a racemate and its S-enantiomer has the highest affinity for TSPO and better brain uptake specificity and good clearance than the R- one [19]. Enantiomers have different pharmacokinetic and pharmacodynamic properties compared to the racemic form and distomers (less active enantiomers) can be present. Racemic drugs have differences between each other since they can have a single bioactive enantiomer or equally active enantiomers or, again, chiral inversion. All these features influence the safety, toxicity, and efficacy of a molecule. Thus, it is vital to separate the different enantiomers and to radiolabel the useful and effective ones [20,21]. For example, GE180 vibrational circular dichroism spectroscopy was used to separate S- and R-isomers in a study by Freedman et al. in 2003 [22]. This systematic review aims to summarize the novel available radiopharmaceuticals used for TSPO detection and explore the steps forward that scientific research has made in TSPO detection in dementia and neuroinflammation.

2. Materials and Methods

2.1. Search Strategy and Study Selection

This systematic review was written in line with PRISMA guidelines [23]. An online search of the literature was conducted in the PubMed, Scopus, Medline, Central (Cochrane Library), and Web Of Science databases. Papers published from January 2004 to December 2022 were searched. The following keywords were used in each database: "radiolabeled" AND "TSPO" AND "brain". Eligible studies had considered the radiopharmaceutical synthesis of radiolabeled TSPO to develop new radiopharmaceuticals for nuclear medicine imaging in dementia and neuroinflammation. Reviews, studies not related to dementia, studies on brain injury, and studies not related to the subject of the research were excluded. Studies being written in the English language was mandatory.

2.2. Quality of the Selected Studies

General data such as the author(s), journal, year of publication, country, and study design were retrieved for each article. The selected studies were analyzed through the Quality Assessment of Diagnostic Accuracy Studies-2 (QUADAS2) tool. Data extraction and quality assessment were separately conducted by two reviewers. Any disagreements were resolved through discussion among researchers.

3. Results

3.1. Search Results

The research produced a total of 50 articles. A total of 36 articles was identified from PubMed and 14 from Scopus, while no studies were detected in the Cochrane Library, Medline, or Web of Science. Five duplicate records were removed. The references of the selected studies were examined to check for any additional relevant articles and 12 papers were identified. A total of 57 articles was screened by examining each abstract to identify studies with potential relevance. From the overall group of 57, 29 articles were excluded because they did not satisfy the inclusion criteria. Among them, 15 were reviews, two articles were related only to traumatic injury and 12 studies were not related. The remaining 28 articles were included and selected for quality assessment. The search strategy and selection criteria applied are represented in a flowchart (see Figure 1).

3.2. Study Characteristics

The main thematic areas of the selected studies can be summarized as follows: (1) studies of new radiopharmaceuticals with an animal model; (2) biodistribution studies conducted on animals and humans; and (3) studies performed only on humans. The majority of the selected articles were conducted by European researchers, followed by Asian, American, and Australian. Among the selected studies, 15 experiments on tracer biodistribution were performed on animals, six both on humans and animals, and seven on humans only.

3.3. Methodological Quality Assessment

The methodological quality of the papers included studies of a very high quality. Of the 28 selected studies, only two did not satisfy all the QUADAS-2 domains (see Table 1). From analyzing the results within each bias assessment domain independently (see Table 2), almost all studies obtained a low concern of bias and no more than one study showed high risk in one domain. Considering all four bias assessment domains, three studies reported unclear results, while one study had high-risk results. Regarding patient selection, only two studies reported unclear results, inflow, and timing: one study had unclear results, while in the references and standards used, only one study was high-risk, due to an insufficient number of details. A low concern of applicability of was found for all the studies (See Figure 2).

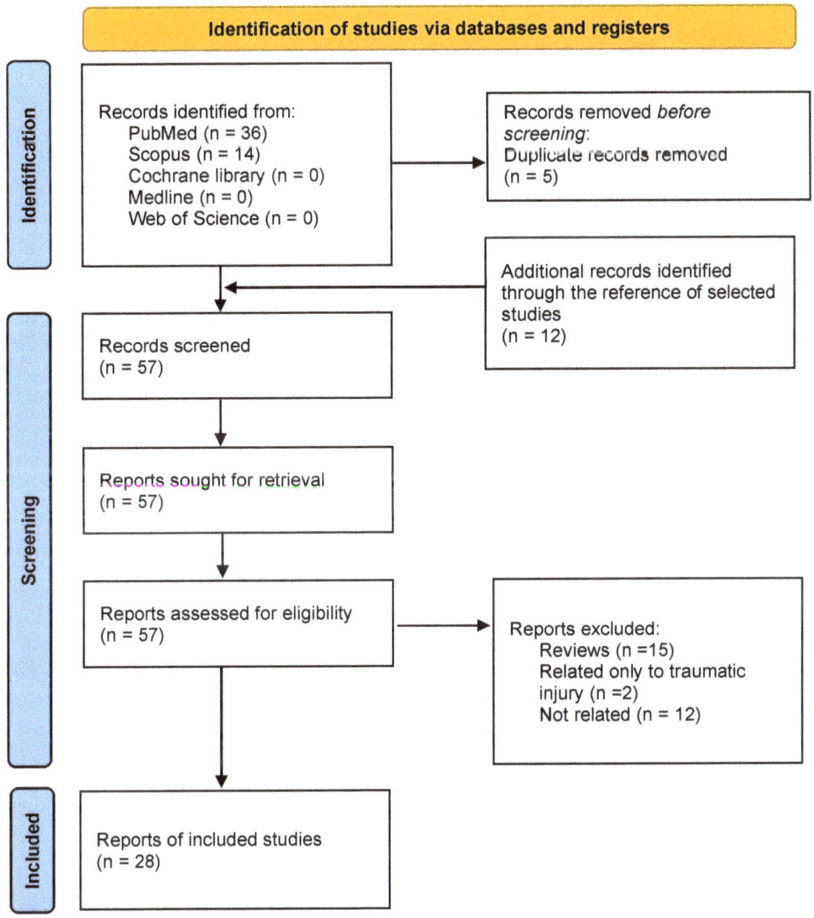

Figure 1. PRISMA workflow for the selection of articles.

Table 1. Quality assessment results.

Study	Risk of Bias				Applicability Concerns		
	P	I	R	FT	P	I	R
Donat et al., 2018 [9]	?	✓	✓	✓	✓	✓	✓
Chauveau et al., 2011 [24]	✓	✓	✓	✓	✓	✓	✓
Chauveau et al., 2009 [25]	✓	✓	✓	✓	✓	✓	✓
Vignal et al., 2018 [26]	✓	✓	✓	✓	✓	✓	✓
Solingapuram et al., 2015 [27]	✓	✓	✓	✓	✓	✓	✓
Mattner et al., 2015 [28]	✓	✓	✓	✓	✓	✓	✓
Tran et al., 2019 [29]	✓	✓	✓	✓	✓	✓	✓
Quiao et al., 2019 [16]	✓	✓	✓	✓	✓	✓	✓
Pike et al., 2011 [30]	✓	✓	✓	✓	✓	✓	✓

Table 1. *Cont.*

Study	Risk of Bias				Applicability Concerns		
	P	I	R	FT	P	I	R
Kumata et al., 2018 [31]	✓	✓	✓	✓	✓	✓	✓
Zhang et al., 2007	✓	✓	✓	✓	✓	✓	✓
Tredwell et al., 2016 [32]	✓	✓	✓	✓	✓	✓	✓
Zischler et al., 2016 [33]	✓	✓	✓	✓	✓	✓	✓
Zischler et al., 2017 [34]	✓	✓	✓	✓	✓	✓	✓
Wadsworth et al., 2012 [35]	✓	✓	✓	✓	✓	✓	✓
Arlicot et al., 2011 [36]	✓	✓	✓	✓	✓	✓	✓
Wadsworth et al., 2012 [37]	✓	✓	✓	✓	✓	✓	✓
Chau et al., 2015 [19]	✓	✓	✓	✓	✓	✓	✓
Nag et al., 2019 [38]	✓	✓	✓	✓	✓	✓	✓
Damont et al., 2015 [39]	✓	✓	✓	✓	✓	✓	✓
Okello et al., 2009 [40]	?	✓	X	?	✓	✓	✓
Cagnin et al., 2001 [41]	✓	✓	✓	✓	✓	✓	✓
Cagnin et al., 2004	✓	✓	✓	✓	✓	✓	✓
Feneey et al., 2016 [42]	✓	✓	✓	✓	✓	✓	✓
Zanotti-Fregonara et al., 2018 [43]	✓	✓	✓	✓	✓	✓	✓
Gulyás et al., 2009 [44]	✓	✓	✓	✓	✓	✓	✓
Fan et al., 2016 [45]	✓	✓	✓	✓	✓	✓	✓
Preshlock et al., 2016 [46]	✓	✓	✓	✓	✓	✓	✓

P = patient selection; I = index test; R = reference standard; FT = flow and timing. ✓ indicates low risk; X indicates high risk; ? indicates unclear risk.

Table 2. QUADAS-2—Number of studies at low, high, or unclear risk of bias and concerns regarding applicability.

	Risk of Bias				Concerns Regarding Applicability		
	Patient Selection	Index Test	Reference Standard	Flow and Timing	Patient Selection	Index Test	Reference Standard
Low	26	28	27	27	28	28	28
High	0	0	1	0	0	0	0
Unclear	2	0	0	1	0	0	0
Total	28	28	28	28	28	28	28

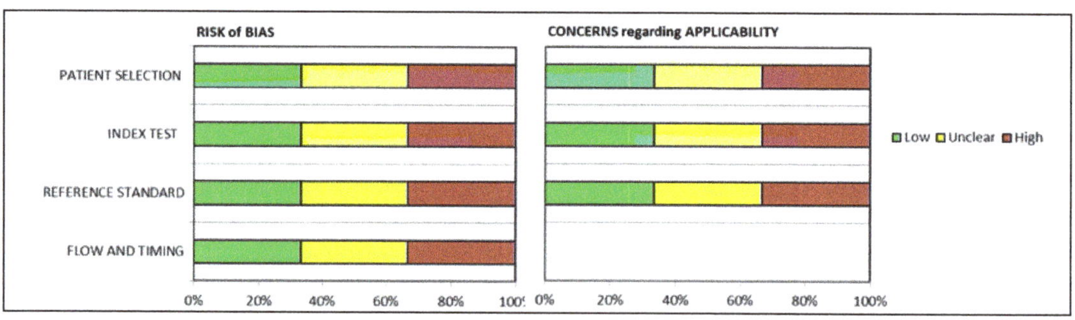

Figure 2. Bias risk results assessed through QUADAS-2.

4. Discussion

4.1. Studies of New Radiopharmaceuticals Using an Animal Model

In 2004, TSPO "cold" ligands were developed by Okubo et al. from FIGN-1-27 (compound **3**) [47]. Their binding affinity was proven on crude mitochondrial fractions derived from a rat cerebral cortex using a competitive assay with [3H]PK11195. The compound

12a [(±)-6,11-Dihydro-5-thia-11-aza-benzo[a]fluorene-6-carboxylic acid dihexylamide] had a high affinity for the translocator protein with an IC_{50} = 3.8 nM, while compounds **12e** [(±)-6,11-Dihydro-5-thia-11-aza-benzo [a]fluorene-6-carboxylic acid diethylamide] and **12f** [(±)-6,11-Dihydro-5-thia-11-aza-benzo[a]fluorene-6-carboxylic acid dipropylamide] had values of 0.44 and 0.97 nM for IC_{50}, respectively, resulting in being deemed more efficient ligands. The compounds **34a** [(±)-5,6-Dihydrobenzo [4,5]imidazo[2,1-a]isoquinoline-6-carboxylic acid dihexylamide] and **34c** [(±)-5,6-Dihydrobenzo[4,5]imidazo[2,1-a]isoquinoline-6-carboxylic acid dipropylamide], meanwhile, exhibited a moderate affinity for TSPO with their IC_{50} equal to 42 and 74 nM, respectively.

Chauveau et al. radiolabeled SSR180575 with ^{11}C (Figure 3) and performed in vivo and in vitro imaging using a model of rodent acute neuroinflammation comparing with [^{11}C](R)-PK11195 [24]. They injected 0.5 μL of (R,S)-α-amino-3-hydroxy-5-methyl-4-isoxazolepropionic (AMPA) into the right striatum of anaesthetized rats. Higher uptake of [^{11}C]SSR180575 in the AMPA-lesioned striatum was observed. In the right striatum, the uptake was higher for [^{11}C]SSR18057 than [^{11}C](R)-PK11195 while the uptake in the contralateral striatum was lower for [^{11}C]SSR18057 than [^{11}C](R)-PK11195. This result demonstrated that [^{11}C]SSR18057 had a better capacity for the discrimination of healthy tissue compared to [^{11}C](R)-PK11195.

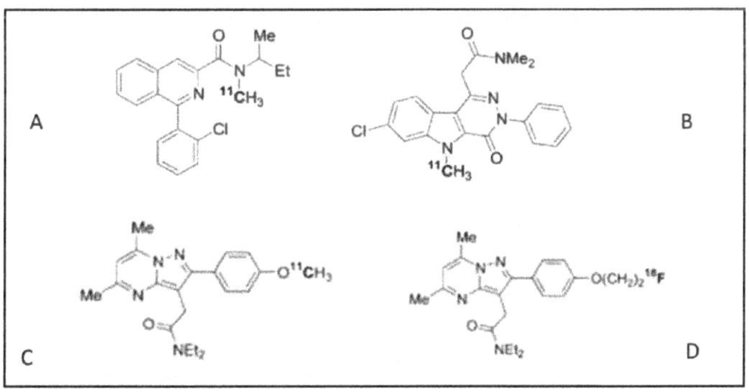

Figure 3. (**A**) ^{11}C-PK11195 (or [^{11}C]1), a component of the isoquinoline family; (**B**) indoleacetamide-derived 7-chloro-N,N,5-trimethyl-4-oxo-3-phenyl-3,5-dihydro-4H-pyridazino [4,5-b]indole-1-acetamide labeled with ^{11}C; (**C**) ^{11}C-labeled N,N-diethyl-2- [2-(4-methoxyphenyl)-5,7-dimethylpyrazolo [1,5-a]pyrimidin-3-yl] acetamide [**^{11}C-DPA-713**]; (**D**) ^{18}F-labeled N,N-diethyl-2-(2-(4-(2-fluoroethoxy)phenyl)-5,7-dimethylpyrazolo [1,5-a]pyrimidin-3-yl) acetamide (^{18}F-DPA-714 or ^{18}F [2]). DPA-713 and DPA-714 belong to the pyrazolopyrimidine acetamide family.

Another study by Chauveau et al. [25], compared ^{11}C-labeled N,N-diethyl-2- [2-(4-methoxyphenyl)-5,7-dimethylpyrazolo [1,5-a]pyrimidin-3-yl] acetamide (^{11}C-DPA-713) and ^{18}F-labeled N,N-diethyl-2-(2-(4-(2-fluoroethoxy)phenyl)-5,7-dimethylpyrazolo [1,5-a]pyrimidin-3-yl) acetamide (^{18}F-DPA-714), two molecules that belong to the pyrazolopyrimidine class (see Figure 3). The study was conducted in vivo and in vitro with a murine model of neuroinflammation. In vitro, ^{11}C-DPA-713 and ^{18}F-DPA-714 had a similar capability in the detection of neuroinflammatory areas and a similar signal-to-noise ratio. In vivo, the fluorinated compound had the highest ratio of ipsilateral to contralateral uptake and a grated greater binding affinity compared to ^{11}C-DPA-713 and ^{11}C-PK11195.

Another intriguing study is the research of Vignal et al. [26]. They applied the optimized [^{18}F]FEPPA (see Figure 4) in a mouse model of cerebral inflammation elicited by the intraperitoneal injection of Salmonella enterica serovar typhimurium lipopolysaccharides preceding a 24 h PET scan. The tracer was synthesized through nucleophilic substitution with a tosylated precursor. The tracer distribution was dependent on TSPO mouse expression as well as in the heart and kidneys, as observed during small animal PET acquisition.

Western blotting showed a 2.2-fold greater expression of TSPO in the brain of treated mice than in the control.

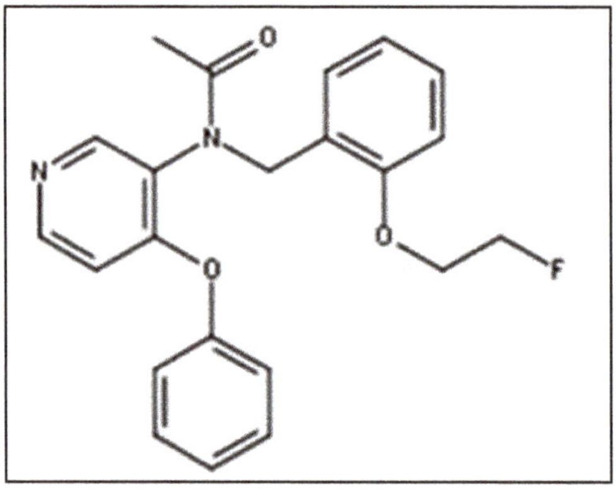

Figure 4. Chemical structure of (2-(2-((*N*-4-phenoxypyridin-3-yl)acetamido)methyl)phenoxy)ethyl 4-methylbenzenesulfonate) radiolabeled with F-18 ([^{18}F]FEPPA) which belongs to the family of phenoxypyridinylacetamides.

Solingapuram et al. undertook a biodistribution study of [^{11}C]PBR28 (Figure 5) in the male rhesus monkey [27]. A dynamic PET scan of 100 min was executed. After 10 min, cerebellum and basal ganglia uptake were evaluated with a complete washout in 2 h.

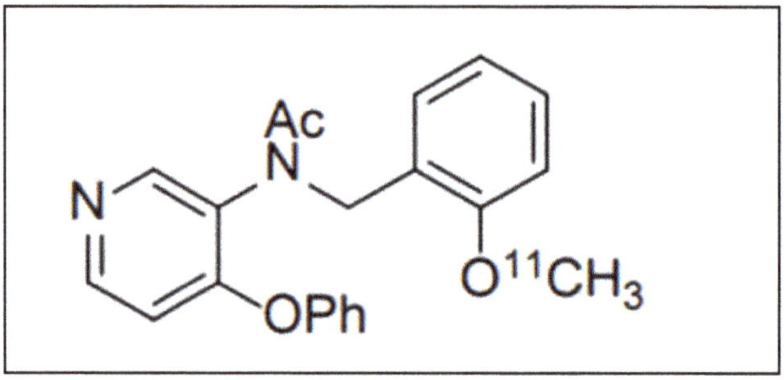

Figure 5. Chemical structure of the aryloxyanilide-based tracer N-(2-[11C]methoxybenzyl)-N-(4-phenoxypyridin-3-yl)acetamide, ([11C]PBR28) belonging to the phenoxypyridinylacetamides family.

Another fascinating tracer with possible applications in SPECT imaging is [^{123}I]-CLINME. Mattener et al. studied the biodistribution of [^{123}I]-CLINME (Figure 6) in an AMPA-induced excitotoxic murine model through SPECT imaging [28]. After 30 min to 6 h post-injection, a plateau in the adrenal glands was seen. Maximal activity was reached in the heart, lungs, liver, spleen, and kidneys at 5–15 min post tracer administration, while blood concentration was very low during the exam. Higher activity was detected in the olfactory bulbs compared to the rest of the brain. Thyroid uptake rose from 5 min to 1 h post-injection, then diminished at 3 h to 4%, remaining steady until 6 h post-injection. Competition studies conducted of PK11195 and Ro 5-4854 confirmed the specificity of CLINME binding to TSPO linking sites.

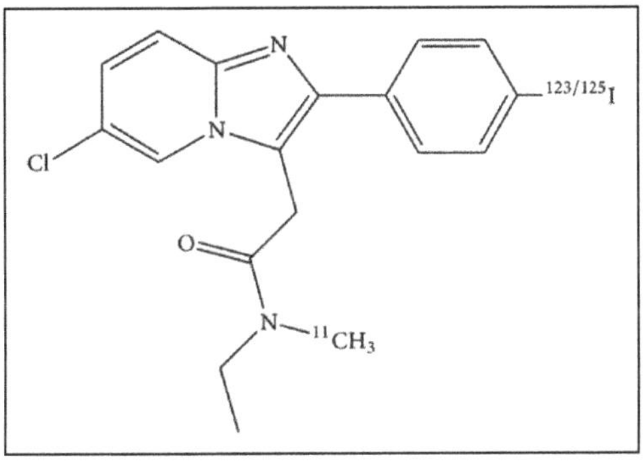

Figure 6. Chemical structure of 6-chloro-2-(4′-iodophenyl)-3-(N,N-methylethyl)imidazo[1,2-a]pyridine-3-acetamide (CLINME) radiolabeled with iodine-123, a compound of the 2-arylimidazo[1,2-a]pyridine-3-acetamide family.

Tran et al. tested a variety of TSPO ligands (11a–c and 13a–d) with a 2-phenylpyrazolo[1,5-a]pyrimidin-3-yl acetamide structure by conducting an in vitro binding assay [29]. A major part of the compound had a better affinity than DPA-714, with particular affinity and lipophilicity in compound **11a**. It was radiolabeled with ^{18}F and PET imaging was conducted in an LPS-induced neuroinflammatory murine model. [^{18}F]11a demonstrated a high affinity for inflammation sites, confirmed by an immunohistochemical exam conducted on the dissected brain, which showed how the PET uptake was in correspondence with areas of activated microglia.

Qiao et al. conducted a synthesis study on compound **6** (GE387) radiolabeled with ^{18}F [16]. The racemic [^{18}F]6 was studied in male Wistar rats: the tracer showed low sensitivity to the polymorphism rs6971 in human embryonic kidney cell lines and modest affinity in murine brains because the animals were healthy.

Pike et al. synthesized a library of new 2-phenylindol-3-ylglyoxylamide derivatives of the general formula II (compounds **19–31**), attributing a methyl group on the indole nitrogen. Then, they tested ligand 31 (Figure 7) which, thanks to its methyl group, was feasible to label with carbon-11 for positron emission tomography (PET) imaging in monkeys [30]. Peak radioactivity was observed in a region rich in TSPO 40 min after tracer injection, with a maximal accumulation in putamen between 12 and 32 min. After the pre-blocked administration of [^{11}C]PK 1119519 ([^{11}C]1, the rapid uptake in all the examined regions and rapid washout demonstrated the specific binding of [^{11}C]31. PET images at 4–100 min post-injection demonstrated a high uptake in the putamen and cerebellum and intermediate levels of accumulation in the cortical regions. Ligand 31 exhibited high-affinity binding in HABs and MABs, which was lower for low-affinity binders even if the authors suggested that further studies in the larger group were needed to establish if differences between binding toward MABs and HABs could be revealed.

DAA1106 (Figure 8), a molecule with high binding selectivity for cerebral rat and monkey mitochondria with a weak affinity for CBR, GABA, and Kappa1 receptors, was tested by Kumata et al. [31]. They labeled DAA1106 with fluorine-18 in a previous study in 2007 through the fluorination of diphenyliodonium salt used as a precursor, but, due to the instability of the diphenyl iodonium precursor, the clinical use of the compound was not feasible. This was until the possibility of adding nucleophilic ^{18}F into an electron-rich benzene ring, Cu-mediated radiofluorination of pinacol arylboronic ester, or the use of arylstannane precursors with [^{18}F]F− and high radiochemical yields were introduced [33–35,46]. In their later paper, Kappal et al. ventured into the synthesis of [^{18}F]DAA1106 via the [^{18}F]fluorination of a spirocyclic iodonium ylide (SCIDY), compound (1), used as a precursor with [^{18}F]F−. The SCIDY precursor was demonstrated to be stable at room temperature for 60 days. DAA1106 was stable after radiolabeling with [^{18}F] for the duration of at least one PET scan. Biodistribution studies executed on rats al 30 and 60 min after tracer injection showed high uptake in the heart, lungs, spleen, and kidneys and, interestingly, a low uptake in bone, which indicated that no defluorination had occurred thanks to the direct introduction of fluorine-18 in the benzene ring. The polar metabolite generated in rats did not pass the blood–brain barrier or, if penetrated, did not remain in brain tissue due to its hydrophilicity. For the PET study, an ischemic rat brain model was used and demonstrated that [^{18}F]DAA1106 had a high binding specificity for TSPO-expressing ischemic areas.

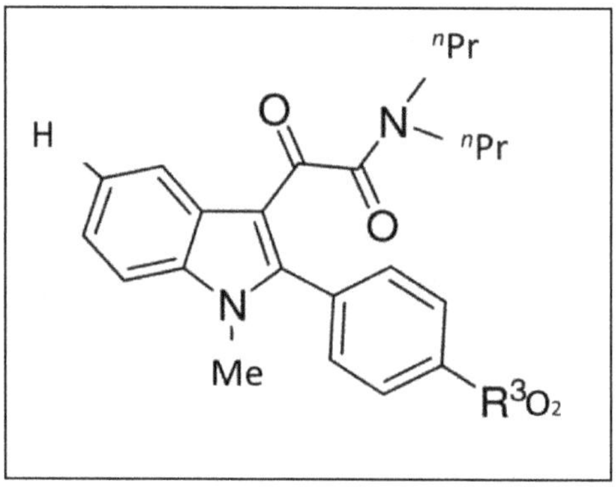

Figure 7. Chemical structure of N1-methylated derivative N,N-di-n-propyl-(N1-methyl-2-(4′-nitrophenyl)indol-3-yl)glyoxylamide (ligand 31).

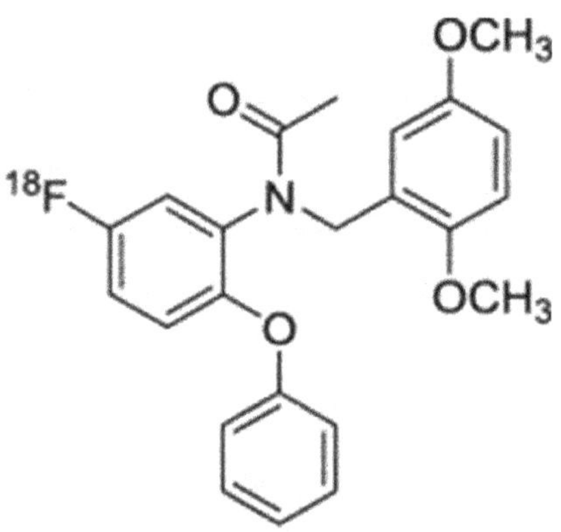

Figure 8. Chemical structure of N-(2,5-Dimethoxybenzyl)-N-(5-fluoro-2-phenoxyphenyl)acetamide (DAA1106) labeled with ^{18}F belonging to the phenoxyphenylacetamide family.

In 2012, in another study by Wadsworth et al., fluorine-18 analogues of DAA1106 were developed [35]. DAA1106 had a high affinity and less lipophilicity with better brain biodistribution. They valued the structure–activity relationships of the analogues by varying their aromatic rings. The reduction of their nitro groups was performed, followed by the reduction of their alanine. Acetylation or fluoroacetylation allowed them to obtain the desired molecule which was tested for affinity screening. Less lipophilicity was obtained by the substitution of phenyl groups with aliphatic groups. The aromatic rings, instead, gave rise to compounds with a higher affinity. The biodistribution of the compounds was studied in naïve Wistar rats. They had a higher whole brain uptake when compared to PK11195 but had a good capacity in discriminating high TSPO-expressing areas and lower ones. In vivo metabolic stability studies stated that compounds **27** and **30** reached the target. In a platelet binding assay however, the question of the affinity of the two binding sites (high affinity, low affinity) of TSPO ligands was not answered.

4.2. Studies Conducted on Animals and Humans

Some of the selected studies were performed on animals but also on humans for in vitro examination, in vivo biodistribution, and metabolism studies or post-mortem binding examination.

^{11}C-PBR28 was demonstrated to be a valid radiopharmaceutical for the detection of glial response in a study by Donat et al. [9]. A transgenic mouse model of AD was used. The tracer was injected in the tail vein through a cannula and 60 min of dynamic PET scanning was executed. The activity of 370 MBq was also administered to AD patients and then 90 min PET images were acquired. CT scans were obtained for attenuation, scatter correction, and to add morphostructural information to the PET images. To improve the spatial resolution of the PET results, autoradiography with 3H-PBR28 and immunochemistry were executed to provide a tracer density of PBR28 binding sites in mouse brains. During the in vivo characterization of TSPO density through autoradiography, higher microglial representation was associated with higher 3H-PBR28 binding, with major representation in the cortical and hippocampal regions in the AD mouse model.

Arlicot et al. conducted a 90 min dynamic PET scan in seven healthy patients post 245 ± 45 MBq injection of [(^{18})F]DPA-714 [36]. Arterial and venous samples were taken while two volunteers underwent whole-body acquisition 1 h after the tracer's administration. PET imaging post-processing showed a maximum cerebral uptake in the pons and cerebral radioactive peak within 5 min with a rapid clearance between 5 and 30 min. In whole-body scans, high uptake was seen in the vertebral bodies, gallbladder, heart wall, spleen, intestinal wall, kidneys, and adrenal gland. They also conducted biodistribution studies on mice highlighting similar results.

Starting with the tetracyclic indole-based pharmacophore described in the previously reported by Okubo et al., Wadsworth et al. developed a new tracer, the tricyclic derivative [^{18}F]GE180 ([^{18}F]5, flutriciclamide) [37]. It demonstrated in rats a high specific binding, a relevant brain uptake with significant uptake and retention in the olfactory bulb, a region rich in TSPO expression, and a good clearance from the region with low TSPO density, that is, the striatum.

Chau et al. tested the higher affinity of the S-enantiomeric tricyclic indole compound, [^{18}F]GE-180, a third-generation tracer for the detection of TSPO [19]. In the Wistar rat heart, the S-enantiomer was demonstrated as having a higher affinity with a Ki of 0.87 nM, while the R-GE180 enantiomer had a Ki equal to 3.87 nM. Analogous results were demonstrated in humans (human colonic cell membranes). S-enantiomer radiolabeled with ^{18}F had a rapid clearance from blood followed by racemate and R-[^{18}F]GE-180. Brain retention in rats was higher in S- radiolabeled enantiomers than R-enantiomers at 10 and 30 min post-injection. S-[^{18}F]GE-180 also had a greater lung uptake than the R-enantiomer and racemate and was demonstrated to be stable in vivo without conversion to a distomer.

Nag et al. reported the radiosynthesis of [^{18}F]fluorovinpocetine, the fluorinate analogue of [^{11}C]vinpocetine [38]. During the autoradiography assay, homogeneous binding between the cortical and subcortical regions was seen in whole hemisphere human brain slices of healthy subjects. Two cynomolgus monkeys underwent a PET scan and a rapid tracer accumulation was observed in the first 4 min with a decrease in around 20 min. [^{18}F]fluorovinpocetine had a good brain penetration and similar brain distribution pattern to [^{11}C]vinpocetine (high values in the thalamus and striatum, low in the cerebellum), but smaller regional differences.

Damont et al. synthesized a group of pyrazolo[1,5-a]pyrimidines, related to 2, DPA-714 (compound **2** in Figure 3 is represented labeled with ^{18}F) to test the in vitro binding affinity of TSPO [39]. Fluroalakyl- and fluoroalakynyl- analogues were created via Sonogashira coupling reactions. In competition experiments against [3H]1 ([3H]-PK11195) in a membrane rat heart, all compounds had a subnanomolar affinity compared to compound **2**, but compound **12** of the fluoroalkyl series and compound **23** from the alkynyl series had the lowest Ki. The specificity of linking was exquisite in all compounds except for **29** and **30**, which showed a modest affinity for central benzodiazepine receptor (CBR). New compounds (**12**, **21–24**, and **28–30**) have undergone oxidative metabolism investigation in humans, rats, and mouse hepatic microsome assays. In murine microsomes, 90% of biotransformation occurred in 20 min, while in human microsomes, the process was variable in different molecules (31% in 2, 91% in 29). In general, alkynyl derivatives were less metabolized than alkyl compounds. Oxidation did not generate fluoroacetate and so non-brain penetrant species could degrade PET image quality. On the basis of these results, compounds **12** and **23** were chosen for fluorine-18 radiolabeling. Wistar rats underwent PET scans with [^{18}F]12 and [^{18}F]23 7 days after AMPA-induced brain inflammation in the right striatum. Selective tracer uptake in the right striatum was seen 2 min post tracer injection and maintained for 60 min even though it became slightly lower. A better contrast was obtained with [^{18}F]23 due to the rapid washout in the non-lesioned striatum.

4.3. Studies Conducted on Humans

In this paragraph, studies conducted on humans are reported, including an in vivo biodistribution evaluation and post-mortem examination. Okello et al. conducted different studies on microglial activation and PET imaging. In their study in 2009, they used the fibrillar amyloid tracer ^{11}C-PIB (Pittsburgh compound B) and the peripheral benzodiazepine binding site ligand ^{11}C-(R)-PK11195 to explore the possible correlation between microglial activation and Aβ plaque in amnestic mild cognitive impairment (MCI) [40]. A total of 50% of the included 14 subjects with amnestic MCI had increased amyloid deposition but no correlation between the regional C-11-labeled PK-11195 binding and PIB uptake was seen in this group. The authors interpreted this result as due to the multifactorial activation of microglia as a response to amyloid deposition.

^{11}C-(R)-PK11195 was used also by Cagnin et al. for Alzheimer's type dementia (AD), including mild and early forms [41]. The in vivo detection demonstrated high levels of tracer binding in these patients, hinting that the microglial response is an event that could happen in the precocious phase of pathogenesis. In 2004, the same group of researchers demonstrated the presence of higher levels of ^{11}C-(R)-PK11195 in people with frontotemporal lobar degeneration, suggesting microglial activation that went hand-in-hand with neuronal loss. Both in AD and frontotemporal dementia, the tracer uptake was independent of the augmented formation of amyloid plaque [41].

Interesting estimations of ([^{18}F]5 were conducted in human brains by Feeney et al. [42] and previously, by Zanotti-Fregonara et al. [43], which stated ^{18}F-GE180 as an unfavorable tracer for TSPO brain imaging compared with ^{11}C-PBR28 because of ^{18}F-GE180's lower brain penetration.

Gulyás et al. [44] performed an in vitro autoradiography on human post-mortem brains of patients who suffered from Alzheimer's disease by evaluating the binding on TSPO of N-(5-[125I]Iodo-2-phenoxyphenyl)-N-(2,5-dimethoxybenzyl) acetamide ([^{125}I]desfluoro-DAA1106) and N-(5-[^{125}I]Fluoro-2-phenoxyphenyl)-N-(2-[^{125}I]Iodo-5-methoxybenzyl)acetamide ([^{125}I] desmethoxy-DAA1106). Their research demonstrated effective binding for both tracers with higher uptake in the hippocampus, temporal and parietal cortex, basal ganglia, and thalamus. Through a comparison with healthy age-matched controls, the tracer had high specificity in microglia-activated areas in the Alzheimer's brain, confirmed by immunohistochemical examination.

[^{18}F]GE180 was also studied in a pharmacokinetics study by Fan et al. in older healthy adults, which stated the two-tissue compartment model was the better model to describe brain kinetics and that 90 min was the ideal scan length for a good assessment [45].

5. Conclusions

TSPO represents a hallmark of neuroinflammation. The data from our systematic review demonstrate the huge effort in developing more and more specific tracers. The long half-life of ^{18}F (110 min) makes this isotope a preferable choice compared to the shorter life of ^{11}C (20 min). Moreover, new chemical synthesis technologies have permitted researchers to improve radiotracer production and allowed research to increase the use of radiolabeling procedures. However, an observed limitation to this research was the huge widespread neuroinflammation involvement that hampers the possibility of detecting a slight inflammation status change in patients. This issue finds a partial solution using the cerebellum as a reference region, as observed in the mentioned studies on ^{11}C-PBR28-TSPO. Improvement in the radiosynthesis and detection of molecules with higher TSPO affinity has permitted research to overcome this issue. To cite some results, [^{11}C]SSR18057 has a better capacity in discriminating healthy tissue compared to [^{11}C](R)-PK11195, while the choice of fluorinated compounds such as ^{18}F-DPA-714 has made it possible to obtain a greater binding affinity compared to ^{11}C-DPA-713 and ^{11}C-PK11195, as demonstrated by Chauveau et al.

Different synthesis strategies have been also exploited to ensure a better stability of the compound: the fluorination of diphenyliodonium salt used as a precursor was demonstrated to not be a feasible approach while the synthesis of [18F]DAA1106 via [18F]fluorination of the SCIDY precursor compound **1** was demonstrated to guarantee a more stable compound for a considerable time. Moreover, [18F]DAA1106 could be used to detect ischemic areas due to its high binding specificity for TSPO. DAA1106 also has a better brain distribution. In particular, the substitution of its phenyl groups with aliphatic groups permits obtaining compounds with less lipophilicity while its aromatic ring, instead, is responsible for higher affinity. Similar consideration could be directed at ^{18}F-GE180, which is an unfavorable tracer for TSPO brain imaging for lower brain penetration compared with ^{11}C-PBR28.

The introduction of [123I]-CLINME has been interesting in opening the possibility of conducting TSPO-SPECT imaging. The competition studies conducted with PK11195 and Ro 5-4854 have confirmed the specificity of TSPO binding even if thyroid uptake represents a relevant aspect to consider for dosimetry in humans.

Finally, the necessity to consider the presence of distomers or racemic compounds in the synthesis process should be emphasized, which could interfere with the pharmacodynamic and pharmacokinetic effects of the tracer, generating metabolites incapable of overcoming the blood–brain barrier and/or that could increase the noise ratio in PET or SPECT images. The complexity of these subjects was not evaluated in this review and requires further efforts.

Author Contributions: Conceptualization, V.F.; methodology, L.F.; validation, M.S.D.F. and F.C.; formal analysis, J.G.; investigation, M.M.A.S.; data curation, M.M.A.S. and F.D.C.; writing—original draft, M.C.; preparation supervision, M.R., V.F. and G.D.V. All authors have read and agreed to the published version of the manuscript.

Funding: This research received no external funding.

Institutional Review Board Statement: Not applicable.

Informed Consent Statement: Not applicable.

Data Availability Statement: Not applicable.

Conflicts of Interest: The authors declare no conflict of interest.

References

1. Shoshan-Barmatz, V.; Pittala, S.; Mizrachi, D. VDAC1 and the TSPO: Expression, Interactions, and Associated Functions in Health and Disease States. *Int. J. Mol. Sci.* **2019**, *20*, 3348. [CrossRef] [PubMed]
2. Tallman, J.F.; Gallager, D.W.; Mallorga, P.; Thomas, J.W.; Strittmatter, W.; Hirata, F.; Axelrod, J. Studies on benzodiazepine receptors. *Adv. Biochem. Psychopharmacol.* **1980**, *21*, 277–283. [PubMed]
3. Lacapère, J.J.; Papadopoulos, V. Peripheral-type benzodiazepine receptor: Structure and function of a cholesterol-binding protein in steroid and bile acid biosynthesis. *Steroids* **2003**, *68*, 569–585. [CrossRef] [PubMed]
4. Anholt, R.R.; Pedersen, P.L.; De Souza, E.B.; Snyder, S.H. The peripheral-type benzodiazepine receptor. Localization to the mitochondrial outer membrane. *J. Biol. Chem.* **1986**, *261*, 576–583. [CrossRef] [PubMed]
5. Braestrup, C.; Albrechtsen, R.; Squires, R.F. High densities of benzodiazepine receptors in human cortical areas. *Nature.* **1977**, *269*, 702–704. [CrossRef] [PubMed]
6. Ricci, M.; Cimini, A.; Chiaravalloti, A.; Filippi, L.; Schillaci, O. Positron Emission Tomography (PET) and Neuroimaging in the Personalized Approach to Neurodegenerative Causes of Dementia. *Int. J. Mol. Sci.* **2020**, *21*, 7481. [CrossRef]

7. Jacobs, A.H.; Tavitian, B. Noninvasive molecular imaging of neuroinflammation. *J. Cereb. Blood. Flow. Metab.* **2012**, *32*, 1393–1415. [CrossRef]
8. Rupprecht, R.; Papadopoulos, V.; Rammes, G.; Baghai, T.C.; Fan, J.; Akula, N.; Groyer, G.; Adams, D.; Schumacher, M. Translocator protein (18 kDa) (TSPO) as a therapeutic target for neurological and psychiatric disorders. *Nat. Rev. Drug. Discov.* **2010**, *9*, 971–988. [CrossRef]
9. Donat, C.K.; Mirzaei, N.; Tang, S.P.; Edison, P.; Sastre, M. Imaging of Microglial Activation in Alzheimer's Disease by [(11)C]PBR28 PET. *Methods. Mol. Biol.* **2018**, *1750*, 323–339. [CrossRef]
10. Heneka, M.T.; Sastre, M.; Dumitrescu-Ozimek, L.; Dewachter, I.; Walter, J.; Klockgether, T.; Van Leuven, F. Focal glial activation coincides with increased BACE1 activation and precedes amyloid plaque deposition in APP[V717I] transgenic mice. *J. Neuroinflammation.* **2005**, *2*, 22. [CrossRef]
11. Solito, E.; Sastre, M. Microglia function in Alzheimer's disease. *Front. Pharmacol.* **2012**, *3*, 14. [CrossRef] [PubMed]
12. Frantellizzi, V.; Conte, M.; De Vincentis, G. Hybrid Imaging of Vascular Cognitive Impairment. *Semin. Nucl. Med.* **2021**, *51*, 286–295. [CrossRef]
13. Frantellizzi, V.; Pani, A.; Ricci, M.; Locuratolo, N.; Fattapposta, F.; De Vincentis, G. Neuroimaging in Vascular Cognitive Impairment and Dementia: A Systematic Review. *J. Alzheimers. Dis.* **2020**, *73*, 1279–1294. [CrossRef] [PubMed]
14. Guo, Q.; Colasanti, A.; Owen, D.R.; Onega, M.; Kamalakaran, A.; Bennacef, I.; Matthews, P.M.; Rabiner, E.A.; Turkheimer, F.E.; Gunn, R.N. Quantification of the specific translocator protein signal of 18F-PBR111 in healthy humans: A genetic polymorphism effect on in vivo binding. *J. Nucl. Med.* **2013**, *54*, 1915–1923. [CrossRef] [PubMed]
15. Vivash, L.; O'Brien, T.J. Imaging Microglial Activation with TSPO PET: Lighting Up Neurologic Diseases? *J. Nucl. Med.* **2016**, *57*, 165–168. [CrossRef] [PubMed]
16. Qiao, L.; Fisher, E.; McMurray, L.; Milicevic Sephton, S.; Hird, M.; Kuzhuppilly-Ramakrishnan, N.; Williamson, D.J.; Zhou, X.; Werry, E.; Kassiou, M.; et al. Radiosynthesis of (R,S)-[(18) F]GE387: A Potential PET Radiotracer for Imaging Translocator Protein 18 kDa (TSPO) with Low Binding Sensitivity to the Human Gene Polymorphism rs6971. *ChemMedChem* **2019**, *14*, 982–993. [CrossRef]
17. Owen, D.R.; Gunn, R.N.; Rabiner, E.A.; Bennacef, I.; Fujita, M.; Kreisl, W.C.; Innis, R.B.; Pike, V.W.; Reynolds, R.; Matthews, P.M.; et al. Mixed-affinity binding in humans with 18-kDa translocator protein ligands. *J. Nucl. Med.* **2011**, *52*, 24–32. [CrossRef]
18. Lyoo, C.H.; Ikawa, M.; Liow, J.S.; Zoghbi, S.S.; Morse, C.L.; Pike, V.W.; Fujita, M.; Innis, R.B.; Kreisl, W.C. Cerebellum Can Serve as a Pseudo-Reference Region in Alzheimer Disease to Detect Neuroinflammation Measured with PET Radioligand Binding to Translocator Protein. *J. Nucl. Med.* **2015**, *56*, 701–706. [CrossRef]
19. Chau, W.F.; Black, A.M.; Clarke, A.; Durrant, C.; Gausemel, I.; Khan, I.; Mantzilas, D.; Oulie, I.; Rogstad, A.; Trigg, W.; et al. Exploration of the impact of stereochemistry on the identification of the novel translocator protein PET imaging agent [(18)F]GE-180. *Nucl. Med. Biol.* **2015**, *42*, 711–719. [CrossRef]
20. Slováková, A.; Hutt, A.J. Chiral compounds and their pharmacologic effects. *Ceska. Slov. Farm.* **1999**, *48*, 107–112.
21. Nguyen, L.A.; He, H.; Pham-Huy, C. Chiral drugs: An overview. *Int. J. Biomed. Sci.* **2006**, *2*, 85–100. [PubMed]
22. Freedman, T.B.; Cao, X.; Dukor, R.K.; Nafie, L.A. Absolute configuration determination of chiral molecules in the solution state using vibrational circular dichroism. *Chirality.* **2003**, *15*, 743–758. [CrossRef] [PubMed]
23. Page, M.J.; McKenzie, J.E.; Bossuyt, P.M.; Boutron, I.; Hoffmann, T.C.; Mulrow, C.D.; Shamseer, L.; Tetzlaff, J.M.; Akl, E.A.; Brennan, S.E.; et al. The PRISMA 2020 statement: An updated guideline for reporting systematic reviews. *BMJ* **2021**, *372*, n71. [CrossRef] [PubMed]
24. Okubo, T.; Yoshikawa, R.; Chaki, S.; Okuyama, S.; Nakazato, A. Design, synthesis, and structure-activity relationships of novel tetracyclic compounds as peripheral benzodiazepine receptor ligands. *Bioorg. Med. Chem.* **2004**, *12*, 3569–3580. [CrossRef]
25. Chauveau, F.; Boutin, H.; Van Camp, N.; Thominiaux, C.; Hantraye, P.; Rivron, L.; Marguet, F.; Castel, M.N.; Rooney, T.; Benavides, J.; et al. In vivo imaging of neuroinflammation in the rodent brain with [11C]SSR180575, a novel indoleacetamide radioligand of the translocator protein (18 kDa). *Eur. J. Nucl. Med. Mol. Imaging* **2011**, *38*, 509–514. [CrossRef] [PubMed]
26. Chauveau, F.; Van Camp, N.; Dollé, F.; Kuhnast, B.; Hinnen, F.; Damont, A.; Boutin, H.; James, M.; Kassiou, M.; Tavitian, B. Comparative evaluation of the translocator protein radioligands 11C-DPA-713, 18F-DPA-714, and 11C-PK11195 in a rat model of acute neuroinflammation. *J. Nucl. Med.* **2009**, *50*, 468–476. [CrossRef]
27. Vignal, N.; Cisternino, S.; Rizzo-Padoin, N.; San, C.; Hontonnou, F.; Gelé, T.; Declèves, X.; Sarda-Mantel, L.; Hosten, B. [(18)F]FEPPA a TSPO Radioligand: Optimized Radiosynthesis and Evaluation as a PET Radiotracer for Brain Inflammation in a Peripheral LPS-Injected Mouse Model. *Molecules* **2018**, *23*, 1375. [CrossRef] [PubMed]
28. Solingapuram Sai, K.K.; Gage, D.; Nader, M.; Mach, R.H.; Mintz, A. Improved Automated Radiosynthesis of [(11)C]PBR28. *Sci. Pharm.* **2015**, *83*, 413–427. [CrossRef]
29. Mattner, F.; Quinlivan, M.; Greguric, I.; Pham, T.; Liu, X.; Jackson, T.; Berghofer, P.; Fookes, C.J.; Dikic, B.; Gregoire, M.C.; et al. Radiosynthesis, In Vivo Biological Evaluation, and Imaging of Brain Lesions with [123I]-CLINME, a New SPECT Tracer for the Translocator Protein. *Dis. Markers.* **2015**, *2015*, 729698. [CrossRef]
30. Hieu Tran, V.; Park, H.; Park, J.; Kwon, Y.D.; Kang, S.; Ho Jung, J.; Chang, K.A.; Chul Lee, B.; Lee, S.Y.; Kang, S.; et al. Synthesis and evaluation of novel potent TSPO PET ligands with 2-phenylpyrazolo[1,5-a]pyrimidin-3-yl acetamide. *Bioorg. Med. Chem.* **2019**, *27*, 4069–4080. [CrossRef]

31. Pike, V.W.; Taliani, S.; Lohith, T.G.; Owen, D.R.; Pugliesi, I.; Da Pozzo, E.; Hong, J.; Zoghbi, S.S.; Gunn, R.N.; Parker, C.A.; et al. Evaluation of novel N1-methyl-2-phenylindol-3-ylglyoxylamides as a new chemotype of 18 kDa translocator protein-selective ligand suitable for the development of positron emission tomography radioligands. *J. Med. Chem.* **2011**, *54*, 366–373. [CrossRef] [PubMed]
32. Kumata, K.; Zhang, Y.; Fujinaga, M.; Ohkubo, T.; Mori, W.; Yamasaki, T.; Hanyu, M.; Xie, L.; Hatori, A.; Zhang, M.R. [(18)F]DAA1106: Automated radiosynthesis using spirocyclic iodonium ylide and preclinical evaluation for positron emission tomography imaging of translocator protein (18 kDa). *Bioorg. Med. Chem.* **2018**, *26*, 4817–4822. [CrossRef] [PubMed]
33. Tredwell, M.; Preshlock, S.M.; Taylor, N.J.; Gruber, S.; Huiban, M.; Passchier, J.; Mercier, J.; Génicot, C.; Gouverneur, V. A general copper-mediated nucleophilic 18F fluorination of arenes. *Angew. Chem. Int. Ed. Engl.* **2014**, *53*, 7751–7755. [CrossRef]
34. Preshlock, S.; Calderwood, S.; Verhoog, S.; Tredwell, M.; Huiban, M.; Hienzsch, A.; Gruber, S.; Wilson, T.C.; Taylor, N.J.; Cailly, T.; et al. Enhanced copper-mediated (18)F-fluorination of aryl boronic esters provides eight radiotracers for PET applications. *Chem. Commun.* **2016**, *52*, 8361–8364. [CrossRef] [PubMed]
35. Zischler, J.; Kolks, N.; Modemann, D.; Neumaier, B.; Zlatopolskiy, B.D. Alcohol-Enhanced Cu-Mediated Radiofluorination. *Chemistry* **2017**, *23*, 3251–3256. [CrossRef]
36. Zischler, J.; Krapf, P.; Richarz, R.; Zlatopolskiy, B.D.; Neumaier, B. Automated synthesis of 4-[(18)F]fluoroanisole, [(18)F]DAA1106 and 4-[(18)F]FPhe using Cu-mediated radiofluorination under "minimalist" conditions. *Appl. Radiat. Isot.* **2016**, *115*, 133–137. [CrossRef] [PubMed]
37. Wadsworth, H.; Jones, P.A.; Chau, W.F.; Durrant, C.; Morisson-Iveson, V.; Passmore, J.; O'Shea, D.; Wynn, D.; Khan, I.; Black, A.; et al. Exploration of the structure-activity relationship of the diaryl anilide class of ligands for translocator protein–potential novel positron emitting tomography imaging agents. *Bioorg. Med. Chem. Lett.* **2012**, *22*, 5795–5800. [CrossRef] [PubMed]
38. Arlicot, N.; Vercouillie, J.; Ribeiro, M.J.; Tauber, C.; Venel, Y.; Baulieu, J.L.; Maia, S.; Corcia, P.; Stabin, M.G.; Reynolds, A.; et al. Initial evaluation in healthy humans of [18F]DPA-714, a potential PET biomarker for neuroinflammation. *Nuclear. medicine. and. biology.* **2012**, *39*, 570–578. [CrossRef]
39. Wadsworth, H.; Jones, P.A.; Chau, W.F.; Durrant, C.; Fouladi, N.; Passmore, J.; O'Shea, D.; Wynn, D.; Morisson-Iveson, V.; Ewan, A.; et al. [^{18}F]GE-180: A novel fluorine-18 labelled PET tracer for imaging Translocator protein 18 kDa (TSPO). *Bioorg. Med. Chem. Lett.* **2012**, *22*, 1308–1313. [CrossRef]
40. Nag, S.; Krasikova, R.; Airaksinen, A.J.; Arakawa, R.; Petukhovd, M.; Gulyas, B. Synthesis and biological evaluation of [(18)F]fluorovinpocetine, a potential PET radioligand for TSPO imaging. *Bioorg. Med. Chem. Lett.* **2019**, *29*, 2270–2274. [CrossRef]
41. Damont, A.; Médran-Navarrete, V.; Cacheux, F.; Kuhnast, B.; Pottier, G.; Bernards, N.; Marguet, F.; Puech, F.; Boisgard, R.; Dollé, F. Novel Pyrazolo[1,5-a]pyrimidines as Translocator Protein 18 kDa (TSPO) Ligands: Synthesis, in Vitro Biological Evaluation, [(18)F]-Labeling, and in Vivo Neuroinflammation PET Images. *J. Med. Chem.* **2015**, *58*, 7449–7464. [CrossRef] [PubMed]
42. Okello, A.; Edison, P.; Archer, H.A.; Turkheimer, F.E.; Kennedy, J.; Bullock, R.; Walker, Z.; Kennedy, A.; Fox, N.; Rossor, M.; et al. Microglial activation and amyloid deposition in mild cognitive impairment: A PET study. *Neurology* **2009**, *72*, 56–62. [CrossRef] [PubMed]
43. Cagnin, A.; Brooks, D.J.; Kennedy, A.M.; Gunn, R.N.; Myers, R.; Turkheimer, F.E.; Jones, T.; Banati, R.B. In-vivo measurement of activated microglia in dementia. *Lancet* **2001**, *358*, 461–467. [CrossRef] [PubMed]
44. Feeney, C.; Scott, G.; Raffel, J.; Roberts, S.; Coello, C.; Jolly, A.; Searle, G.; Goldstone, A.P.; Brooks, D.J.; Nicholas, R.S.; et al. Kinetic analysis of the translocator protein positron emission tomography ligand [(18)F]GE-180 in the human brain. *Eur. J. Nucl. Med. Mol. Imaging* **2016**, *43*, 2201–2210. [CrossRef] [PubMed]
45. Zanotti-Fregonara, P.; Pascual, B.; Rizzo, G.; Yu, M.; Pal, N.; Beers, D.; Carter, R.; Appel, S.H.; Atassi, N.; Masdeu, J.C. Head-to-Head Comparison of (11)C-PBR28 and (18)F-GE180 for Quantification of the Translocator Protein in the Human Brain. *J. Nucl. Med.* **2018**, *59*, 1260–1266. [CrossRef]
46. Gulyás, B.; Makkai, B.; Kása, P.; Gulya, K.; Bakota, L.; Várszegi, S.; Beliczai, Z.; Andersson, J.; Csiba, L.; Thiele, A.; et al. A comparative autoradiography study in post mortem whole hemisphere human brain slices taken from Alzheimer patients and age-matched controls using two radiolabelled DAA1106 analogues with high affinity to the peripheral benzodiazepine receptor (PBR) system. *Neurochem. Int.* **2009**, *54*, 28–36. [CrossRef] [PubMed]
47. Fan, Z.; Calsolaro, V.; Atkinson, R.A.; Femminella, G.D.; Waldman, A.; Buckley, C.; Trigg, W.; Brooks, D.J.; Hinz, R.; Edison, P. Flutriciclamide (18F-GE180) PET: First-in-Human PET Study of Novel Third-Generation In Vivo Marker of Human Translocator Protein. *J. Nucl. Med.* **2016**, *57*, 1753–1759. [CrossRef]

MDPI AG
Grosspeteranlage 5
4052 Basel
Switzerland
Tel.: +41 61 683 77 34

Biomolecules Editorial Office
E-mail: biomolecules@mdpi.com
www.mdpi.com/journal/biomolecules

www.ingramcontent.com/pod-product-compliance
Lightning Source LLC
Chambersburg PA
CBHW061437210926
51850CB00008B/1533

* 9 7 8 3 7 2 5 8 6 1 9 3 4 *